Elements of Engineering Probability and Statistics

Elements of Engineering Probability and Statistics

Rodger E. Ziemer

Electrical and Computer Engineering Department
University of Colorado at Colorado Springs

Prentice Hall
Upper Saddle River, New Jersey 07458

Original edition, entitled ELEMENTS OF ENGINEERING PROBABILITY AND STATISTICS, 1st Edition, by ZIEMER, RODGER E., published by Pearson Education, Inc, publishing as Prentice Hall, Copyright © 1997.

ISBN 978-89-450-4517-1
ISBN 89-450-4517-1

Contents

Preface

This book covers in a single volume the essentials of probability and statistics and certain of their engineering applications. Many undergraduate engineering curricula include a required course in probability. It is important that this course include not only the elements of probability and its engineering applications, but statistics as well. The reasons for this are at least twofold. First, the graduate engineer is served well in the work environment by these subject areas, since many aspects of modern engineering analysis and design require the tools of probability and statistics. Second, the engineering accreditation agency, Accreditation Board for Engineering and Technology, requires *probability and statistics* as part of the criteria for accreditation in many undergraduate specialties. It is doubtful that most undergraduate engineering curricula will include two courses devoted to these subject areas, so it seems most prudent to include the essentials of both in one course. This book has been written with this as one of its features—an abbreviated treatment of both that can be fit into a one-semester three-credit-hour course.

An important feature of the book is the inclusion of computer exercises in addition to the practice problems at the end of each chapter. It is recommended that a computer mathematics package such as MATLAB® or MATHCAD® be used in working with these computer exercises. It is felt that inclusion of computer exercises with the student's homework will go a long way toward better understanding of the subject matter, which appears at first sight to be deceptively simple.

The layout of the book is as follows. After an introductory chapter, fundamental concepts of probability are dealt with in the second chapter. This includes various approaches to probability culminating in the axiomatic approach. Various derived probability relationships are then introduced and used in problem solving through several examples of varying difficulty.

Chapter 3 introduces the concepts of random variable, cumulative distribution function (cdf), and probability density function (pdf). Probability mass functions are introduced, but the main vehicles for describing random variables are the cdf and pdf, even for discrete random variables through the use of unit impulse functions. Several commonly occurring random variables and their distribution functions are summarized in the chapter. Chapter 3 also includes transformations of random variables, expectation, Chebyshev's inequality, and generation of pseudorandom numbers by means of computer software.

Chapter 4 essentially repeats the content of Chapter 3, but for two random variables. In addition, this chapter includes the important concepts of statistical independence for random variables, conditional cdf's and pdf's, the central limit theorem, the weak law of large numbers, and the extension of joint random variable ideas to more than two random variables.

At Chapter 5, a break is taken from probability topics, and some elementary statistical ideas are introduced. These include the sample mean, sample variance, regression techniques, empirical distribution functions, Monte Carlo simulation, and process control.

Chapter 6 includes an introduction to estimation theory and engineering applications. After discussing desirable properties of point estimators, the maximum likelihood technique is introduced. Next, confidence intervals for various estimators, including the sample mean, sample variance, and relative frequency, are discussed. The chapter closes with discussions of the conditional mean and the othogonality principle.

In Chapter 7, decision-making approaches are discussed. These include Bayes' decision rules, maximum a posteriori and maximum likelihood decision strategies, and classical decision theory.

Chapter 8 provides an introduction to reliability theory, including time-dependent reliability, reliability analysis for various system structures, and the Weibull distribution in reliability modeling.

Chapter 9 gives an introduction to random processes. Included are several examples of random processes, their statistical description, the autocorrelation function and its properties, cross-correlation functions, and the Gaussian process and its importance. The Weiner–Khinchine theorem is mentioned as a footnote in conjunction with a property of the autocorrelation function. Although deserving much more attention to this, its treatment is limited to this mention in keeping with the desire to keep the book understandable to students early in their college educations. The chapter closes with a presentation of the first-order autoregressive discrete-time process as an illustration of a random process whose second-order statistics are relatively easy to derive.

The last chapter, Chapter 10, is a consideration of the effect of systems on random processes. After a short review of system descriptions and properties, characterization of the output of zero-memory nonlinear systems and fixed, linear systems in response to stationary random processes that have been present for the infinite past (i.e., the output is also stationary) is considered.

Each chapter includes several example problems, a set of homework problems, a set of computer exercises, a detailed point-by-point summary, and a further reading section. Also included are several appendices, including a summary of some important formulas from Fourier analysis and linear system analysis that will provide some information on these ideas for the few places in the text that they appear, an appendix on programming hints for using MATLAB in statistical analysis and simulation of random phenomena, probability tables, mathematical tables, and answers to selected problems. A solutions manual for all end-of-chapter problems and computer exercises is available from the publisher for instructors who adopt the text.

Although the author does not presume to tell potential users of this book how to teach a course on the subject matter, the following is one suggestion for coverage of material from it in a 15-week session (30 twice-per-week class sessions).

Chapter	Topics	Time (75-min class sessions)
1	Introduction	1
2	Probability concepts	4
3	Single random variables	5
4	Two random variables	5
	Midterm	1
5	Elementary statistics	3
6 or 7	Estimation theory or engineering decisions	3
8	Reliability[a]	3
9 and 10	Random processes[a]	5
	Final Exam	Week 16

[a]Possible deletions or topics to scale back for a quarter system.

I have taught probability and random processes to engineering students at both the undergraduate and graduate levels several times over the span of 30 years of teaching. I have yet to teach the ideal course. When as much is crammed into a course as one taught from this book, it may seem that one is facing an insurmountable task. However, I have also found that if the students are reasonably ambitious, it is possible to cover the material listed above. To motivate them, I assign problems weekly and give a short 10-minute quiz each week. Computer exercises are assigned sporadically throughout the semester, with several coming at the front of the course to get students familiarized (or reacquainted) with programming (we require a course on MATLAB of freshman). During the heavy-duty coverage of probability, the computer exercises are suspended. Toward the end of the course, two or three of the more challenging exercises are assigned as projects.

My thanks are extended to my students, who sometimes willingly and sometimes grudgingly accepted my teaching methods. Without their feedback, this book would not have been possible. I also acknowledged my wife, Sandy and her acceptance of my writing venture. Apparently, we have struck a happy medium between my computer and time with each other, and it is to her patience that I dedicate this book. Finally, I wish to thank the reviewers who improved the quality of the final product through their criticisms and suggestions. Those who have contributed include: Zoran Siveski of New Jersey Institute of Technology; Stephen Horan of New Mexico State University; Abraham H. Haddad of Northwestern University; Tim Healy of Santa Clara University; Roger Conant of University of Illinois-Chicago and James A. Bucklow of University of Wisconsin.The shortcomings that remain are mine alone; any constructive feedback would be welcomed. My email address is as follows; Ziemer@.ucs.edu.

Rodger E. Ziemer

1

Introduction

1–1 INTRODUCTION

Purpose of This Book

The underlying ideas of probability and statistics, and certain of their engineering applications are introduced in this book. Probability and statistics are important in many engineering applications, including characterizing the reliability of systems,[1] estimation of quantities from random data, making engineering decisions based on imperfect and/or incomplete data, and analyzing the effects of noise in components and systems. Probability and statistics can be studied on a purely mathematical basis; it is the intent of this book not only to introduce the student to the mathematical theory of these two subject areas, but also to show how they can be applied in typical engineering applications.

It is always a good idea to define what one is studying. *Probability* theory deals with averages of mass, unpredictable phenomena, whereas *statistics* is concerned with the collection and representation of engineering data so that practical conclusions can be drawn. Phenomena occurring in nature can be classified as either deterministic or random. Deterministic phenomena happen in the same way each time that an experiment giving rise to them is performed under the same conditions. Random phenomena, on the other hand, do not happen in the same way each time the underlying experiment is performed, even though the underlying conditions are the same insofar as we are able to

[1] A *system* can be defined as a combination and interconnection of components or subsystems to perform a desired task.

make them.[2] For the most part, the student up to this time has probably dealt with deterministic phenomena, although it can be argued that nothing in nature is truly deterministic. For example, Newtonian mechanics in physics is taught and learned as a deterministic theory: the mechanics of a pool game is idealized in that the angle of incidence of a ball bouncing off a bumper is equal to the angle of return; the striking of a ball by the cue ball is idealized as conserving momentum; and so on. These physical phenomena obey fixed laws and are expected to happen consistently time after time. The student should not have to think too hard, however, to realize that not every experiment, either naturally occurring or otherwise, can be treated as deterministic. The birth of male and female human beings is a random phenomenon where, on the average, the ratio of male to female births appears to be about 51% to 49%, respectively. That this appears to be the case is estimated by using statistics.

Word of Warning

Probability and its applications to engineering problems is perhaps one of the more difficult bodies of knowledge for the undergraduate engineering student to grasp. This is true also of the study of statistics, although many of the initial ideas of either body of knowledge are intuitive in nature. There are probably several reasons for the difficulty in becoming proficient in these two subject areas. First, probabilistic and statistical concepts run counter to the systematic way of thinking in terms of the fixed laws of physics that have been drilled into the student from the start. Second, the basic ideas about probability and statistics appear *deceptively simple*. Thus, the student is tempted to gloss over the definitions and theorems and gain only a cursory understanding of the subject matter. Third, these can be somewhat dry subjects and it is difficult for the student to maintain interest until enough knowledge has been assimilated to make them useful. A fourth reason may be that it is difficult to perform experiments involving probabilistic concepts so that the student can be convinced that it really works. Finally, there is a gap between the popular manifestation of the subject (say, sports or gambling) and the theory to be covered in this book. Like any subject, there is a certain minimum amount of knowledge that must be assimilated about probability and statistics to make them minimally useful, and the student is urged to be diligent in getting a firm grasp on the fundamental ideas about these two subject areas.

Role of Models

It is assumed that the student is reasonably familiar with the idea of a model of an actual situation, even though we sometimes become so engrossed in the mathematical solution of a problem that we forget that the mathematical description is simply a model of the real, physical thing we are analyzing. For example, the motion of an object, such as a billiard

[2]We refer to an experiment in this discussion as though it can always be performed, when indeed it often cannot. For example, if we are studying the average rate of occurrence of lightning strikes in a given location, we obviously cannot perform the underlying experiment at will (i.e., an electrical storm). Rather, we must wait for the opportunity that nature affords us to make our measurements. Quite often, the underlying experiment must be performed conceptually or through computer simulation. An example of this is flipping a coin 1 million times. Although possible, we probably do not have the patience and stamina to do so.

ball, is modeled by certain laws of physics, such as conservation of momentum. Its motion is then predicted with the aid of mathematical solution of the equations of motion based on the laws of physics. Finally, if we are so inclined, we can observe how well the model predicts the actual motion of the billiard ball by playing a game of eightball. In summary, the process just described is composed of four steps: identification of a model; solution for certain quantities of interest using mathematics appropriate for the modeling method; comparison of the predicted results with the outcomes of an actual physical experiment for verification of the model; and modification of the model, if necessary, to get closer correspondence to the behavior of the actual physical phenomenon being modeled. Having been assured of the accuracy of the modeling method by the final verification, one can then go on to predict the outcomes of other situations using a similar physical setup. Or, one can come up with alternative solutions for a given problem. The latter is referred to as *design,* whereas predicting the response of a given physical setup is referred to as *analysis.* Unfortunately, much of the undergraduate engineering curriculum deals with analysis because it is usually necessary to be able to carry out reliable analysis before meaningful design can be accomplished. In the light of our earlier discussion about deterministic and probabilistic models, it goes without saying that either one or both approaches to modeling may be useful for representing natural phenomena.

In recent years, the use of computer simulation for modeling physical phenomena has become popular. The reasons for this are (1) that personal computers have become sufficiently powerful to allow comprehensive simulations to be developed and run conveniently, and (2) that special-purpose software is available and is reasonably priced and allows fairly rapid development of accurate simulations. We make use of such simulations in this book to illustrate probabilistic modeling.

Why Probabilistic Approaches to Engineering Problems?

Why use probabilistic methods in the solution of engineering problems? The reasons are varied. In the example of the billiard game just discussed, we may not be able to model accurately the friction of the table felt or the imperfect elasticity of the bumpers. A probabilistic approach can be used to reflect our inability to model perfectly some aspects of the actual situation. Another reason is that we are unable to make perfect measurements of physical quantities—for example, random motion of air molecules may make the needle of a sensitive meter vibrate sporadically, or we may not be able to make accurate measurements because of our inability to read the measurement instrument consistently. Finally, certain phenomena in nature are "random," and we are only able to describe them in probabilistic terms. One might argue that random-appearing phenomena are the result of our inability to observe them to the finest detail, such as the motion of electrons in atomic structure, or it may be that we are unable to observe them closely because of their inaccessibility such as the occurrence of sunspots on the sun. Whatever the reason, the fact remains that probabilistic modeling methods are particularly applicable for certain phenomena where our state of knowledge, our inability to observe sufficiently closely, or certain phenomena in nature being chaotic leave an element of uncertainty. For these reasons, it is helpful, and in many cases necessary, to use probabilistic methods in the analysis and design of certain engineering problems.

1–2 TWO EXAMPLES

To illustrate further the difference between deterministic and probabilistic models, we consider two simple examples in this section. These examples also illustrate some ways of summarizing random data.

Measurement of Voltage Levels. Consider the measurement of a number of voltages with a meter having a certain amount of internal noise. The internal root-mean-square (rms) noise level of the meter is 1 microvolt (μV),[3] and the voltages to be measured are constants of 1 V, 0.02 millivolt (mV) and 5μV. To get an accurate estimate of the voltage level, 50 measurements (V_i, $i = 1, 2, \ldots, 50$) are taken and averaged to get the sample mean, defined for this specific example as:

$$V_{sm} = \frac{1}{50}\sum_{i=1}^{50} V_i \tag{1-1}$$

When plotted versus sample number, they appear as shown in Figure 1-1a-c (note that the vertical scales are different on each graph). This series of figures points out that it is easy to obtain an accurate measurement, despite the noise, if the voltage level being measured is large compared with the rms noise level. In fact, for the 1-V level, a sufficiently accurate model would be to ignore the noise and let the voltage be the constant 1 V itself.

When the sample means are computed according to (1-1), the results are 1 V, 0.0197 mV, and 4.97 μV, respectively. It is seen that even for the voltage of 5 μV, for which the noise can have a significant effect on a single measurement, the use of the sample mean of 50 measurements provides a fairly accurate estimate of the true voltage level. We explore the theoretical basis for this in a subsequent chapter.

Examination Grades in a Class of Students. Consider a class of 25 students wherein an examination is given and the following grades result: 97, 93, 92, 87, 85, 84, 83, 80, 80, 77, 76, 76, 71, 71, 68, 65, 63, 62, 55, 53 52, 49, 43, 40, and 16. Typically, when the examination is returned, the students ask for the maximum, the average (sample mean), and the minimum score. These are 97, 68.7, and 16, respectively. Often, the instructor needs more information in order to assign letter grades. This can be provided by a *dot diagram,* as shown in Figure 1-2. Yet another type of diagram that summarizes the data and will be discussed in a subsequent chapter is the *histogram,* which divides the range of the data into contiguous bins. This then forms the abscissa of the plot, with the number of occurrences of the data within a given bin shown on the ordinate, as presented in Figure 1-3. The histogram plot entails some loss of information since the raw scores are not available from it, whereas the dot diagram does not. Another representation that contains all the information available is a *stem-and-leaf display.* This is given in Table 1-1 where ranges of 10 are given in the left-most column for the test scores, and the tally figures in the right-hand column are the units' digits of the test scores. The appearance of the stem-and-leaf display is not too unlike that of

[3]Certain abbreviations will be convenient for later use: giga = $\times 10^9$; mega = $\times 10^6$; kilo = $\times 10^3$; milli = $\times 10^{-3}$; micro = $\times 10^{-6}$; nano = $\times 10^{-9}$; pico = $\times 10^{-12}$. The root-mean-square value of a waveform is the square root of the average (over one period, if periodic) of the square of the waveform.

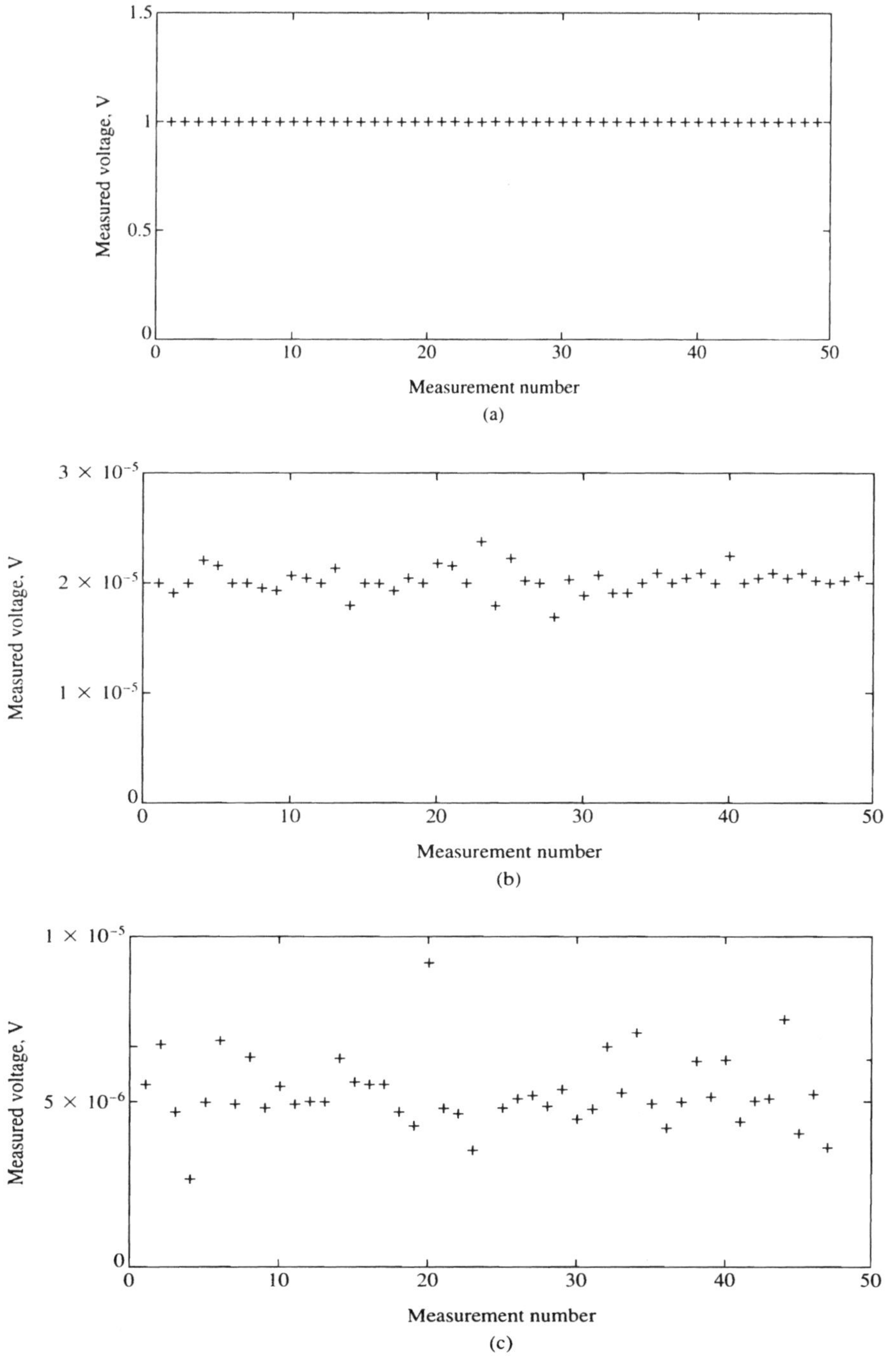

Figure 1-1 Measurement of three noisy voltages: (a) 1 V; (b) 20 μV; (c) 5 μV. The rms noise level is 1 μV.

TABLE 1-1 STEM-AND-LEAF DISPLAY OF 25 EXAMINATION SCORES

Decade	Unit
0– 9	
10–19	6
20–29	
30–39	
40–49	0 3 9
50–59	2 3 5
60–69	2 3 5 8
70–79	1 1 6 6 7
80–89	0 0 3 4 5 7
90–99	2 3 7

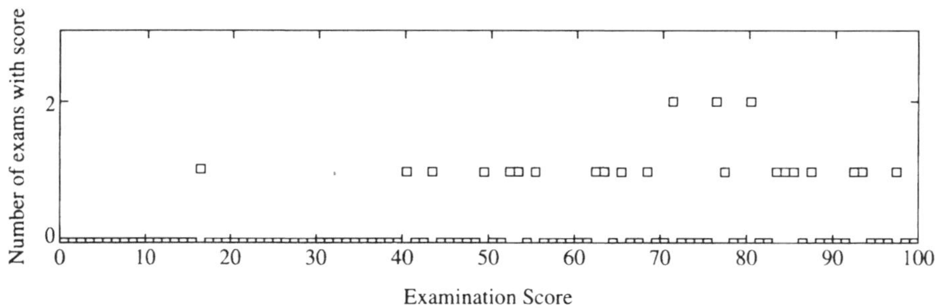

Figure 1-2 Dot diagram for scores achieved on an examination.

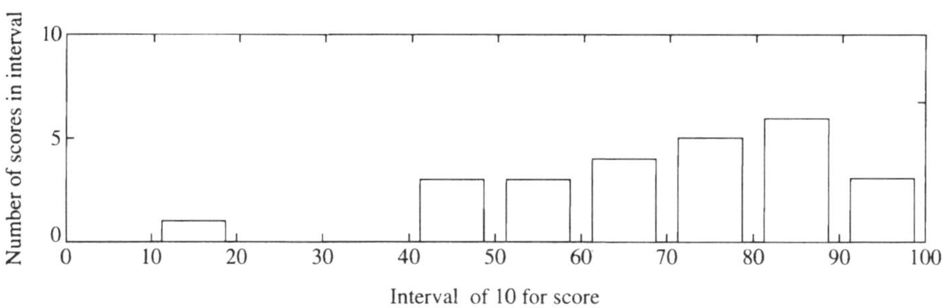

Figure 1-3 Histogram of examination scores.

the histogram tipped on its side, but instead of bars we have the actual units' digits for the scores. Both the histogram and stem-and-leaf displays allow us to assess the grouping of scores about the mean, whether or not this distribution of scores is skewed to one side or the other, and the fact that one score lies far outside the grouping of the other scores. Called an *outlier,* we may wish to exclude this extreme-value score before computing such quantities as the sample mean. This is done, for example, in figure skating competitions where the high and low scores of the judges are excluded before computing the final tally.

1–3 SIMULATION OF RANDOM PHENOMENA

The data used to plot Figure 1-1 was generated by means of a *pseudorandom number generator*. Pseudorandom numbers have properties that appear to be random but are really generated by a deterministic algorithm, often by making use of the overflow properties of integer arithmetic in computers. Specifically, a good pseudorandom number generator should produce sequences of numbers (usually with values between 0 and 1) with the following properties:

1. Different seeds or starting values for the algorithm will produce different sequences of numbers.
2. The numbers will be evenly (uniformly) distributed between 0 and 1 (or whatever other interval the generator is designed to operate within).
3. Any specific number x_i will bear little dependence (called *uncorrelated*) with other numbers generated in the chain.

We will not go into the mechanics of pseudorandom number generation here because such number generators are available in most high-level programming languages, such as C and Pascal. If, as recommended in this book, a mathematics package such as MATLAB[4] or Mathcad[5] is used, pseudorandom numbers are readily generated. For example, in MATLAB, the following statement will generate a row vector of 1000 random numbers uniformly distributed in the interval (0, 1):

```
X=rand(1,1000);
```

The semicolon suppresses the output appearing on the screen. To compute and plot a histogram with 10 intervals showing the distribution of these pseudorandom numbers, the following statement would be used:

```
hist(X, 10)
```

With Mathcad, 1000 pseudorandom numbers can be generated with the statements

```
i=0. . . 999        X_i=rnd(1)
```

Note that the indices now run from 0 to the last digit minus 1, and the argument of the random number generator is the length of the interval (starting at 0) over which the numbers are uniformly distributed.

[4]Copyright by The Mathworks, Inc., Natick, MA.

[5]Copyright by MathSoft, Inc., Cambridge, MA.

If we have a pseudorandom number generator that generates uniformly distributed random numbers in the interval $[0, 1)$, any other interval of uniform distribution can be created, say $[a, b)$, very simply by using the relationship

$$Y_i = a + (b - a) X_i \tag{1-2}$$

For example, if we wish to simulate the widths of keys for mounting gears on shafts that have nominal widths of 5 mm but can vary from this uniformly by ± 0.1 mm, we would follow the generation of a set of uniform random numbers (however many key widths we wish to simulate) by the statement

$$W = 4.9 + 0.2X \tag{1-3}$$

where the notation suggests a MATLAB implementation, with W and X being vectors of the proper size. Note that the elements of the W vector will range from a minimum of 4.9 mm (for an element of X that is 0) to 5.1 mm (corresponding to an element of X that is 1). An example MATLAB program is given in Table 1-2, and the results of a typical run are provided in Table 1-3. A histogram plot is provided in Figure 1-4. For more information on MATLAB, a brief overview is provided in Appendix B, and the manual for the Student Edition of MATLAB is highly recommended.

We take up the subject of the use of pseudorandom numbers for simulation of probabilistic phenomena in later chapters.

TABLE 1-2 SAMPLE MATLAB PROGRAM FOR GENERATING PSEUDORANDOM KEY WIDTHS AND COMPUTING CERTAIN STATISTICS

```
% Program for generating N key widths
%
N=input('Enter the number of key widths to generate:');
w=input('Enter the nominal width (mm) for the keys:');
p=input('Enter the + or - uniform variation (mm) about the nominal width:');
flag1=input('Enter 1 if all widths are to be displayed; otherwise enter 0:');
flag2=input('Enter 1 if histogram is to be displayed; otherwise enter 0:');
X=rand(1,N);
W=w-p+2*p*X;
mu=mean(W);
std_dev=std(W);
disp('Sample mean and standard deviation of key widths:');
disp(mu)
disp(std_dev)
if flag1 == 1
disp(' The widths in mm are: ');
disp(W')
end
if flag2 == 1
hist(W), xlabel('Key width, mm'), ylabel('Number of counts'), ...
title(['Distribution of ', num2str(N), ' key widths ']), grid
end
```

TABLE 1-3 COMMAND WINDOW FOR A RUN OF THE MATLAB PROGRAM IN TABLE 1-2

```
»prog_ex1

Enter the number of key widths to generate: 100
Enter the nominal width (mm) for the keys: 5
Enter the + or - uniform variation (mm) about the nominal width: .1
Enter 1 if all widths are to be displayed; otherwise enter 0: 1
Enter 1 if a histogram is to be displayed; otherwise enter 0: 1

Sample mean and standard deviation of key widths:
4.9928
0.0558

The widths in mm are:
4.9036    5.0035    4.9433    4.9128
5.0187    4.9210    5.0604    4.9695
5.0444    5.0683    5.0055    4.9585
4.9993    4.9737    5.0338    5.0019
4.9107    4.9848    5.0727    5.0652
4.9883    4.9654    4.9488    5.0669
5.0038    5.0427    5.0423    5.0453
5.0544    5.0154    4.9527    4.9659
4.9131    5.0744    4.9766    5.0604
4.9886    5.0516    4.9162    5.0264
5.0954    5.0673    4.9920    4.9091
4.9935    5.0400    4.9466    4.9083
4.9658    4.9778    5.0591    4.9307
4.9892    5.0870    4.9983    5.0534
5.0487    5.0211    4.9006    4.9123
4.9411    4.9724    4.9532    5.0039
4.9343    5.0079    5.0329    5.0858
4.9745    4.9910    4.9850    4.9441
5.0879    4.9803    4.9374    4.9409
5.0930    4.9481    5.0927    4.9178
4.9510    4.9166    4.9268    5.0399
4.9141    4.9843    5.0592    5.0413
4.9248    4.9945    5.0892    4.9989
5.0053    4.9580    4.9342    4.9524
4.9320    4.9210    5.0212    5.0423
```

1–4 SCOPE OF THIS BOOK

The intent of this book is to provide an *introduction* to probability and statistics and their applications to engineering problems. In the next three chapters we provide much of the background theory for probability. In Chapter 2 we define what is meant by probability

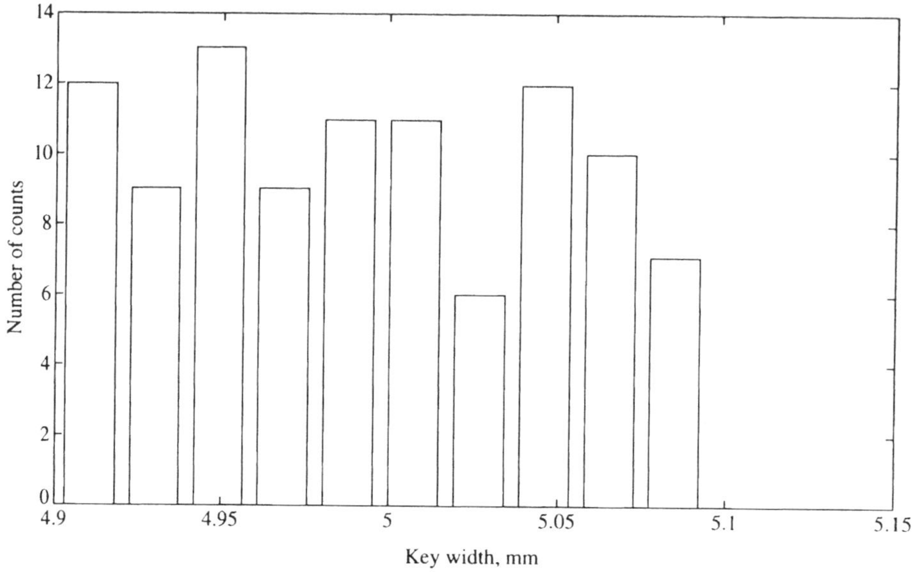

Figure 1-4 Histogram of 100 key widths uniformly distributed between 4.9 and 5.1 mm.

after illustrating several intuitive approaches that, although useful, do not provide a firm mathematical basis on which to build. The idea of a chance experiment is described in Chapter 2, and the description of the results from chance experiments in terms of outcomes and events is covered. It is seen that these concepts can be represented in terms of a sample space and combinations of subsets within the sample space. Some elementary set theory ideas are therefore summarized next and are then used to develop derived relationships for probabilities. The chapter ends with a discussion of counting methods. In Chapter 3 we develop the concept of a random variable and the probability functions that describe it. Also discussed in this chapter are the notions of transformation of a random variable and averages of a random variable. The purpose of Chapter 4 is to redo the concepts of Chapter 3 for two or more random variables. In Chapter 5 some very basic ideas from statistics are introduced, and Monte Carlo simulation is described. In Chapter 6 we look at estimation theory and its engineering applications. In Chapter 7 we ask the question: How can reasonable engineering decisions be made when the observables are subject to randomness? The concept of reliability for components and interconnection of components is considered in Chapter 8. In Chapter 9 we look at random time series or random processes; how probabilistic methods can be used in analyzing the response of systems to random forcing functions is taken up in Chapter 10. It is assumed that the student has not taken a previous course on linear system analysis, in particular the ideas of convolution and the Fourier transform. These ideas are used only a few times in the book. A review of these concepts is provided in Appendix A along with linear system ideas needed for Chapter 10. In this book we make use of MATLAB to illustrate probabilistic ideas and for computation. In Appendix B we give some MATLAB programing hints. Mathcad is another possible option. Appendix C contains MATLAB programs for and tables of certain probability distributions used in the book. Tables of mathematical formulas are included in Appendix D, and answers to selected problems are given in Appendix E.

1–5 FURTHER READING

There are many books on probability and statistics. Two will be cited here as being representative of the level of treatment given here. The book by Johnson (1994) is a very readable revision of the fifth edition of a book authored by Miller and Freund. It has a very lucid presentation of the basic theory of probability and statistics with good examples. It does not treat random processes, which is given an introductory treatment in Chapter 9 of this book. Another somewhat older book written in a vein similar to the one by Johnson is Ross (1987). For those students desiring more examples than given here (and no matter how many examples are provided students seem to want more), the following two resources are recommended: *The Probability Tutoring Book* by Ash (1993) and *Probability and Statistics* by Spiegel (1975).

1–6 PROBLEMS

1-1. Given the data sets below, compute their sample means.

i	1	2	3	4	5	6	7	8	9	10
x_i	3	6	8	9	9	4	4	4	0	8
y_i	16	20	38	61	16	15	−2	24	21	25

1-2. Prepare a dot diagram for each data set of Problem 1-1.

1-3. Prepare a histogram for each data set of Problem 1-1. Use 5 bins for the x_i's and 9 bins for the y_i's.

1-4. Prepare a stem-and-leaf display for the $\{y_i\}$ data set of Problem 1-1.

1–7 COMPUTER EXERCISES[6]

1-1. **(a)** Using the rnd(1) function of Mathcad or the rand(1, N) of MATLAB, generate 100 integer-valued random numbers between 0 and 9. Plot a histogram in each case using 10 bins equally spaced between 0 and 10 (i.e., 0 to 1, 1 to 2, etc.). [*Hint:* For example, if using MATLAB, use the fix (·) function rather than the round (·) function. The latter will produce a set of integers that are biased against 1 and 10. The student should give the reason for this.]

[6]Appendix B gives an overview of MATLAB®

(b) Repeat for 1000 random numbers generated. Is the histogram more "uniform" than for that of 100 random numbers?

(c) Repeat for 5000 random numbers generated. Is the histogram more "uniform" than when 100 or 1000 random numbers were generated?

1-2. Simulate the widths of 5000 shims from a production line that are punched from stock nominally 3 mm thick but which varies uniformly from this nominal width by ± 0.005 mm. Plot a histogram of the shim thicknesses utilizing 20 intervals.

1-3. Resistors are produced by a production line with nominal values of 1000 ohms $\Omega \pm 20\%$. Generate 1000 sample values for resistance and plot a histogram utilizing 10 intervals.

2 Fundamental Concepts of Probability

2-1 INTRODUCTION

The purpose of this chapter is to define the meaning of probability and to develop various relationships among probabilities based on the definition and using mathematical reasoning. The concept of a chance, or random, experiment is discussed along with its outcomes. An example is the flipping of a coin. At each performance of this chance experiment, two outcomes are possible—heads or tails—unless we also include the highly unlikely outcome of the coin standing on edge. Before defining the probability of the occurrence of an outcome of a chance experiment, various approaches to probability used in the past are discussed, including the personal, relative frequency, and equally likely approaches. As will be seen, these are often useful in assigning probabilities to outcomes of chance experiments but do not provide a satisfactory basis for a theory of probability. Thus the axiomatic approach to probability was developed rather recently by the Russian mathematician Kolmogorov, and it is this framework that provides the basis for the entire theory of probability. The axiomatic approach gives properties that the probability of an outcome of a chance experiment must satisfy but does not provide a means for determining the probabilities of actual experimental outcomes. To assign probabilities to experimental outcomes, some other means must be used (i.e., the personal, relative frequency, or equally likely approaches). It is comforting to note that these other approaches to probability actually satisfy the axioms of probability.

2–2 APPROACHES TO PROBABILITY

What is probability? This question is not as easy to answer as might first appear. Each of us has an intuitive notion of what probability is which might be referred to as the *personal* definition of probability. For example, we might have the personal feeling that the longer we wait in a line at the supermarket, the better are our chances of being served within the next 30 seconds. Yet, why is it that when we choose the line that appears to be decreasing the fastest, it seems to slow to a crawl the minute we enter it?

Historically, mathematical analysis was first applied to probability in France in the mid-seventeenth century when games of chance were popular among the nobles (Barr and Zehna, 1983). Discrepancies between the basis for betting and the actual outcomes of a certain popular game at the time led one of the noblemen, Chevalier de Meré, to consult the mathematician Blaise Pascal, who in conjunction with fellow mathematician Pierre de Fermat, developed the first formal analysis of gambling odds in terms of mathematics. Over the years, attempts to define probability mathematically have met with one failure or another; we will examine some of these attempts shortly. As a result, rather than *define* probability, the accepted approach to probability theory nowadays is almost without exception an *axiomatic* approach—that is, to give a list of properties that probability must satisfy. Although the axiomatic approach to probability was developed by Kolmogorov in 1933, it was not widely used until the 1950s, when an English translation of his work was made available.

Before getting into the axiomatic approach to probability, we discuss some of the other approaches and ideas about probability so that the student will have an appreciation of why the axiomatic approach came about. Also, some of these other approaches give us a means for attaching probability to real happenings in the world around us. We first give several definitions that will provide us with a common language in discussing probability.

Any discussion of probability is centered around the idea of a *chance experiment.* Such an experiment is not necessarily actually performed, but is often conceptual; they are referred to as chance or random experiments because, when performed, the results obtained do not obey deterministic[1] laws of nature. The results of a chance experiment are referred to as *outcomes.* Collections of outcomes are referred to as *events.* We denote chance experiments by boldface uppercase letters, the outcomes of chance experiments by lowercase Greek letters, and events by uppercase italic letters. Several examples are given below.

$\mathbf{E}_1$: Two coins are tossed and the number of heads is observed. The possible outcomes are: ζ_1 = tail on coin 1 and tail on coin 2 (*tt*); ζ_2 = head on coin 1 and tail on coin 2 (*ht*); ζ_3 = tail on coin 1 and head on coin 2 (*th*); and ζ_4 = head on coin 1 and head on coin 2 (*hh*). A specified event might be the event E_1 that at least one head occurs, which consists of any of the outcomes ζ_2, ζ_3, or ζ_4.

[1]We are in the process of defining what random phenomena are, so it is difficult to define the term *deterministic* at this point. Perhaps it will suffice for now to define any deterministic law as one where the outcome is predictable once the phenomenon is set in motion. For example, if an apple drops from a tree, we know from experience that it always falls down to the ground, and furthermore, we can predict its rate of fall, time duration of the fall if given the distance, and so on.

E$_2$: A green die and a red die are tossed and the number of spots on the upside of each is observed. The possible outcomes are: $\omega_1 = 1$ spot up on both the green and red dies (G1/R1); $\omega_2 =$ G2/R1, . . . , $\omega_{65} =$ G6/R5; $\omega_{66} =$ G6/R6. An event E_2 of interest might be that the sum of the number of spots up is 3, which would consist of either of the outcomes $\omega_2 =$ G2/R1 or $\omega_7 =$ G1/R2.

E$_3$: A number is drawn from an urn containing slips of paper with the integer numbers zero through 10 written on them. If the number is greater than five, the event E_1 is noted; if five or less, the event E_2 is noted. In this case, the separate outcomes are the numbers zero through 10.

It is convenient to represent events and outcomes of chance experiments in terms of a *sample space, S.* A pictorial representation of a sample space for experiment **E**$_1$ is provided in Figure 2-1. The totality of all possible outcomes is represented by the rectangle. Particular outcomes are represented by points within the rectangle, and events are represented by closed curves, usually circles or ellipses. Such a diagram is also known as a *Venn diagram.*

We now return to our original objective for this section, that of discussing some possible approaches to probability. We have already mentioned the idea of *personal* or *subjective* probability. This is the type of probability that one uses when glancing out the window in the morning to determine whether or not an umbrella should be taken along to work based on the appearance of the sky. One might argue that such an example really does not represent the idea of probability at all, but rather, the art of educated guessing based on past experience, which is really the case.

Another approach to probability can be referred to as the *equally likely outcomes* approach. It is typical of the reasoning used when one is trying to assess the chances of obtaining a certain card when cutting a deck. In this approach, the chance experiment is visualized as consisting of several outcomes that have equal chances of occurrence, such as the experiment of throwing a single die and having the expectation that any number of spots from one to six are equally likely to show on the upside. The probability of a particular event of interest is then the number of outcomes favorable to the event divided by the

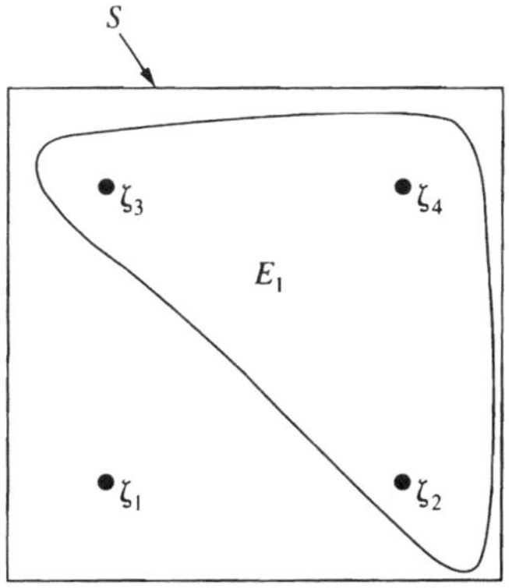

Figure 2-1 Venn diagram for visualizing outcomes and events for a chance experiment.

total number of possible equally likely outcomes. In the defining statement of this approach, note that the phrase "several outcomes that have equal chances of occurrence" really amounts to using probability to define probability. This is the main flaw with this approach. Another is that there is often no way for some chance experiments to be broken down into a set of equally likely outcomes, such as in the case of determining the probability of a hurricane passing through a certain locality. In a similar vein, in terms of the die-throwing example, there is no way to account for the fact that someone may have supplied us with a loaded die, or that the die has become worn on one side. Nevertheless, this approach to probability is useful in many situations where it is possible to associate equally likely outcomes with a chance experiment.

Example 2-1

Use the equally likely approach to determine the probability of at least one head appearing in the tossing of two fair[2] coins, as described in experiment $\mathbf{E}_1$.

Solution There are four outcomes possible each time the two coins are tossed. There is no reason to believe that any one outcome should be favored over any of the others. (Later, we will give this statement more credence based on the separate coins themselves being "fair.") There are three outcomes (at least one head) out of a total of four favorable to the event E_1. Thus the chance of this occurring, for now called the probability of the event and indicated by $P(\cdot)$, is 3 out of 4, or

$$P(\text{at least 1 head}) = P(E_1) = \tfrac{3}{4} \tag{2-1}$$

The final approach to probability to be discussed before introducing the axioms of probability is referred to as the *relative frequency* approach. One can best envision this approach by imagining a coin-toss chance experiment. If a coin is tossed repeatedly, we anticipate about half heads and half tails. This assumes, of course, that the coin and tosser are fair. However, even if there is some bias on the part of the coin or person tossing it, we can estimate the probability of a head by taking the ratio of the number, n_H, of heads that appear in a long sequence of tossings to the total number, n, of tossings. For an arbitrary event, A, of interest in a chance experiment, the relative frequency definition of the probability of the event is the limit

$$P(A) = \lim_{n\to\infty} \frac{n_A}{n} \tag{2-2}$$

where n_A is the number of occurrences of event A in n repetitions of the experiment. The problem with this definition, of course, is that the limit can never be reached. An example is provided in Figure 2-2, where 1000 tossings of a hypothetical coin are illustrated. The quantity on the ordinate is the ratio of the number of heads to the the total number of tossings up to that point. These ratios are connected by straight lines to make the curves easier to follow.

[2]The modifier "fair" is often used to denote the fact that no outcome is favored over any other. When applied to a single coin, it means that "head" and "tail" are equally likely outcomes. In the case of this example, the four outcomes "tail/tail," "tail/head," "head/tail," and "head/head" are assumed equally likely to occur. An example of an "unfair" experiment is to flip thumbtacks.

Plots are given for scales on the abscissa showing 100 tosses and 1000 tosses. This particular coin appears to be biased against heads if we perform about 25 tosses or less; at about 25 tosses, the outcome heads seems to be favored. Over the long term of 1000 tosses, the ratio of heads to total tosses appears to be approaching the limit of $\frac{1}{2}$; we will never be able to reach this limit with our chance experiment. Where do we stop? Obviously 100 tossings is not enough. Increasing to 1000 tossings is better. The dilemma with the relative frequency approach to probability is that we can never reach the limit defined by (2-2).

2-3 AXIOMS OF PROBABILITY

It is because of difficulties like those pointed out in Section 2-2 that we adopt a different approach to probability. As already mentioned, instead of stating what probability *is,* we give three properties that it must satisfy. Both the equally likely outcomes and relative frequency approaches to probability will fit within the axiomatic framework given in this section. To state the axioms of probability, we return to the idea of a sample space as illustrated by Figure 2-3. The probability of an event A in this sample space is by definition a number $P(A)$ that satisfies the following three axioms[3]:

1. $P(A)$ is a nonnegative number; that is,[4]

$$P(A) \geq 0 \tag{2-3}$$

2. The probability of the event S (i.e., the certain event) is unity:

$$P(S) = 1 \tag{2-4}$$

3. If two events A and B have no common outcomes (such events are said to be *disjoint* or *mutually exclusive*) as shown in Figure 2-3, the probability of the event $\{A \text{ or } B\}$ is the sum of their probabilities[5]:

$$P(A \text{ or } B) = P(A) + P(B) \tag{2-5}$$

The axioms of probability do not connect probability with physical events. We simply state at this point that if probabilities satisfy the axioms, a self-consistent theory of probability can be built. The connection with physical events must come about through one of the other approaches to probability discussed in Section 2-2. It is therefore necessary that the probabilities defined there satisfy the axioms of probability just given. For example, consider the relative frequency approach to probability. Imagine that we perform a chance experiment with two outcomes of interest, A and B, among others. In n repetitions

[3]Criteria that might be used to adopt a set of axioms are the following: (1) Are they consistent? (2) Are they useful? (3) Are they inclusive of other approaches to probability?

[4]Note that an event A is not impossible if its probability $P(A) = 0$. To see this, consider the relative frequency approach and a dart-throwing experiment; hitting a specific point has probability 0 but is not impossible.

[5]This axiom may be generalized in the case of a countably infinite number of events to

$$P(A_1 \text{ or } A_2 \text{ or } \ldots) = \sum_{i=1}^{\infty} P(A_i)$$

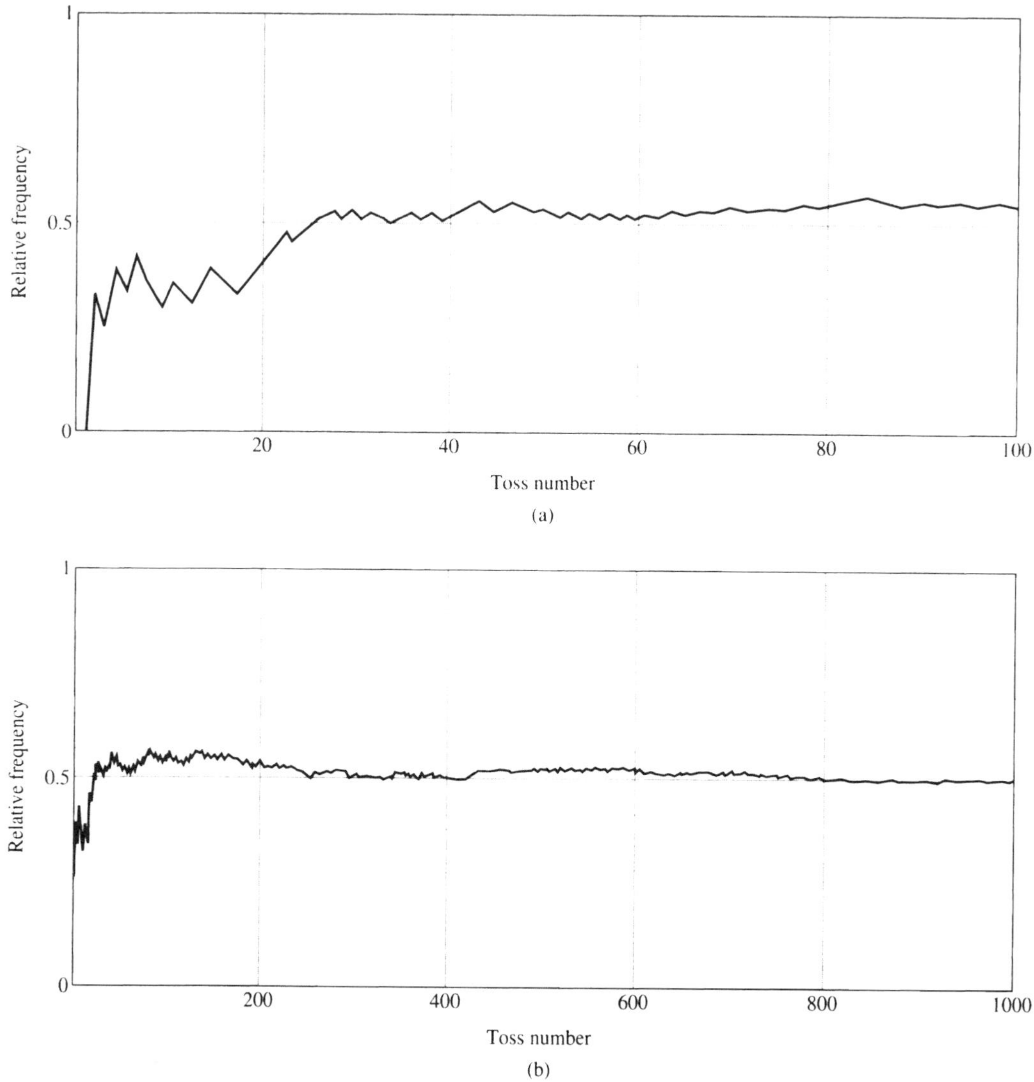

Figure 2-2 Relative frequency for a sequence of 1000 coin tossings: (a) first 100; (b) entire 1000.

of the chance experiment, suppose that event A occurs n_A times and event B occurs n_B times. Also suppose that A and B are mutually exclusive (i.e., cannot happen together). If n is sufficiently large, we have intuitive confidence that the respective probabilities of A and B can be approximated by n_A/n and n_B/n [i.e., the limit in (2-2) is dropped]. Both of these ratios are nonnegative, no matter what n is, so Axiom 1 is satisfied. If one of the events, say A, happens every time the experiment is performed, it is the sample space (i.e., it is the certain event), and the ratio n_A/n is unity, so Axiom 2 is satisfied. If we are interested in the

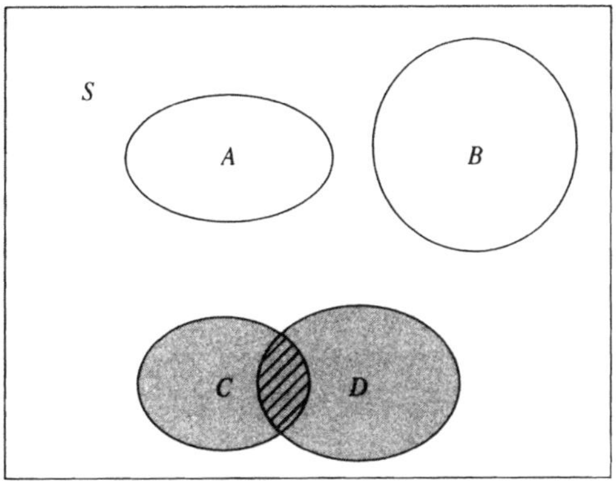

Figure 2-3 Sample space S illustrating two mutually exclusive events A and B and two events C and D which are not mutually exclusive.

event $\{A \text{ or } B\}$, we will observe it $n_A + n_B$ times since A and B cannot happen together by assumption. Therefore,

$$\begin{aligned} P(A \text{ or } B) &\simeq \frac{n_A + n_B}{n} \\ &= \frac{n_A}{n} + \frac{n_B}{n} \\ &\simeq p(A) + P(B) \end{aligned} \tag{2-6}$$

and Axiom 3 is satisfied. Thus the relative frequency approach to probability fits within the framework of the axiomatic approach. A similar demonstration could be carried out for the equally likely outcomes approach.

Example 2-2

Consider the rolling of a fair die. Event A is a six up on the die, and event B is a three up on the die. We take the probabilities of events A and B to both be $\frac{1}{6}$. (Which approach to probability are we using to infer this?) What is the probability of a three or a six?

Solution Since the events "three up" and "six up" are mutually exclusive (i.e., cannot happen together) the probability of a three or a six up is $\frac{1}{6} + \frac{1}{6} = \frac{1}{3}$ by Axiom 3. Since we really used the equally likely approach to infer the probabilities of events A and B as being $\frac{1}{6}$, we again use it to check our calculation of the event $\{3 \text{ or } 6 \text{ up}\}$. There are two outcomes favorable to this event out of six equally likely ones for a probability of $\frac{2}{6} = \frac{1}{3}$ for the event $\{3 \text{ or } 6 \text{ up}\}$.

2-4 SET THEORY

To develop further relations involving probabilities, it is convenient to introduce the notation of set theory. Sets are collections of objects, with the objects in the sets called *elements*. The elements of a set are often placed within braces. The following are examples of sets:

1. $S_1 = \{1, 3, 5, \ldots\}$ is the set of all nonnegative, odd integers.
2. $S_2 = \{0, 2, 4, \ldots\}$ is the set of all nonnegative, even integers.
3. $S_3 =$ {ace of clubs, ace of spades, ace of diamonds, ace of hearts} is the set of all aces in a standard 52-card deck.
4. $S_4 =$ {head, tail} is the set of possible outcomes in tossing a coin.
5. $S_5 = \{hh, ht, th, tt\}$ indicates the set of possible outcomes in tossing a pair of coins.

In keeping with our notation introduced for sample spaces, an arbitrary element of a set will be denoted by a lowercase Greek letter. Sets themselves will be denoted by uppercase Roman italic letters. We indicate that the element ζ_i is an element of the set A (or belongs to A) by the notation

$$\zeta_i \in A \tag{2-7}$$

The notation

$$\zeta_j \notin A \tag{2-8}$$

means that ζ_j is not contained in, or is not a member of, the set A.

To say that a set B is a *subset* of a set A means that every element of B is contained in A and is indicated by the notation

$$B \subset A \tag{2-9}$$

Two sets are *equal* if they contain the same elements. Equality of sets, A and B, can be established by showing that every element of B is an element of A, and vice versa. If B is a subset of A but is not equal to A, we say that B is a *proper* subset of A.

The set containing no elements is called the *null set* and is denoted by $\varnothing$. We will usually be dealing with a set of objects of interest called the *universal set*, or *universe*, which will be denoted by S. An example is the set of all nonnegative integers of which the sets S_1 and S_2 defined in the first part of this section are subsets.

A convenient way of pictorially representing sets and the relationships between them is by means of a Venn diagram, as introduced in Figure 2-1. It should be clear at this point that the set notation discussed here is identical to the idea of a sample space, with the events and outcomes of the sample space corresponding to subsets and their elements, respectively.

Operations with sets include *union*, *intersection*, and *complement*. The union of two sets C and D is a set E that contains the elements of C, or D, or both, and is indicated by the notation

$$E = C \cup D \quad \text{or} \quad E = C + D \tag{2-10}$$

For example, in Figure 2-3 the union of the sets C and D is indicated by the total shaded area of the two ellipses denoting C and D. In the subset examples given above, the union of S_1 and S_2 is the set of all nonnegative integers.

The *intersection* of two sets C and D is a set F that contains only elements that are in *both* C and D, and is indicated by the notation

$$F = C \cap D \quad \text{or} \quad F = CD \tag{2-11}$$

The intersection of the sets C and D in Figure 2-3 is indicated by their common area, which is crosshatched. The intersection of the sets S_1 and S_2 given previously is the null set (i.e., there are no common integers between the set of nonnegative odd integers and the set of nonnegative even integers).

The *complement* of a set A, denoted A^c (also sometimes denoted $\bar{A}$ or A'), consists of all the elements in the universe not in A. For example, if the universal set is the set of all nonnegative integers, S_1 and S_2 defined previously are complements of each other.

Two sets A and B are said to be *disjoint* if their intersection is the null set. That is,

$$A \cap B = \varnothing \text{ (disjoint sets)} \tag{2-12}$$

The sets S_1 and S_2 defined at the beginning of this section are disjoint. Similarly, the sets A and B of Figure 2-3 are disjoint.

Sets can be countable or uncountable. A set is *countable* if its elements can be put in one-to-one correspondence with the positive integers; otherwise, it is uncountable. Examples of countable sets are {all even integers} and {all rational numbers}. The set of real numbers within a finite interval, no matter how small the interval, is *uncountable.* A random experiment with a countably infinite set of outcomes is to flip a fair coin until the first head appears. The universal set in this case is $\{h, th, tth, ttth, \ldots\}$, with the sequence of probabilities being $\{\frac{1}{2}, \frac{1}{4}, \frac{1}{8}, \frac{1}{16}, \ldots\}$. Note that the sum of the probabilities of all possible outcomes is 1 by using the fact that $p/(1 - p) = p + p^2 + \cdots$. (To see that these are the probabilities, consider tth: there are eight possible outcomes if the coin is tossed three times, only one of which is tth, so the probability of tth is $\frac{1}{8}$. Similar reasoning can be used for the remaining probabilities.)

Certain relationships involving the universal set and the null set follow from the definitions of union and intersection. They are

$$A \cup S = S, \quad A \cup \varnothing = A, \quad A \cap S = A, \quad A \cap \varnothing = \varnothing \tag{2-13}$$

Also, if $B \subset A$, it follows that

$$A \cup B = A \quad \text{and} \quad A \cap B = B \tag{2-14}$$

which is obvious by use of a Venn diagram as sketched in Figure 2-4a. It is also clear by use of a Venn diagram that if $A \subset B$ and $B \subset C$, then $A \subset C$, as illustrated in Figure 2-4b.

If union is viewed as addition and intersection is viewed as multiplication, set operations obey the same commutative, associative, and distributive laws as do the corresponding arithmetic operations. In particular, these properties are:

$$\begin{aligned} A \cup B &= B \cup A \\ A \cap B &= B \cap A \end{aligned} \quad \text{(commutative property)} \tag{2-15}$$

$$\begin{aligned} (A \cup B) \cup C &= A \cup (B \cup C) \\ (A \cap B) \cap C &= A \cap (B \cap C) \end{aligned} \quad \text{(associative property)} \tag{2-16}$$

$$(A \cup B) \cap C = (A \cap C) \cup (B \cap C) \quad \text{(distributive law)} \tag{2-17}$$

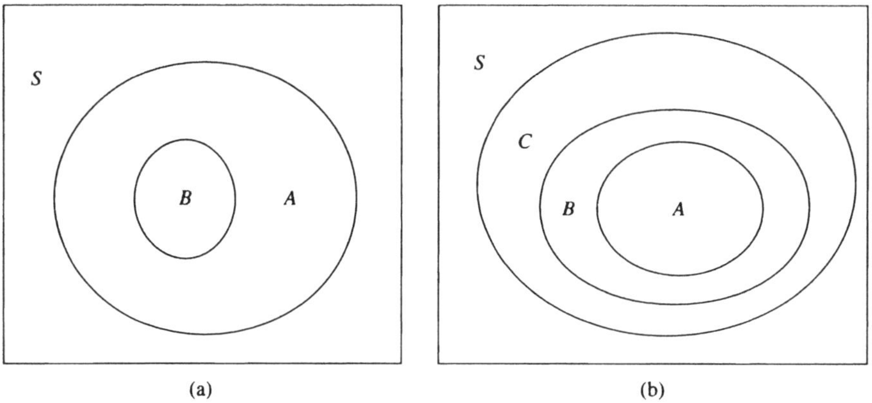

Figure 2-4 Venn diagrams for illustrating set relationships: **(a)** $B \subset A \subset S$; **(b)** $A \subset B \subset C \subset S$.

The student should construct a pair of Venn diagrams, one for each side of the equation, to demonstrate the validity of (2-17).

Example 2-3

Consider the tossing of two coins as in Example 2-1. Let E_1 be defined as in that example (i.e., $E_1 = \{h_1h_2, h_1t_2, t_1h_2\}$), and let E_2 denote a match, or $E_2 = \{h_1h_2, t_1t_2\}$. Describe the events E_1^c, $E_1 \cup E_2$, $E_1 \cap E_2$, and $(E_1 \cup E_2)^c$.

Solution $E_1^c = \{t_1t_2\}$, $E_1 \cup E_2 = \{h_1h_2, h_1t_2, t_1h_2, t_1t_2\}$, $E_1 \cap E_2 = \{h_1h_2\}$, and $(E_1 \cup E_2)^c = \varnothing$.

A *partition* of S is a collection of subsets $A_1, A_2, \ldots, A_m$ with the properties that all are disjoint and their union equals S. That is,

$$A_1 \cap A_2 = \varnothing \qquad (\text{all } i \text{ and } j,\ i \neq j) \tag{2-18}$$

$$A_1 \cup A_2 \cup \ldots \cup A_m = S$$

A partition of a sample space S is illustrated by the Venn diagram of Figure 2-5. The members of a partition are often said to be *mutually exclusive* and *collectively exhaustive.*

The *Cartesian product* of two sets A and B with elements α_i and β_j, respectively, is a new set C whose elements are all possible pairs $\alpha_i\ \beta_j$ and is denoted by $C = A \times B$.

Example 2-4

Consider the tossing of two coins. Separately, the possible outcomes are the sets $A = \{h_1, t_1\}$ and $B = \{h_2, t_2\}$, where the subscripts refer to the first and second coins, respectively. Their Cartesian product is the set $C = \{h_1h_2, h_1t_2, t_1h_2, t_1t_2\}$. Note that the outcomes listed for C form a partition.

Finally, a useful pair of relationships is *De Morgan's laws,* which may be stated as

$$\overline{A \cup B} = \overline{A} \cap \overline{B} \quad \text{and} \quad \overline{A \cap B} = \overline{A} \cup \overline{B} \tag{2-19}$$

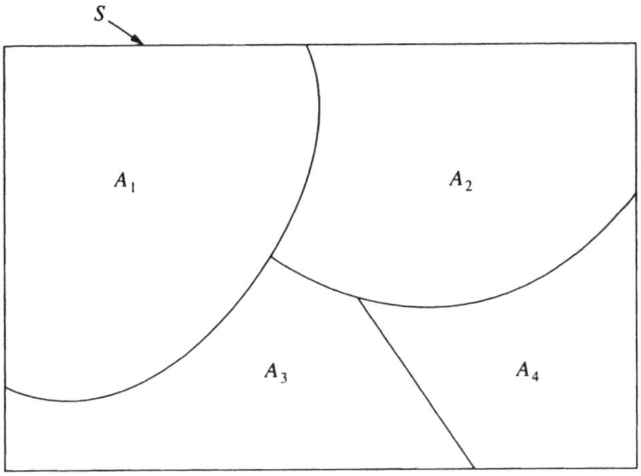

Figure 2-5 Partition of a sample space S into four events.

A demonstration of the validity of these relationships may be carried by the use of Venn diagrams (Problem 2-8).

2-5 DERIVED PROBABILITY RELATIONSHIPS

With the introduction of the set theory notation in Section 2-4, we are now in a position to derive some probability relationships from the axioms stated in Section 2-3. First, from a Venn diagram, it follows that for any event A,

$$A \cup A^c = S \quad \text{and} \quad A \cap A^c = \varnothing \tag{2-20}$$

Thus, since A and A^c are disjoint and their union is S (i.e., form a partition),

$$1 = P(S) = P(A \cup A^c) = P(A) + P(A^c) \qquad \text{(Axioms 2 and 3)}$$

or

$$P(A^c) = 1 - P(A) \tag{2-21}$$

From Axiom 1, it follows that $P(A) \leq 1$. A special case of (2-21) occurs with $A = S$, which results in (note that $S^c = \varnothing$)

$$P(\varnothing) = 1 - P(S) \quad \text{or} \quad P(\varnothing) = 0 \tag{2-22}$$

which follows from Axiom 2.

Finally, we generalize Axiom 3 to the case where A and B are not disjoint. Since Axiom 3 applies only to disjoint sets, we write the union of A and B as

$$A \cup B = A \cup (B \cap A^c) \tag{2-23}$$

This follows by sketching a Venn diagram, as shown in Figure 2-6a, or using the distributive relationship on the right-hand side of (2-23).

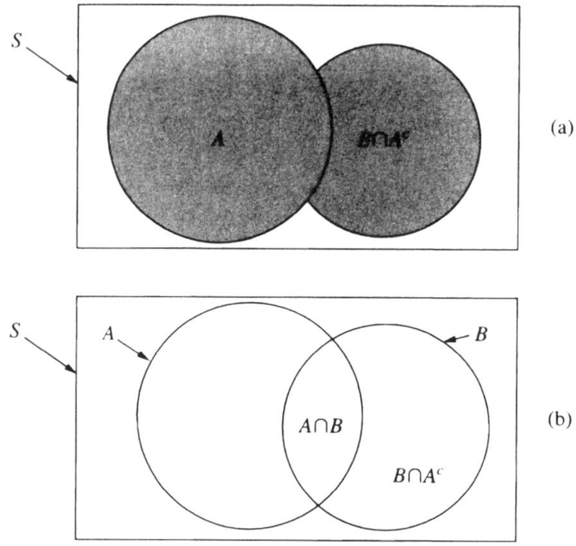

Figure 2-6 Venn diagrams for generalizing the probability relationship for the union of two events.

It is also clear from the Venn diagram of Figure 2-6a that the sets A and $B \cap A^c$ are disjoint. Thus

$$P(A \cup B) = P(A) + P(B \cap A^c) \tag{2-24}$$

Note also from the Venn diagram of Figure 2-6b that the sets $A \cap B$ and $B \cap A^c$ are disjoint and that their union equals B. Thus, from Axiom 3, it follows that

$$P(A \cap B) + P(B \cap A^c) = P(B) \tag{2-25}$$

Solving (2-25) for $P(B \cap A^c)$, substituting into (2-24), and rearranging, we obtain the relationship

$$P(A \cup B) = P(A) + P(B) - P(A \cap B) \tag{2-26}$$

which holds for any pair of events A and B, whether or not they are disjoint. The following examples illustrate the use of (2-26).

Example 2-5

Consider the tossing of two fair coins as in Example 2-1. Let E_1 be the event that at least one head shows (i.e., h_2t_2, t_1h_2, or h_1h_2), and E_4 be the event of a match (i.e., t_1t_2 or h_1h_2). Compute the probability of at least one head and a match or both.

Solution The probability of E_1 was computed in Example 2-1 to be $\frac{3}{4}$. The probability of a match or E_4 is $\frac{1}{2}$ since there are two equally likely outcomes (t_1t_2 or h_1h_2) of the experiment favorable to this event. The intersection of the events E_1 and E_4 is the outcome h_1h_2, so its probability is $\frac{1}{4}$. Thus (2-23) becomes

$$P(E_1 \cup E_4) = P(E_1) + P(E_4) - P(E_1 \cap E_4) = \tfrac{3}{4} + \tfrac{1}{2} - \tfrac{1}{4} = 1 \tag{2-27}$$

which is what it should be since the events "at least one head" and "match" encompass all the possible outcomes of the chance experiment.

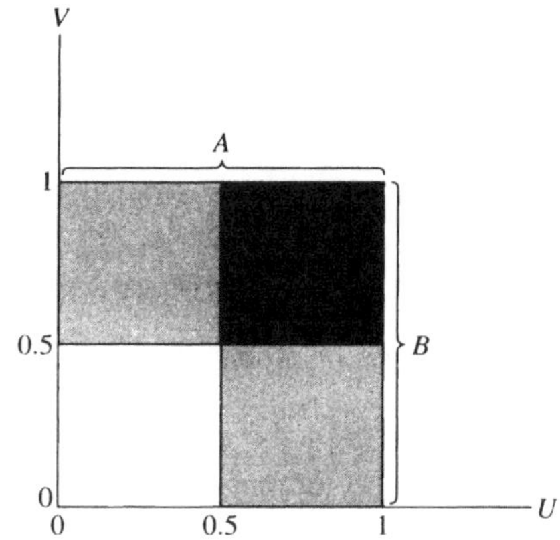

Figure 2-7 Venn diagram illustrating the calculation of the probability of the union of two events, A and B, that are not disjoint.

Example 2-6

Consider the joint occurrence of two events defined as follows. A random number generator yields a number, denoted as U, equally likely to be anywhere in the interval $[0, 1]$. Similarly, a second random number generator yields a number, denoted as V, equally likely to be any place in the interval $[0, 1]$ and independently (i.e., the numbers generated for U have no influence on those generated for V) of the other number generator. Two events A and B are defined as follows:

$$A = \{0 \le U \le 1, 0.5 \le V \le 1\}$$
$$B = \{0.5 \le U \le 1, 0 \le V \le 1\} \qquad (2\text{-}28)$$

Find $P(A \cup B)$.

Solution A Venn diagram showing the events A and B is given in Figure 2-7. From this Venn diagram it follows that

$$P(A) = \tfrac{1}{2}, \quad P(B) = \tfrac{1}{2}, \quad P(A \cap B) = \tfrac{1}{4} \qquad (2\text{-}29)$$

It therefore follows from (2-26) that

$$P(A \cup B) = \tfrac{1}{2} + \tfrac{1}{2} - \tfrac{1}{4} = \tfrac{3}{4} \qquad (2\text{-}30)$$

This is obvious from the Venn diagram. Since the ordered pair of numbers (U, V) is equally likely to be any place in the unit square (this is a result of the independence assumption), the area of the shaded region (S minus the lower left-hand corner of the unit square) is equal to the desired probability. Its area of $\frac{3}{4}$ is equal to the desired probability computed according to (2-26).

2–6 CONDITIONAL PROBABILITIES AND STATISTICAL INDEPENDENCE

Consider two events, A and B, with $P(A)$ and $P(B) > 0$. We define the *conditional probability* of event A given that event B has occurred as

$$P(A|B) = \frac{P(A \cap B)}{P(B)} \qquad (2\text{-}31)$$

and the conditional probability of event B given that event A has occurred as

$$P(B|A) = \frac{P(A \cap B)}{P(A)} \tag{2-32}$$

Note that by solving both (2-31) and (2-32) for $P(A \cap B)$ and equating, we obtain the relationship

$$P(A \cap B) = P(B|A)P(A) = P(A|B)P(B) \tag{2-33}$$

Conditional probabilities also satisfy the axioms of probability and the derived relationships given in Section 2-3 and 2-5.

A glance at a Venn diagram, as given in Figure 2-8, shows that the conditional probability of event A given B is really the probability of that event conditioned on a smaller sample space, in particular that corresponding to event B. Thus $P(A \mid B)$ is illustrated by the crosshatched area in Figure 2-8 divided by the area of B. A similar statement holds for $P(B \mid A)$.

Example 2-7

What is the conditional probability of obtaining two heads when flipping a coin twice given that at least one head was obtained?

Solution The sample space is $\{t_1t_2, t_1h_2, h_1t_2, h_1h_2\}$. Let at least one head be denoted by event A, and two heads be denoted by event B. Using equal likelihood, $P(A) = \frac{3}{4}$ and $P(A \cap B) = \frac{1}{4}$. Thus

$$P(2 \text{ heads given at least 1 head}) = P(B|A) = \frac{P(A \cap B)}{P(A)} = \frac{1/4}{3/4} = \frac{1}{3}$$

Given that the event A has occurred, we now have a universe of three events, one of which is favorable.

Example 2-8

As in Example 2-6, consider the two random numbers U and V. Now, however, in addition to A we consider events C and D, defined as $C = \{U > V\}$ and $D = \{0 \le U \le 1, 0 \le V \le 0.5\}$. Find the conditional probabilities $P(A|C)$ and $P(D|C)$.

Solution A Venn diagram showing the events A, C, and D is given in Figure 2-9. From this figure it follows that

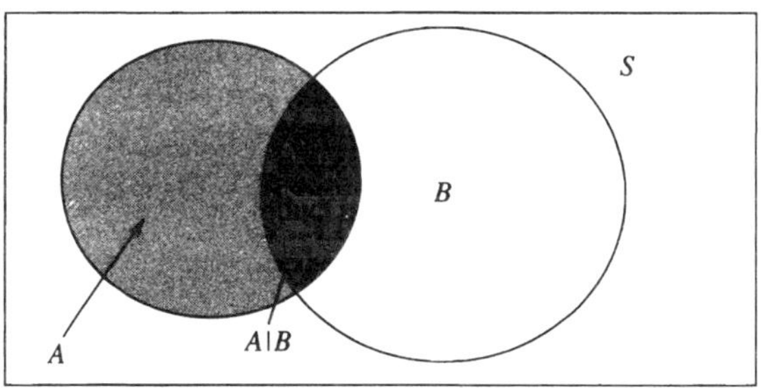

Figure 2-8 Venn diagram illustrating conditional probability.

$$P(A) = P(C) = P(D) = 0.5$$

$$P(A \cap C) = 0.125 \tag{2-34}$$

$$P(D \cap C) = 0.375$$

From (2-31) with B replaced by C, it follows that

$$P(A|C) = \frac{P(A \cap C)}{P(C)} = \frac{0.125}{0.5} = 0.25 \tag{2-35}$$

and again from (2-31) with A replaced by D and B replaced by C, it follows that

$$P(D|C) = \frac{P(D \cap C)}{P(C)} = \frac{0.375}{0.5} = 0.75 \tag{2-36}$$

In the first case $P(A) = 0.5 > P(A \mid C)$, and in the second case $P(D) < P(D \mid C)$. It could, of course, happen that conditioning on a second event leaves the probability of a given event the same, as in the case of Example 2-5, where $P(A \mid B) = P(A)$. This leads to the next topic of statistical independence.

Definition. Two events A and B are said to be *statistically independent* if

$$P(A \cap B) = P(A)P(B) \qquad \text{(statistically independent events)} \tag{2-37}$$

From (2-33) it follows that for statistically independent events

$$P(A|B) = P(A) \qquad \text{and} \qquad P(B|A) = P(B) \tag{2-38}$$

In simple terms, two events are statistically independent if the occurrence or nonoccurrence of one has no influence over the occurrence or nonoccurrence of the other. Events A and B of Example 2-6 are statistically independent, whereas events A and C of Example 2-8 are not.

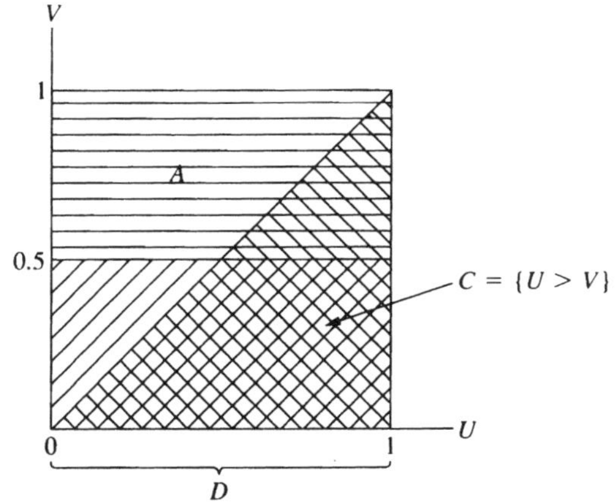

Figure 2-9 Venn diagram for computing the probabilities of Example 2-8.

Example 2-9

Reconsider Example 2-5. Are the events that at least one head shows (E_1) and a match (E_4) statistically independent?

Solution From Example 2-4 we have that $P(E_1) = \frac{3}{4}$, $P(E_4) = \frac{1}{2}$, and $P(E_1 \cap E_4) = \frac{1}{4}$. Note that $P(E_1)\,P(E_4) = \frac{3}{8} \neq P(E_1 \cap E_4)$. Thus E_1 and E_4 are not statistically independent. As a further check, the conditional probabilities can be computed using (2-31) and (2-32), with appropriate events substituted, as

$$P(E_1|E_4) = \frac{P(E_1 \cap E_4)}{P(E_4)} = \frac{1/4}{1/2} = \frac{1}{2}$$

and (2-39)

$$P(E_4\,|\,E_1) = \frac{P(E_1 \cap E_4)}{P(E_1)} = \frac{1/4}{3/4} = \frac{1}{3}$$

Since $P(E_1) \neq P(E_1|E_4)$ and $P(E_4) \neq P(E_4|E_1)$, this again shows that these events are not statistically independent.

2-7 TOTAL PROBABILITY AND BAYES' THEOREM

Let the events $A_1, A_2, \ldots, A_m$ be a partition of a sample space S, where it is recalled that to be a partition, the events are mutually exclusive and their union equals the sample space. Consider an arbitrary event B also in the sample space S. It follows that (recall that $B \cap S = B$)

$$\begin{aligned} B = B \cap S &= B \cap (A_1 \cup A_2 \cup \cdots \cup A_m) \\ &= (B \cap A_1) \cup (B \cap A_2) \cup \cdots \cup (B \cap A_m) \end{aligned} \tag{2-40}$$

It is clear from Figure 2-10 that since the events A_i are all mutually exclusive (disjoint), the events $B \cap A_i$ are also mutually exclusive. Thus

$$P(B) = P(B \cap A_1) + P(B \cap A_2) + \cdots + P(B \cap A_m) \tag{2-41}$$

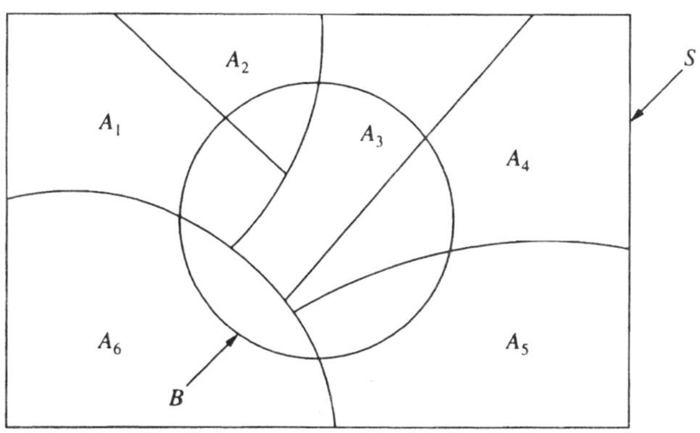

Figure 2-10 Partition of a sample space S and an arbitrary event B for the purpose of obtaining the theorem on total probability.

Since the probability of B is obtained by summing over all the joint probabilities of B with each of the A_i's, $P(B)$ is often referred to as a *marginal probability.*

In accordance with (2-33), each of the probabilities $P(B \cap A_i)$ can be expressed as

$$P(B \cap A_i) = P(B|A_i)P(A_i) \tag{2-42}$$

where $P(B|A_i)$ is the conditional probability of event B given that event A_i has occurred. Thus (2-41) becomes

$$P(B) = \sum_{i=1}^{m} P(B \cap A_i) = \sum_{i=1}^{m} P(B|A_i)P(A_i) \tag{2-43}$$

which is known as the *theorem of total probability.* The use of this theorem is illustrated by the following example.

Example 2-10

Two boxes of coins contain nickels and quarters; box 1 has eight nickels and two quarters and box 2 has 20 quarters and five nickels. A box is selected at random, and one of its coins is selected at random. Find the probability that the coin selected is a quarter.

Solution Let $P(B_1) = P(B_2) = \frac{1}{2}$ be the probabilities of selecting box 1 or box 2, respectively. Using the equally likely approach to probability, the probabilities of selecting a quarter from box 1 and box 2 are

$$P(Q|B_1) = \tfrac{1}{5} \quad \text{and} \quad P(Q|B_2) = \tfrac{4}{5} \tag{2-44}$$

respectively. Note that $Q \cap B_1$ and $Q \cap B_2$ are mutually exclusive events. Therefore, using (2-43), we find that the probability of a quarter being selected is

$$\begin{aligned} P(Q) &= P(Q|B_1)P(B_1) + P(Q|B_2)P(B_2) \\ &= \tfrac{1}{5} \cdot \tfrac{1}{2} + \tfrac{4}{5} \cdot \tfrac{1}{2} = \tfrac{1}{2} \end{aligned} \tag{2-45}$$

That is, the probability of picking at random a coin from a randomly selected box results in a quarter 50% of the time. Note that this is a different answer than if one considers the probability of randomly selecting a coin from either box (by the relative frequency approach to probability, this is the ratio of the number of quarters to the total number of coins, or $\frac{22}{35}$).

Exercise. Repeat Example 2-10 for the case of $P(B_1) = p$ and $P(B_2) = q = 1 - p$. Plot the probability of getting a quarter versus p.

We now turn to another very useful relationship which makes use of the theorem on total probability. This relationship is known as *Bayes' theorem.* To obtain a special case of Bayes' theorem, we solve (2-33) for $P(B|A)$ to obtain

$$P(B|A) = \frac{P(A|B)P(B)}{P(A)} \tag{2-46}$$

To illustrate the use of (2-46), consider the following example.

Example 2-11

Assume that a certain class is given a midterm examination. The probability of a student studying for the examination is 0.7. Of those students who study, the probability of their passing the examination is 0.9; if a student does not study, his or her probability of passing is 0.05. Given that a student did not pass the examination, what is the probability that he or she studied?

Solution Let S denote the event that a student studied and P denote the event that a student passed the examination. Then, by Bayes' theorem,

$$P(S|P^c) = \frac{P(P^c|S)P(S)}{P(P^c)} \tag{2-47}$$

From the data given for the problem, we have $P(S) = 0.7$ and $P(P^c|S) = 1 - P(P|S) = 1 - 0.9 = 0.1$. To obtain the denominator of (2-47), we use the theorem on total probability, which is

$$\begin{aligned} P(P^c) &= P(P^c|S)P(S) + P(P^c|S^c)P(S^c) \\ &= (0.1)(0.7) + (0.95)(0.3) = 0.355 \end{aligned} \tag{2-48}$$

Thus the probability that a student studied given that he or she did not pass is

$$P(S\,|\,P^c) = \frac{0.07}{0.355} = 0.197. \tag{2-49}$$

The generalization of Bayes' theorem to the partition of a sample space into m disjoint events, $A_1, A_2, \ldots, A_m$, is

$$P(A_i\,|\,B) = \frac{P(B\,|\,A_i)P(A_i)}{\sum_{j=1}^{m} P(B\,|\,A_j)P(A_j)}, \quad 1 \leq i \leq m \tag{2-50}$$

where (2-43) has been used to express $P(B)$ in terms of the conditional probabilities $P(B|A_j), j = 1, 2, \ldots, m$.

Example 2-12

Manufactured devices are tested for defects. Hopefully, the devices should fail (F) if a defect (D) is identified. Similarly, a device should not fail (F^c) if the test indicates no defect (D^c). However, it is possible for a given device having no defect to fail the test (called a *type I error*), and for a device having a defect to pass the test (called a *type II error*). Given the following probabilities,

$$\begin{aligned} P(D^c) &= 0.95 \\ P(F|D) &= 0.98 \\ P(F|D^c) &= 0.03 \end{aligned} \tag{2-51}$$

what are the probabilities that a device is defective if it failed the test, and that a device is not defective if it did not fail the test?

Solution We apply Bayes' theorem. Using (2-21), we compute

$$\begin{aligned} P(D) &= 1 - P(D^c) = 1 - 0.95 = 0.05 \\ P(F^c|D) &= 1 - P(F|D) = 1 - 0.98 = 0.02 \\ P(F^c|D^c) &= 1 - P(F|D^c) = 1 - 0.03 = 0.97 \end{aligned} \tag{2-52}$$

By the theorem on total probability (2-43), we obtain

$$\begin{aligned} P(F) &= P(F|D)P(D) + P(F|D^c)P(D^c) \\ &= (0.98)(0.05) + (0.03)(0.95) = 0.078 \end{aligned} \tag{2-53}$$

Bayes' Theorem gives us

$$\begin{aligned} P(D|F) &= \frac{P(F|D)P(D)}{P(F)} \\ &= \frac{(0.98)(0.05)}{0.078} = 0.632 \end{aligned} \tag{2-54}$$

Similarly, we calculate

$$\begin{aligned} P(D^c|F^c) &= \frac{P(F^c|D^c)P(D^c)}{P(F^c)} \\ &= \frac{(0.97)(0.95)}{0.922} = 0.999 \end{aligned} \tag{2-55}$$

where we have used $P(F^c) = 1 - P(F) = 1 - 0.078 = 0.922$. The probabilities of type I and II errors are, respectively,

$$P(F|D^c) = 1 - P(F^c|D^c) = 1 - 0.999 = 0.001$$

and (2-56)

$$P(F^c|D) = 1 - P(F|D) = 1 - 0.98 = 0.02$$

Example 2-13

A good application of Bayes' Theorem is the *Monte Hall problem,* which may be stated as follows. A game show contestant is to choose among three curtains behind two of which are goats, with a car being behind the third. The contestant makes a selection whereupon the game show host opens one of the remaining two curtains to reveal a goat. Should the contestant stay with his or her choice, or switch to the other unopened curtain?

Solution Before observing the goat behind the curtain opened by the game show host, the probability of selecting the car is $\frac{1}{3}$ and the probability of selecting a goat is $\frac{2}{3}$. After seeing the goat behind the curtain opened by the game show host, the contestant has additional information. The probability of having selected a car (SC), given the revealed goat (GR) is, by Bayes' Theorem,

$$P(SC|GR) = \frac{P(GR|SC)P(SC)}{P(GR)}$$

By the theorem on total probability, the denominator is given by

$$P(GR) = P(GR|SC)P(SC) + P(GR|SG)P(SG)$$

where SG stands for "goat selected." The conditional probabilities above depend on whether or not the game show host makes a random selection between the unchosen curtains or uses prior knowledge so as always to reveal a goat. They are:

$$P(GR|SC) = 1, \quad P(GR|SG) = \tfrac{1}{2} \quad \text{(host randomly selects)}$$

$$P(GR|SC) = 1 = P(GR|SG) \quad \text{(host uses prior knowledge)}$$

Thus

$$P(GR) = 1 \times \tfrac{1}{3} + \tfrac{1}{2} \times \tfrac{2}{3} = \tfrac{2}{3} \text{ (random selection)}, \quad P(GR) = 1 \text{ (prior knowledge)}$$

$$P(\mathrm{SC}|\mathrm{GR}) = \begin{cases} \dfrac{1 \times \frac{1}{3}}{\frac{2}{3}} = \dfrac{1}{2} & \text{(random selection)} \\ \dfrac{1 \times \frac{1}{3}}{1} = \dfrac{1}{3} & \text{(prior knowledge)} \end{cases}$$

Therefore, the contestant is as well off staying with the original choice as with switching if the host selects the unchosen curtain randomly, but should switch curtains if the host uses prior knowledge so as always to reveal a goat.

2–8 COUNTING TECHNIQUES

In using the equally likely outcomes approach to probability, it is necessary to determine the ratio of the number of favorable outcomes corresponding to the event of interest to the total number of outcomes of the chance experiment. Various ways of counting these outcomes are necessary, and these are discussed in this section.

Multiplication Principle

Often, we can imagine filling slots with objects as illustrated by the following example.

Example 2-14

How many different security codes are there for a locking briefcase with a combination lock having three thumbwheels, with each thumbwheel numbered 0 to 9?

Solution There are 10 ways we can choose the first number (i.e., the digits 0, 1, . . . , 9), times 10 ways we can choose the second number, times 10 ways we can choose the third number—for a total of $10 \times 10 \times 10 = 1000$ different combinations.

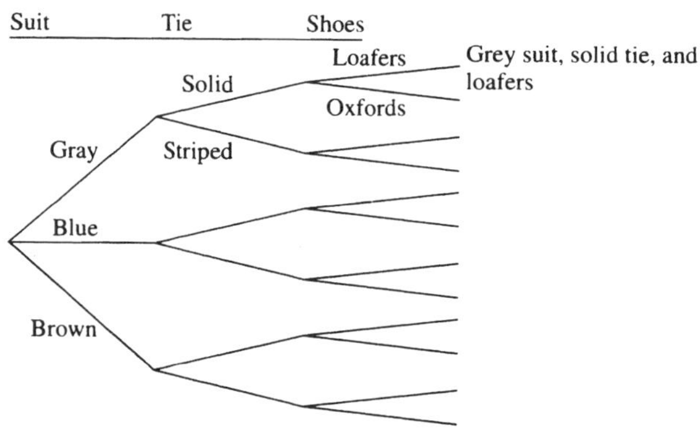

Figure 2-11 Possible clothing selections.

More complex situations arise if the number of possibilities at each selection vary, for example, the number of different clothes selections with three different-colored suits, two different designs of tie, and two selections of shoes. For such cases, a tree diagram is useful as illustrated in Figure 2-11. When diagrammed in this way, it is obvious that there are $3 \times 2 \times 2 = 12$ different selections.

The next problem is that of making selections from a set of objects if a selection means taking one object away. An example of this is the number of ways in which we can draw a five-card hand containing all hearts *in a given order* from a 52-card deck. The first heart can be any of 13 possibilities (13 hearts in a standard deck), the second heart can be any of 12 possibilities (one of the 13 hearts has already been taken out of the deck), and so on, for a total of

$$13 \times 12 \times 11 \times 10 \times 9 = \frac{13!}{(13\text{–}5)!} = \frac{13!}{8!} \tag{2-57}$$

Permutations

A *permutation* is the number of ways we can line up a set of objects. We imagine filling slots or boxes. Once we have filled a slot with an object, we cannot use it again. Thus, if we have n objects, we can permute them $n!$ ways. We can have a permutation of size $m < n$, where n is the total number of objects from which we make our selection. The number of possible permutations of n objects taken m at a time is

$$P_m^n = \frac{n!}{(n - m)!} \tag{2-58}$$

Example 2-15

How many different five-letter computer passwords are there if the same letter cannot be repeated?

Solution This is the number of permutations of size five from a 26-object set, or

$$\frac{26!}{(25 - 5)!} = 26 \times 25 \times 24 \times 23 \times 22 \times 21 = 165{,}765{,}600 \tag{2-59}$$

Combinations

A *combination* is a permutation in which order is of no consequence. For example, for each of the permutations of five-letter passwords of Example 2-15, there are $5 \times 4 \times 3 \times 2 \times 1 = 5!$ containing the same letters in a different order. Thus, if we are not interested in the order of the letters, we divide (2-59) by 5!. This is the number of combinations of 26 letters taken five at a time. The number of combinations of n objects taken k at a time is

$$C_k^n = \binom{n}{k} = \frac{n!}{(n - k)k!} \tag{2-60}$$

This is the *binomial coefficient of algebra*[6]; in a counting context it is called *n choose k*.

Example 2-16

What is the probability of drawing a flush (i.e., five cards all of the same suit) in five-card stud poker?

Solution There are four possible suits, with each suit having the number of possible five-card hands given by

$$\binom{13}{5} = \frac{13!}{5! \times 8!} \tag{2-61a}$$

The total number of possible five-card hands in a 52-card deck is

$$\binom{52}{5} = \frac{52!}{5! \times 47!} \tag{2-61b}$$

Thus, using the equally likely definition of probability, we have

$$\begin{aligned} P(\mathit{flush}) &= \frac{4 \times \dfrac{13!}{5! \times 8!}}{\dfrac{52!}{5! \times 47!}} = \frac{4 \times 13! \times 5! \times 47!}{5! \times 8! \times 52!} \\ &= \frac{4 \times 13 \times 12 \times 11 \times 10 \times 9}{52 \times 51 \times 50 \times 49 \times 48} = 0.001981 \end{aligned} \tag{2-62}$$

Several useful properties of the binomial coefficient are

$$\binom{n}{k} = \binom{n}{n-k}, \quad \binom{n}{k} = \binom{n-1}{k} + \binom{n-1}{k-1}, \quad \binom{n}{0} = \binom{n}{n} = 1, \quad \binom{n}{1} = n \tag{2-63}$$

If we flip a coin n times in a row and count the number of ways that k heads can occur without regard to order, this is given by the binomial coefficient. If the probability of a head is p, the probability of a particular sequence of k heads and $n - k$ tails is $p^k(1 - p)^{n-k}$. Thus the probability of *any sequence* of k heads and $n - k$ tails is

$$\binom{n}{k} p^k (1 - p)^{n-k} \tag{2-64}$$

This is called the *binomial probability distribution* and is discussed further in Chapter 3.

[6]Recall that a handy way to compute the binomial coefficients is with *Pascal's triangle.* Begin with a 1 and place below it a 1 on either side. The next row is computed by placing 1's on the outside and adding the two numbers above to produce the middle one. Each row is constructed is a similar way to produce the array

$$\begin{array}{ccccccc} & & & 1 & & & \\ & & 1 & & 1 & & \\ & 1 & & 2 & & 1 & \\ 1 & & 3 & & 3 & & 1 \end{array}$$

MATLAB will construct Pascal's triangle with the statement pascal(n), where n is the size of the desired array. It is tipped 60 degrees counterclockwise, with 1's along the first row and down the first column.

Finally, consider the *multinomial coefficient,* which arises when dividing n distinct objects into k classes of sizes $n_1, n_2, \ldots, n_k$, where $\sum n_i = n$. For the first class there are $\binom{n}{n_1}$ possible choices. For each choice in the first class, there are $\binom{n-1}{n_2}$ possible choices, and so on. Using the multiplication principle, the total number of divisions is

$$\binom{n}{n_1}\binom{n-n_1}{n_2}\binom{n-n_1-n_2}{n_3}\cdots\binom{n-n_1-\cdots-n_{k-1}}{n_k} = \frac{n!}{n_1!\,n_2!\cdots n_k!} \tag{2-65}$$

Example 2-17

In a class of 25 students, the instructor wants a group of 10 to do project 1, a group of eight to do project 2, and a group of seven to do project 3. How many possible ways can the instructor assign the students to the three projects?

Solution Using the multinomial coefficient justified above, the total number is

$$\frac{25!}{10!\,8!\,7!} = 956{,}112{,}300$$

As a final example, we present one that combines the multiplication principle and the idea of combinations.

Example 2-18

What is the probability of obtaining three aces and two kings in a five-card poker hand?

Solution The number of possible hands containing three aces is

$$\binom{4}{3} = \frac{4!}{3! \times 1!}$$

The number of possible hands containing two kings is

$$\binom{4}{2} = \frac{4!}{2! \times 2!}$$

By the multiplication principle, the number of hands containing three aces and two kings is

$$\binom{4}{3}\binom{4}{2} = \frac{4!}{3! \times 1!}\,\frac{4!}{2! \times 2!}$$

The number of possible five-card hands in a 52-card deck is

$$\binom{52}{5} = \frac{52!}{5! \times 47!}$$

Thus, the probability of being dealt three aces and two kings in a five-card poker hand is

$$P(3 \text{ aces and } 2 \text{ kings}) = \frac{\dfrac{4!}{3! \times 1!}\,\dfrac{4!}{2! \times 2!}}{\dfrac{52!}{5! \times 47!}} = \frac{4! \times 4! \times 5! \times 47!}{3! \times 1! \times 2! \times 2! \times 52!}$$

$$= \frac{1}{13 \times 17 \times 10 \times 49} = 9.23 \times 10^{-6}$$

2-9 SUMMARY

In this chapter the fundamental concepts of probability were introduced. Several possible approaches to probability, including the personal approach, the equally likely approach, and the relative frequency approach were discussed. Although handy for ascertaining probabilities of certain events, each of these had some flaw in regard to its use for constructing a mathematical theory of probability. Thus an axiomatic approach is chosen as the basis of a mathematical theory of probability. The equally likely and relative frequency definitions of probability satisfy the axioms of probability, however. Once a satisfactory basis for defining probability is established, one can build a consistent theory on it and derive relations useful for solving engineering problems. The following are the main points covered in this chapter.

1. Probability is a real number associated with each outcome of a chance experiment, i.e., an experiment whose outcomes are not known in advance, which reflects the likelihood of that outcome occurring. If the occurrence of the outcome is certain, the probability is 1, whereas if the occurrence is impossible, the probability is zero. Numbers between zero and 1 reflect varying certainty of the occurrence of the outcome.
2. The *equally likely approach* to probability assumes that all the outcomes of a chance experiment are equally likely, one or more of which must occur upon performance of the experiment. The probability of a given *event* (i.e., a collection of outcomes) is then the number of outcomes favorable to the event divided by the total number of possible outcomes. In this definition of probability, we really use probability to define probability because of the assumption that the outcomes are equally likely. It is a handy definition for determining probabilities for the outcomes of games of chance, such as card games, but is not satisfactory for building a mathematical theory of probability.
3. In the *relative frequency approach* to probability, we assume that we can repeat the chance experiment an arbitrary number of times, noting the occurrence or nonoccurrence of an outcome of interest at each trial of the experiment, and at the end divide the number of occurrences of the outcome of interest by the total number of repetitions. The limit of this ratio is then the probability of the desired outcome. Unfortunately, we can never reach this limit experimentally, and the ratio of the number of favorable outcomes to the total number will continue to vary, even though we expect this variation to decrease as the number of repetitions is increased. Again, this definition is a useful way to estimate or assign probabilities but is not a satisfactory definition on which to base a theory.
4. The *axiomatic approach* to probability is based on three axioms or properties that the probability, $P(A)$, of an event A must satisfy:
   - **(a)** $P(A) \geq 0$.
   - **(b)** The probability of the certain event is unity.
   - **(c)** If two events A and B have no common outcomes (i.e., are *mutually exclusive*), the probability of either A or B occurring is the sum of their respective probabilities:

$$P(A \text{ or } B) = P(A) + P(B)$$

5. Set theory is convenient for representing probabilistic concepts. We represent all possible outcomes of a chance experiment as the universal set, called the *sample space* of the experiment. Outcomes are elements of this universal set or sample space, and events are subsets of the universal set. All set operations, properties of these operations, and operations derived can be applied in our theory of probability. Set operations include union, intersection, and complement. The definition of mutually exclusive events of probability theory given above is then that of disjoint sets (i.e., sets that have no elements in common) of set theory. The union of two sets corresponds to either or both of two events occurring, the intersection of two sets corresponds to the simultaneous occurrence of two events, and the complement of a set corresponds to a given event not occurring. The universal set corresponds to the certain event, and the null set corresponds to the impossible event. Commutative, associative, and distributive laws of unions and intersections in set theory apply equally to events in a sample space.
6. A partition of a universal set or sample space is the division of the sample space into a number of disjoint subsets, the union of which is equal to the sample space. The Cartesian product of two sets is a new set whose elements are composed of all possible pairs of elements, one from each of the original sets.
7. Derived probability relationships (i.e., those obtained by reasoning from the axioms stated above) include the following:
   (a) The probability of the null, or impossible, event is zero.
   (b) The probability of the union of two events, disjoint or not, is given by

   $$P(A \cup B) = P(A) + P(B) - P(A \cap B)$$

   where union corresponds to either or both of the events occurring, and intersection corresponds to both or simultaneous occurrence of the events.
8. The *conditional probability* of an event A occurring given that an event B has occurred is defined as

   $$P(A|B) = \frac{P(A \cap B)}{P(B)}$$

   and the conditional probability of event B given A has occurred is

   $$P(B|A) = \frac{P(A \cap B)}{P(A)}$$

   The probabilities $P(A)$ and $P(B)$ are called *unconditional probabilities,* to distinguish them from the conditional probabilities of the respective events.
9. By solving both equations above for $P(A \cap B)$, equating results, and solving for $P(B|A)$, we obtain a simplified version of *Bayes' Theorem:*

   $$P(A|B) = \frac{P(B|A)P(A)}{P(B)}$$

Similarly,

$$P(B|A) = \frac{P(A|B)P(B)}{P(A)}$$

10. *Statistically independent events* are events for which the occurrence or nonoccurrence of one does not influence the occurrence or nonoccurrence of the others. For such events, the conditional probabilities are equal to the respective marginal probabilities, or their joint probability factors into the product of their respective marginal probabilities.
11. If the events $A_1, A_2, \ldots, A_m$ are a partition of a sample space (i.e., are mutually exclusive and their union is the certain event) and event B is an arbitrary event in the sample space, then the *theorem on total probability* states that

$$P(B) = \sum_{i=1}^{m} P(B|A_i)P(A_i)$$

12. Use of the theorem on total probability in Bayes' Theorem gives

$$P(A_j|B) = \frac{P(B|A_j)P(A_j)}{\sum_{i=1}^{m} P(B|A_i)P(A_i)}$$

where the A_i's form a partition of the sample space.
13. The number of *permutations* (lineups) of n objects taken k at a time is

$$P_k^n = \frac{n!}{(n-k)!}$$

14. The number of combinations of objects taken k at a time is

$$\binom{n}{k} = \frac{n!}{k!(n-k)!}$$

A combination is a permutation without regard to the order in which the objects occur. In algebra, this is called the binomial coefficient; in counting we call it *n choose k*.
15. Flipping a coin n times and counting the number of ways that k heads can occur regardless of order gives the binomial coefficient. If the probability of a head is p, the probability of a particular sequence of k heads and $n - k$ tails is $p^k(1-p)^{n-k}$. Thus the probability of *any sequence* of k heads and $n - k$ tails, called the *binomial probability distribution*, is

$$\binom{n}{k} p^k(1-p)^{n-k}$$

16. The number of divisions of n distinct objects into k classes of size $n_1, n_2, \ldots, n_k$ is given by the *multinomial coefficient*, which is

$$\frac{n!}{n_1!\, n_2! \cdots n_k!}$$

2–10 FURTHER READING

In addition to the books cited in Chapter 1, the following books provide alternative introductions to probability theory oriented to the engineering undergraduate: Barr and Zehma (1983), Breipohl (1970), Helstrom (1992), Leon-Garcia (1994), Papoulis (1990), Peebles (1987), Roberts (1992), Walpole and Myers (1993), and Williams (1991).

2–11 PROBLEMS

Section 2-2

2-1. Which approach to probability is illustrated by the following experiments?

- **(a)** The sky is overcast, leading you to debate in your mind whether you should take your umbrella to class.
- **(b)** You are playing blackjack (also called "twenty-one") with your roommate and wonder if you should take a hit with 16 in your hand and no face cards showing.
- **(c)** You have never had a royal flush (ace, king, queen, jack, ten, all of the same suite) playing five-card draw, and wonder if you will in your lifetime, even though you play poker weekly.
- **(d)** Your roommate complains that you are using loaded dice. You offer to prove that you are not by rolling one 10 times and it comes up with seven spots showing in half of those 10 rolls.
- **(e)** You toss a coin 100 times to find out how many times you get heads.

2-2. Describe the possible outcomes for the following experiments, using notation introduced in this chapter, and find the probabilities asked for using the equally likely outcomes definition.

- **(a)** A numbered disk is drawn from an urn containing 10 red and 20 white disks numbered from 1 through 10 (one red disk has a 1, one red disk has a 2, . . . , two white disks have a 1, two white disks have a 2, . . .). What is the probability of drawing a red disk? A white disk with a 3 on it?
- **(b)** Three fair coins are tossed simultaneously. What is the probability of at least one head?
- **(c)** A six-sided die is tossed with one through six spots on the faces. What is the probability of a 2 or a 3?
- **(d)** A "wheel-of-fortune" is spun with 30 pins, dividing 30 segments marked with amounts of money in increments of \$100 starting at \$100 and going through \$3000 (each amount appears on one and only one segment). What is the probability of landing on a segment marked with an amount greater than or equal to \$500 when the wheel is spun?

2-3. Using the equally likely approach to probability, compute the following probabilities of events where the experiment is drawing cards from a standard 52-card deck of cards (consisting of four suits—clubs, spades, hearts, and diamonds—with 13 cards each of values ace, king, queen, jack, 10, nine, . . . , two):

- **(a)** An ace of spades.
- **(b)** Any ace.
- **(c)** A red (i.e., diamond or heart) ace.
- **(d)** Any face card (i.e., an ace, king, queen, or jack of any suit).
- **(e)** Any black face card.
- **(f)** Any pair of aces when drawing twice without replacement.

2-4. Consider a weather vane, equally likely to stop at angle θ with respect to north. What are the probabilities of the following events? **(a)** $A = \{0 \le \theta \le 45 \text{ degrees}\}$; **(b)** $B = \{22.5 \le \theta < 67.5 \text{ degrees}\}$; **(c)** $A \cap B$; **(d)** $A \cup B$.

Section 2-3

2-5. Extend Axiom 3 to the case of n mutually exclusive events, $A_1, A_2, \ldots, A_n$; that is,

$$P(A_1) + P(A_2) + \cdots + P(A_n) = \sum_{i=1}^{n} P(A_i), \qquad A_i \text{ and } A_j \text{ mutually exclusive, } i \ne j$$

[*Hint:* In (2-5) let $A = A_1$ and $B = A_2 \cup A_3$ (mutually exclusive). Apply Axiom 3 to B, thus proving the relationship for these events. Then use induction, assuming the relationship to be true for $n - 1$, and prove it for n.]

Section 2-4

2-6. Given the sets

$$A = \{1, 3, 5, 7, 9\}$$
$$B = \{0, 2, 4, 6, 8\}$$
$$C = \{10\}$$
$$D = \{3, 4, 5, 6\}$$

assume that the universal set in this case is $\{0, 1, 2, 3, 4, 5, 6, 7, 8, 9, 10\}$. Using Venn diagrams, describe the following sets: **(a)** $A \cup B$; **(b)** $A \cup B \cup C$; **(c)** $A \cup B \cup C \cup D$; **(d)** $A \cup B \cup C \cap D$; **(e)** $A \cap B$; **(f)** $A \cap D$; **(g)** $(A \cup B) \cap (C \cup D)$; **(h)** $A \cap B \cup C$.

2-7. Given that the universal set corresponding to the sets of Problem 2-6 is $\{0, 1, 2, 3, 4, 5, 6, 7, 8, 9, 10\}$, specify the complements of A, B, C, and D as given in Problem 2-6.

2-8. Demonstrate De Morgan's laws by drawing Venn diagrams corresponding to each side of the equation

$$\overline{A \cap B} = \bar{A} \cap \bar{B} \quad \text{and} \quad \overline{\mathrm{A} \cap \mathrm{B}} = \bar{\mathrm{A}} \cup \bar{\mathrm{B}}$$

Section 2-5

2-9. Two random number generators operate independently of each other and generate random numbers, U and V, equally likely to be anywhere in the interval $[0, 1]$. Two events are defined by

$$A = \{0 \le U \le 0.75, 0.5 \le V \le 1\}$$
$$B = \{0.5 \le U \le 1, 0 \le V \le 1\}$$

Find the following probabilities: **(a)** $P(A)$; **(b)** $P(B)$; **(c)** $P(A \cup B)$; **(d)** $P(A \cap B)$.

2-10. Given the numbers of automobile parts of types A, B, and C manufactured by companies X, Y, and Z as shown in the following table. A part is selected at random. Find the following probabilities and tell what they signify in words: **(a)** $P(X)$; **(b)** $P(B)$; **(c)** $P(B \cap X)$; **(d)** $P(B \cup X)$.

Number of Parts from Manufacturer

	Manufacturer		
Part Type	X	Y	Z
A	113	207	342
B	57	116	175
C	202	83	77

2-11. A city of 100,000 persons has three newspapers, A, B, and C. In the city, 10,000 persons read paper A, 30,000 read paper B, and 5,000 read paper C. Also, 8,000 read papers A and B, 2,000 read A and C, 4,000 read B and C, and 1,000 read all three. Deduce the following:

(a) The number of persons reading at least one newspaper.

(b) The number reading two newspapers.

(c) The number reading no newspapers.

2-12. Two numbers, X and Y, are generated independently by random generators with uniform probability in [0, 1]. Compute the following probabilities: **(a)** $P(X > Y)$; **(b)** $P(X + Y > \frac{1}{2})$; **(c)** $P[\max(X, Y) > \frac{1}{2}]$; **(d)** $P(X\,Y < \frac{1}{4})$.

2-13. Given the following probabilities involving two events, A and B: $P(A) = 0.3$; $P(A^c \cap B) = 0.4$; $P(A \cap B) = 0.2$. Compute the following probabilities: **(a)** $P(B)$; **(b)** $P(A \cup B)$; **(c)** $P(B)$; **(d)** $P(A \cap B^c)$. Draw Venn diagrams to help you in your deductions.

Section 2-6

2-14. Given two events A and B that are mutually exclusive. What do you conclude about the probabilities of their occurrences if you are also told they are statistically independent? Give reasoning for your answer.

2-15. Referring to Problem 2-9, find the following probabilities: **(a)** $P(A|B)$; **(b)** $P(B|A)$. **(c)** Are the events A and B statistically independent? Provide proof for your answer.

2-16. Referring to Problem 2-10, find the following conditional probabilities: **(a)** $P(A|X)$; **(b)** $P(X|A)$; **(c)** $P(B|X)$; **(d)** $P(X|B)$; **(e)** $P(C|X)$; **(f)** $P(X|C)$; **(g)** $P(A|Y)$; **(h)** $P(Y|A)$; **(i)** $P(B|Y)$; **(j)** $P(Y|B)$; **(k)** $P(C|Y)$; **(l)** $P(Y|C)$; **(m)** $P(A|Z)$; **(n)** $P(Z|A)$; **(o)** $P(B|Z)$; **(p)** $P(Z|B)$; **(q)** $P(C|Z)$; **(r)** $P(Z|C)$.

2-17. Consider Example 2-5, but with events E_5 = {only one tail} and E_6 = {exactly two tails}. Find the following probabilities and describe the corresponding events in words: **(a)** $P(E_5)$; **(b)** $P(E_6)$; **(c)** $P(E_5 \cap E_6)$; **(d)** $P(E_5 \cup E_6)$.

2-18 The experiment is rolling a pair of dice. Are the events {the sum of the spots up is a 7} and {the sum of the spots up is an 11} statistically independent?

Section 2-7

2-19. Referring to Problem 2-10, a type C part is selected. What is the probability that it came from **(a)** manufacturer X? **(b)** manufacturer Y? **(c)** manufacturer Z?

2-20. Given three boxes containing nickels, dimes, and quarters in the amounts given in the following table:

	1	2	3
Nickels	10	5	5
Dimes	5	20	10
Quarters	30	15	5

A box is selected at random (with equal probability), and one of its coins is selected at random. Find the probability that the coin selected is **(a)** a nickel; **(b)** a dime; **(c)** a quarter.

2-21. A digital communications system sends 1's and 0's in sequence with equal probability. Errors are made, so the probabilities of receiving a digit r_j (equal to a 0 or 1) given that a certain digit s_i (equal to a 0 or 1) was sent are as follows:

	Received	
Sent	0	1
0	0.99	0.01
1	0.005	0.995

Compute the conditional probabilities that a certain digit was sent given that a certain one was received [e.g., $P(s_i = 0 | r_j = 1)$] and similarly for the other three possibilities.

2-22. In a certain locality, the probability that it rains during the day given that the sky is cloudy in the morning is 0.7, while the probability that is does not rain given that the sky is not cloudy in the morning is 0.3. Two-thirds of the days in the year begin as cloudy, and one-third begin as sunny. Find:

(a) The probabilities of rain and no rain irrespective of whether or not the sky is cloudy in the morning.

(b) The probability that if it does not rain during the day, the sky is cloudy in the morning.

(c) The probability that if it rains during the day, the sky is not cloudy in the morning.

2-23. A digital data transmission system selects one of the digits 0, 1, 2, or 3 to transmit through a channel with prior probabilities 0.1, 0.3, 0.3, and 0.3, respectively. The conditional probabilities of receiving the digits 0, 1, 2, or 3 given that a 0, 1, 2, or 3 is transmitted are given in the following table.

$P(Y_j \text{ received} | X_i \text{ transmitted})$

	Y_j			
X_i↓	0	1	2	3
0	0.95	0.02	0.02	0.01
1	0.005	0.98	0.005	0.01
2	0.01	0.01	0.97	0.01
3	0.02	0.03	0.02	0.93

Obtain the following posterior probabilities: **(a)** $P(X = 2 | Y = 2)$; **(b)** $P(X = 1 | Y = 2)$; **(c)** $P(X = 1 | Y = 2)$; **(d)** $P(X = 3 | Y = 1)$. Comment on the reliability of the channel (i.e., the probabilities of the digits being received correctly).

2-24. In a certain lot of personal computers it is known that 1% have some minor defect as they come off the production line. They are put through a test procedure, which detects any defect 98% of the time if a defect is really present, and indicates a defect 1% of the time even though there is none present. What is the probability that a computer is **(a)** defective if it fails the test; **(b)** not defective if it did not fail the test?

2-25. A problem very similar to Example 2-12 is that of testing for controlled substances or testing for diseases. Consider a population that contains 95% nonusers in the case of a controlled substance, or 95% disease-free persons in the case of testing for disease. Consider a test that is 98% accurate when someone is actually using drugs (i.e., 98% of the users test as users) and 1% false positive (i.e., 1% of the nonusers test positive). Find the conditional probability of a person being a nonuser if he or she tested positive (i.e., user). Are you surprised by the result you obtained? Why or why not?

2-26. Reconsider Example 2-13 but generalize it to n curtains, one of which has a car behind it and the rest goats. Consider the cases of the host randomly picking from among the $n - 1$ curtains left after the contestant makes his or her choice, and the case of the host always choosing a curtain with a goat behind it. In each case, tell whether the contestant should make a new choice after the curtain that the host chooses is opened and the goat revealed.

Section 2-8

In these problems, use counting techniques and equal likelihood.

2-27. What is the probability that if n people are present in the same place, at least two will have the same birthday? Compute this probability numerically for $n = 25$.

2-28. Find the probability of getting four aces in a five-card poker hand.

2-29. Find the probability of getting **(a)** all hearts and **(b)** all cards in the same suit in a 13-card bridge hand.

2-30. A committee is picked randomly from a population of 10 men and 12 women. What is the probability that it contains three men and four women?

2-31. To pass a certain class, you must answer m out of n true-false questions correctly.

(a) Suppose that you guess on all n questions.

(i) If $n = 10$ and $m = 8$, what is the probability that you pass?

(ii) Generalize your answer to arbitrary m and $n (m \leq n)$.

(b) Suppose that you know the answer to one question and guess on the rest. Again answer the questions of part (a).

2–12 COMPUTER EXERCISES

2-1. Write a program to take two events A and B that are composed of m and n independent outcomes, respectively, with probabilities given. In addition, there are p outcomes common to events A and B, also with probabilities given. Your program should be able to compute all possible probabilities involving A and B. That is, $P(A)$, $P(B)$, $P(A \cap B)$, $P(A \cup B)$, $P(A|B)$, and $P(B|A)$.

2-2. Design a Monte Carlo simulation to test which of the following is more likely: (1) at least 1 six when six dice are rolled; (2) at least 2 sixes when 12 dice are rolled; (3) at least 3 sixes when 18 dice are rolled. [From Frederick Mosteller, *Fifty Challenging Problems in Probability with Solutions* (New York: Dover Publications, 1965).]

2-3. Design a simulation that will verify the results of the Monte Hall example both for the random choice on the part of the host and for the certain choice of a curtain with a goat behind it (Example 2-13). Attempt to make it the general case where there are n curtains with a car behind one of them.

3 Single Random Variables and Probability Distributions

3–1 WHAT IS A RANDOM VARIABLE?

In our consideration of elementary probability relations in Chapter 2, we had to describe the outcomes and events of our chance experiments in words: for example, one spot up on the die, a tail, the ace of spaces, and so on. Clearly, it is easier to describe and manipulate things using numerical values. That is the purpose of a random variable: to map each point in the sample space of a chance experiment to a point on the real line. Such a mapping from the elements of one set (in this case, the sample space) to the elements of another set (in this case, the real line) in mathematical language is called a *function.* When applied to probability theory, the mapping of sample space to real line is called a *random variable.* In addition to the name change, a different notation is used. It is customary in probability to denote the mapping by capital letters, usually near the end of the alphabet (e.g., X, Y, Z, and so on). An arbitrary point in the sample space is denoted as in Chapter 2 (e.g., in this book, by lowercase Greek letters such as ζ, ω, ν, etc.). The value of the random variable assumed for a given sample space element or point is denoted by the corresponding lowercase letter for the denotation of the random variable. Thus, corresponding to the sample space element ζ, we have the numerical value

$$x = X(\zeta) \tag{3-1}$$

assigned by the random variable X. To emphasize, uppercase English alphabet letters, with some exceptions, denote random variables (the function or rule of assignment), and the corresponding lowercase letter denotes the value it takes on. The usual mathematical notation used, of course, would be something like $y = f(x)$. The mapping of a random variable would be shown as in Figure 3-1. Note that after the experiment is performed, the function X yields a specific value x. Before the experiment is performed, the values for $X(\zeta)$ depend on the particular outcomes of the chance experiment and are described by their probabilities. We will want to carry this probabilistic assignment onto the real line.[1] Since more than one point in the sample space may be mapped into the same value on the real line, this may be more complex than merely determining what probability goes with which value and may include what probabilities correspond to a given mapped point. An example will illustrate the foregoing remarks.

Example 3-1

In certain games involving the tossing of a pair of dice, winning or losing depends on the sum of the spots up on the die. A suitable mapping from sample space elements (or chance experiment outcomes) to numerical values is provided in Table 3-1.

All combinations of spots up on the diagonals going down and to the left would be mapped onto the same point on the real line (i.e., the single 2 onto 2, the two 3's onto 3, the three 4's onto 4, etc.). Assuming equally likely outcomes, each value in the table has a probability occurrence of $\frac{1}{36}$. The probabilities of the various values for the random variable are as follows:

$$\begin{aligned}
&P(X=2)=\tfrac{1}{36}, && P(X=3)=\tfrac{2}{36}=\tfrac{1}{18}, && P(X=4)=\tfrac{3}{36}=\tfrac{1}{12},\\
&P(X=5)=\tfrac{4}{36}=\tfrac{1}{9}, && P(X=6)=\tfrac{5}{36}, && P(X=7)=\tfrac{6}{36}=\tfrac{1}{6},\\
&P(X=8)=\tfrac{5}{36}, && P(X=9)=\tfrac{4}{36}=\tfrac{1}{9}, && P(X=10)=\tfrac{3}{36}=\tfrac{1}{12},\\
&P(X=11)=\tfrac{2}{36}=\tfrac{1}{18}, && P(X=12)=\tfrac{1}{36}, &&
\end{aligned} \tag{3-2}$$

As a simpler example, consider the tossing of a single coin. We might assign the numerical value 1 to the occurrence of a head and a 0 to the occurrence of a tail. For a fair coin, therefore, the probabilities of the two possible values for the random variable in this case are

$$P(X=1)=\tfrac{1}{2}, \quad P(X=0)=\tfrac{1}{2} \tag{3-3}$$

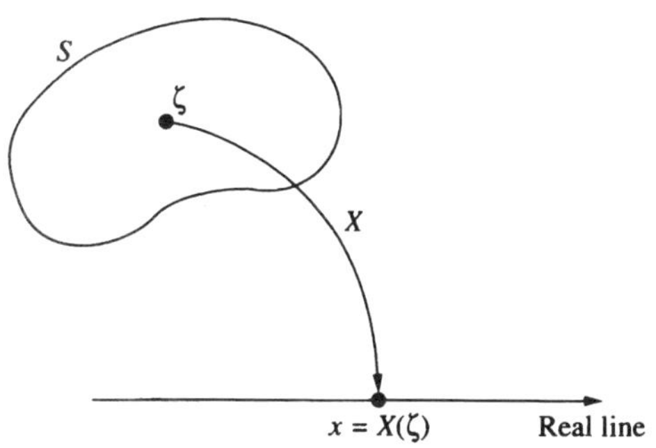

Figure 3-1 A random variable is a mapping of the sample space onto the real line.

[1]Other than mapping the outcomes of a chance experiment onto the real line, the mapping is arbitrary. It should be chosen conveniently, however.

Table 3-1 VALUES FOR A RANDOM VARIABLE REFERRING TO THE SUM OF THE SPOTS UP WHEN A PAIR OF DICE ARE THROWN

			Die 1			
Die 2	1 spot	2 spots	3 spots	4 spots	5 spots	6 spots
1 spot	2	3	4	5	6	7
2 spots	3	4	5	6	7	8
3 spots	4	5	6	7	8	9
4 spots	5	6	7	8	9	10
5 spots	6	7	8	9	10	11
6 spots	7	8	9	10	11	12

The particular assignment used for a random variable is arbitrary, although we wish to have a convenient one. For example, we could have assigned the values π and e to heads and tails, respectively, in the coin-tossing experiment, but it probably would not be done because these values are not particularly convenient.

Finally, we note that any function of a random variable is also a random variable (there are, of course, degenerate exceptions such as a switch that always sticks on or off). Examples are

$$Y = e^x, \qquad W = \ln X, \qquad U = \cos X, \qquad V = X^2 \tag{3-4}$$

We may also be interested in the probabilities of the values of these derived random variables; this is a more complex topic that will be discussed later in the chapter.

Random variables can be *discrete, continuous,* or *mixed.* The example of rolling a pair of dice just discussed provides an example of a discrete random variable (i.e., a random variable that can assume only a countable number of possible values). A continuous random variable can assume a continuum of values, any given value of which occurs with infinitesimal probability. An example of such a random variable is provided by a weather vane that indicates the direction of the wind. It can pivot in any direction, and the angle that it makes with respect to the direction north is a continuous random variable. We usually agree to restrict this angle to the range of zero to 360 degrees, or zero to 2π radians. Mixed random variables can assume a discrete set of values each with finite probability as well as a continuum of values, a specific value of which has infinitesimal probability.

Example 3-2

A wheel of fortune provides an example of a mixed random variable, where the random variable is defined as the angle at which the wheel comes to rest with respect to some reference mark on the wheel, say the $1000 slot. The pegs provide several points around the circumference of the wheel where the probability is nonzero that the wheel will come to rest at those particular angles. Between adjacent pegs, it is equally likely that the stopping point will be any angle in between the angles defined by the peg positions. Of course, the way the game is played is to assign the same prize amount to all angles between an adjacent pair of pegs.

3-2 PROBABILITY DISTRIBUTION FUNCTIONS

Descriptions of random variables using probability can be done in various ways. If a random variable is discrete, we can simply tabulate or plot the probability of it assuming its various values versus these values. This is referred to as the *probability mass distribution.* For the example of a pair of dice being thrown and the spots up summed as given in Table 3-1, the probability mass distribution is given by (3-2). The probability mass function of the numbers up on the two dies separately would consist of 36 vertical lines perpendicular to the x–y plane, each with height $\frac{1}{36}$, located at the integer pairs $(X = i,\ Y = j)$, where i and j take on the values 1, 2, 3, 4, 5, and 6, as considered in Chapter 4.

In this section we are concerned primarily with two other methods of description, however. These are referred to as the *cumulative probability distribution function,* or simply *cumulative distribution function* (cdf), and the *probability density function* (pdf). The latter is the derivative of the former. The cdf works for discrete, continuous, or mixed random variables. If we agree that the derivative of a step discontinuity is an impulse, or delta, function, the pdf can be applied to all three types of random variables as well.

Cumulative Distribution Function

The cdf of a random variable, X, is defined as

$$F_X(x) = P(X \leq x) \tag{3-5}$$

That is, the cdf is the probability that random variable X is less than or equal to the running value x.

Example 3-3

To illustrate the cdf, consider the random variable defined in Table 3-1 with the probability mass distribution given by (3-2).

Solution We consider various values of the running variable, x. For x less than 2, there are zero values of the random variable X less than or equal to x. Therefore,

$$F_X(x) = 0,\ x < 2$$

If $x = 2$, there is one value of the random variable included, namely 2, with probability of occurrence of $\frac{1}{36}$ according to (3-2). Hence the cdf jumps from 0 to $\frac{1}{36}$ at $x = 2$, and it stays at this level until $x = 3$ since no other values of X are included in this range. The random variable X assumes the value 3 with probability $\frac{2}{36}$. Thus

$$F_X(3) = P(X \leq 3) = P(X = 2) + P(X = 3) = \frac{1}{36} + \frac{2}{36} = \frac{3}{36} = \frac{1}{12}$$

and this value is held until $X = 4$, since no other values of the random variable are possible in the interval $3 < X < 4$. We keep on reasoning in this fashion, and eventually end up with the following description of the cdf for the dice-throwing example:

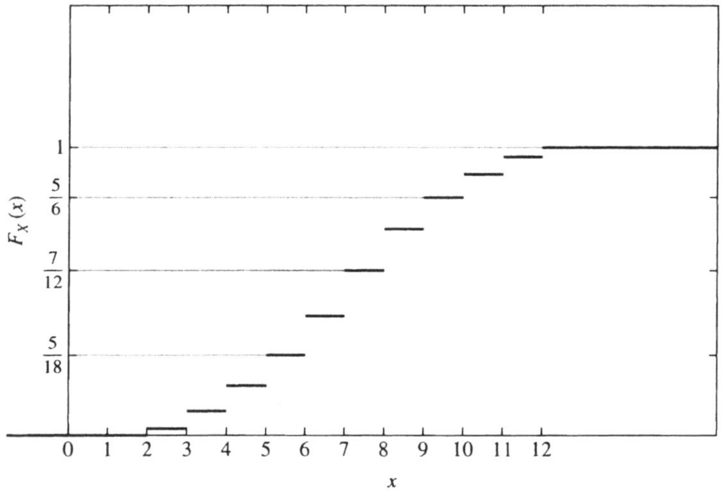

Figure 3-2 Cumulative distribution function for a random variable numerically equal to the sum of the spots up on rolling a pair of dice.

$$F_X(x) = \begin{cases} 0, & x < 2 \\ \frac{1}{36}, & 2 \le x < 3 \\ \frac{3}{36}, & 3 \le x < 4 \\ \frac{6}{36}, & 4 \le x < 5 \\ \frac{10}{36}, & 5 \le x < 6 \\ \frac{15}{36}, & 6 \le x < 7 \\ \frac{21}{36}, & 7 \le x < 8 \\ \frac{26}{36}, & 8 \le x < 9 \\ \frac{30}{36}, & 9 \le x < 10 \\ \frac{33}{36}, & 10 \le x < 11 \\ \frac{35}{36}, & 11 \le x < 12 \\ \frac{36}{36}, & 12 \le x \end{cases} \tag{3-6}$$

A plot of this cdf is shown in Figure 3-2.

From the cdf of Example 3-1, we can surmise several properties of cdf's in general. These are:

1. Limiting values:

$$\lim_{x \to -\infty} F_X(x) = 0$$
$$\lim_{x \to \infty} F_X(x) = 1 \tag{3-7}$$

2. Continuity: cdf's are continuous from the right. That is, if x_0 is a point of discontinuity,

$$\lim_{x \to x_0^+} F_X(x) = F_X(x_0) \tag{3-8}$$

3. Cdf's are monotonically nondecreasing functions of their arguments.
4. The probability that a random variable lies in the range $x_1 < X \le x_2$ is given by

$$P(x_1 < X \le x_2) = F_X(x_2) - F_X(x_1) \tag{3-9}$$

The reasonableness of these properties can be explained as follows. First, if $x \to -\infty$, no possible outcomes of the underlying chance experiment are included in the inequality $X \le x$, whereas if $x \to \infty$, all possible outcomes are included. Thus the probability of the former event is zero, and the probability of the latter event is unity. The continuity from the right follows because we define the cdf as the probability that the random variable X is less than *or equal to* the running value x. The monotonicity property follows because as x increases, we are always including the same number of or more outcomes of the underlying chance experiment. Finally, property (3-9) follows by noting that the events $X \le x_1$ and $x_1 < X \le x_2$ are mutually exclusive events, and their union is the event $X \le x_2$. Thus

$$P(X \le x_1) + P(X_1 < X \le x_2) = P(X \le X_2)$$

or

$$F(x_1) + P(x_1 < X \le x_2) = F(x_2) \tag{3-10}$$

When (3-10) is rearranged, we obtain (3-9).

Example 3-4

Find the probability of the sum of the faces up being between 3 and 7, inclusive when a pair of dice is thrown.

Solution Using (3-9) along with Figure 3-1 or (3-6) and remembering that the cdf is continuous from the right, we obtain

$$P(3 \le X \le 7) = F_X(7) - F_X(3^-) = F_X(7) - F_X(2) = \tfrac{21}{36} - \tfrac{1}{36} = \tfrac{20}{36} = \tfrac{5}{9}$$

Using equal likelihood and Table 3-1, we find 20 outcomes favorable to the event that X is greater than or equal to 3 but less than or equal to 7 out of 36 possibilities, which is the same answer.

Example 3-5

Show that the function

$$F_X(x) = \frac{1}{2}\left(1 + \frac{2}{\pi}\tan^{-1}x\right) \tag{3-11}$$

is a proper cdf.

Solution The limit of the arctangent function as $x \to \pm\infty$ is $\pm\,\pi/2$. Thus, the proper limits are achieved by (3-11) as $x \to \pm\infty$. This function is also continuous (from the right) and monotonically increasing. Therefore, it is a proper cdf. A plot of this function is shown in Figure 3-3.

Probability Density Function

For continuous random variables, the pdf can be defined as the derivative of the cdf; that is,

$$f_X(x) = \frac{dF_X(x)}{dx} \tag{3-12}$$

Using the definition of a derivative, we may write

$$f_X(x) = \lim_{\Delta x \to 0} \frac{F_X(x + \Delta x) - F_X(x)}{\Delta x} \tag{3-13}$$

Thus, for Δx sufficiently small, we can remove the limit in (3-13) and rearrange it to obtain

$$f_X(x)\,\Delta x \simeq F_X(x + \Delta x) - F_X(x) = P(x < X \leq x + \Delta x) \tag{3-14}$$

where (3-10) has been used. Therefore, we may interpret the pdf, when multiplied by a small increment in the independent variable Δx, as the probability that the random variable X lies in the range x to $x + \Delta x$. Since the pdf is the derivative of the cdf, we may deduce certain properties of it. These are:

1. It is a nonnegative function:

$$f_X(x) \geq 0, \qquad \text{all } x \tag{3-15}$$

   (Note that the pdf can exceed 1 in value.)

2. The area under the pdf is unity.

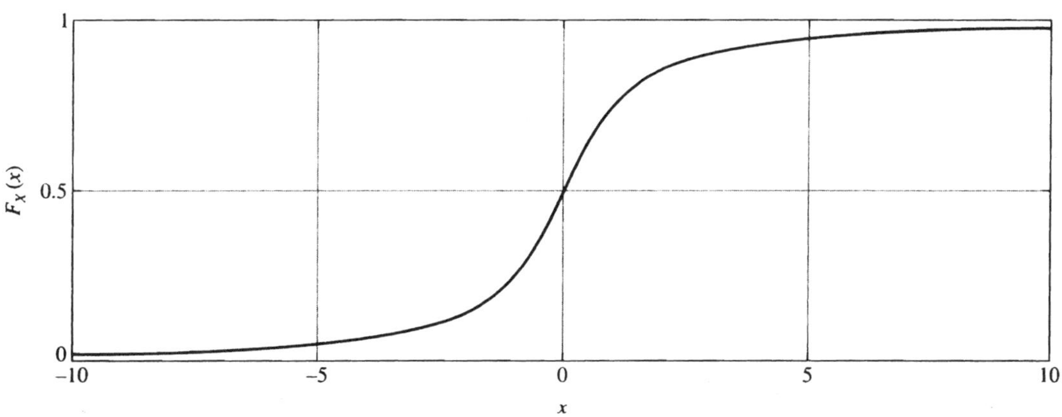

Figure 3-3 Suitable cumulative distribution function.

3. The integral of the pdf from $-\infty$ to x is the cdf:

$$F_X(x) = \int_{-\infty}^{x} f_X(u)\, du \tag{3-16}$$

From (3-16), we obtain the relationship

$$\begin{aligned} P(x_1 < X \le x_2) &= F_X(x_2) - F_X(x_1) \\ &= \int_{-\infty}^{x_2} f_X(u)\, du - \int_{-\infty}^{x_1} f_X(u)\, du = \int_{x_1}^{x_2} f_X(u)\, du \end{aligned} \tag{3-17}$$

That is, to get the probability of a random variable lying in the range x_1 to x_2, we integrate the pdf over this range.

The properties of the pdf stated above can be justified as follows. First, since the cdf is a nondecreasing function, its slope is nonnegative everywhere. Second, from (3-17) with $x_1 = -\infty$ and $x_2 = \infty$, we obtain

$$\int_{-\infty}^{\infty} f_X(u)\, du = F_X(\infty) = 1 \tag{3-18}$$

The last property follows by the antiderivative of (3-12) along with the fact that $F_x(-\infty) = 0$.

Example 3-6

Obtain the pdf corresponding to the cdf (3-11). What is the probability that the random variable X lies between 2 and 5?

Solution Using the derivative formula for the arctangent, we obtain

$$f_X(x) = \frac{1/\pi}{1 + x^2} \tag{3-19}$$

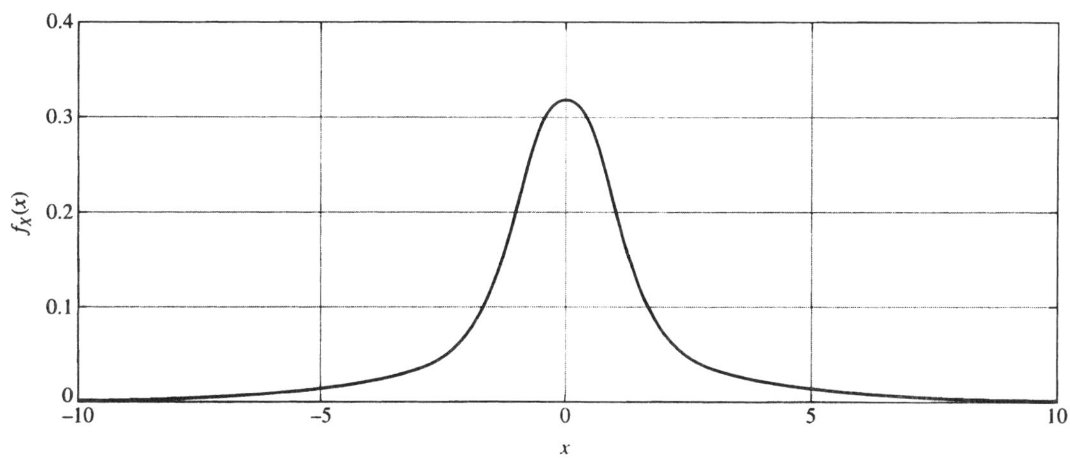

Figure 3-4 Probability density function for Example 3-6.

This pdf is shown in Figure 3-4. Using (3-9) or (3-17), we obtain

$$P(2 < X \leq 5) = \frac{1}{\pi}(\tan^{-1}5 - \tan^{-1}2) = 0.085 \tag{3-20}$$

Discrete random variables can be described by either their *probability mass functions* or their cdf's, as given by (3-2) and (3-6), respectively, for the dice-throwing experiment. The pdf for a discrete random variable also may be expressed mathematically by recalling that the derivative of a unit step function is a unit impulse function:

$$\frac{du(x)}{dx} = \delta(x) \tag{3-21}$$

where $u(x) = 1, x \geq 0$, and is 0 otherwise. Since the cdf of a discrete random variable can be expressed as a sum of step functions, its pdf is a sum of unit impulse functions. Mixed random variables have pdf's that are a combination of a sum of unit impulse functions and a continuous function.[2]

Example 3-7

Express the pdf of the cdf shown in Figure 3-2 mathematically.

Solution In terms of step functions, the cdf of Figure 3-2 is

$$\begin{aligned} F_X(x) = \tfrac{1}{36}[&u(x-2) + 2u(x-3) + 3u(x-4) + 4u(x-5) + 5u(x-6) \\ &+ 6u(x-7) + 5u(x-8) + 4u(x-9) + 3u(x-10) \\ &+ 2u(x-11) + u(x-12)] \end{aligned} \tag{3-22}$$

Taking the derivative, we find the pdf to be

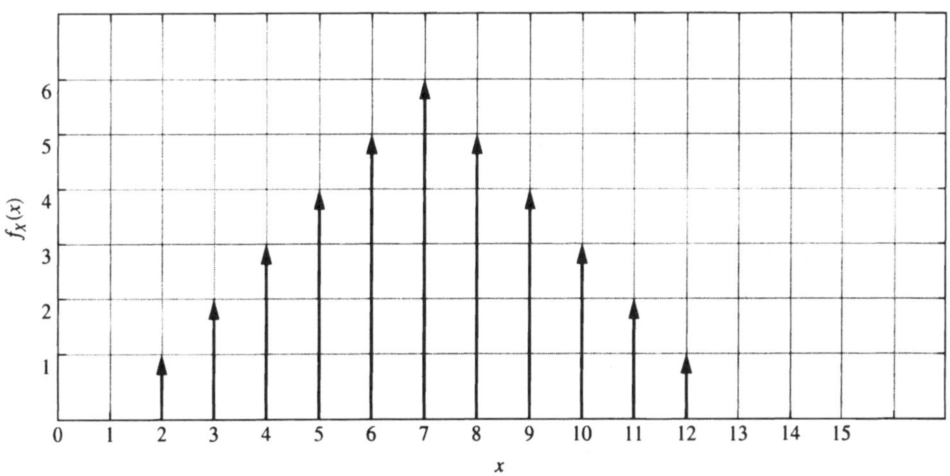

Figure 3-5 Probability density function for Example 3-7. (*Note:* Arrow heights are proportional to weights of impulses.)

[2]We do not deal in this book with certain pathological cases for which this representation in terms of impulses does not work.

$$f_X(x) = \tfrac{1}{36}[\delta(x-2) + 2\delta(x-3) + 3\delta(x-4) + 4\delta(x-5) + 5\delta(x-6) \\ + 6\delta(x-7) + 5\delta(x-8) + 4\delta(x-9) + 3\delta(x-10) + 2\delta(x-11) + \delta(x-12)] \tag{3-23}$$

A plot of this pdf is provided in Figure 3-5.

3-3 COMMON RANDOM VARIABLES AND THEIR DISTRIBUTION FUNCTIONS

Several commonly occurring random variables are discussed in this section. They are described in terms of their pdf's and cdf's. Examples are provided to illustrate their use in modeling of random phenomena. Continuous random variables are considered first, followed by discrete random variables.

Uniform Random Variable

In a situation where we know lower and upper bounds for a measured quantity, the uniform random variable is sometimes useful. It is described in terms of its pdf as

$$f_X(x) = \begin{cases} \dfrac{1}{b-a}, & a \le x \le b,\ b > a \\ 0, & \text{otherwise} \end{cases} \tag{3-24}$$

By integration, we find the cdf to be

$$F_X(x) = \begin{cases} 0, & x \le a \\ \dfrac{x-a}{b-a}, & a < x \le b \\ 1, & x > b \end{cases} \tag{3-25}$$

These are sketched in Figure 3-6 for $a = 0$ and $b = 5$.

Example 3-8

Resistors are manufactured to a tolerance of $\pm$ 10%. Within this tolerance region, they are approximately uniformly distributed. What is the probability that a nominal 1000-Ω resistor has a value between 990 and 1010 Ω?

Solution For a nominal 1000-Ω resistor, the tolerance region is bounded by 900 and 1100 Ω. Thus, from (3-26), the pdf for the resistance value is

$$f_R(r) = \begin{cases} \dfrac{1}{200}, & 900 \le r \le 1100\ \Omega \\ 0, & \text{otherwise} \end{cases} \tag{3-26}$$

From (3-17), the probability that the resistance lies in the range (990, 1010 Ω) is

$$P(990 < R \le 1010\ \Omega) = \int_{990}^{1010} \frac{dr}{200} = 0.1 \tag{3-27}$$

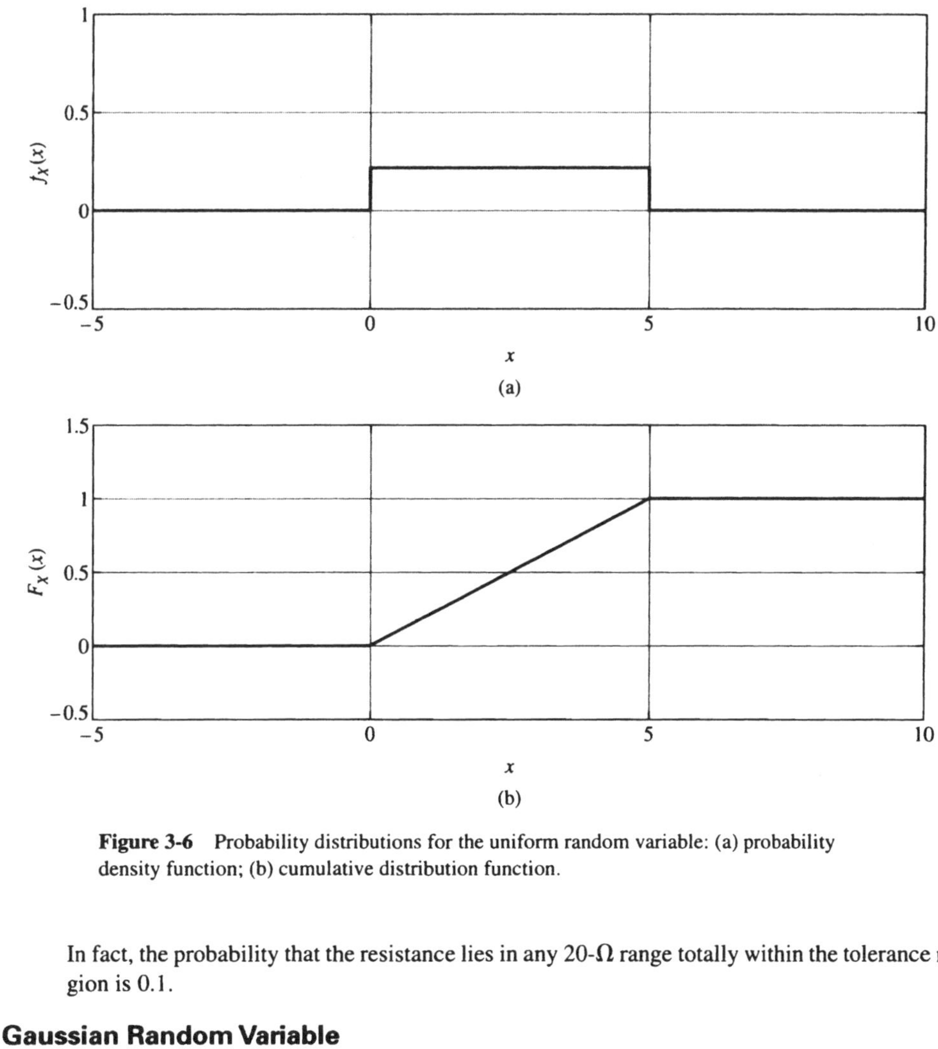

Figure 3-6 Probability distributions for the uniform random variable: (a) probability density function; (b) cumulative distribution function.

In fact, the probability that the resistance lies in any 20-Ω range totally within the tolerance region is 0.1.

Gaussian Random Variable

Often, a random error is the sum of a number of independent component random errors. Later, in Chapter 4, when we discuss the *central limit theorem,* we shall see that this is a situation for which a Gaussian, or normal, random variable is a good model. A Gaussian (capitalized because it is named after a seventeenth-century mathematician, Karl Friedrich Gauss) random variable is described by the pdf

$$f_X(x) = \frac{e^{-(x-m)^2/2\sigma^2}}{\sqrt{2\pi\sigma^2}} \tag{3-28}$$

where m and σ^2 are parameters. The cdf cannot be found as an exact integral; one way of expressing it is in terms of the integral

$$Q(x) = \frac{1}{\sqrt{2\pi}} \int_x^\infty e^{-u^2/2} du \tag{3-29}$$

which is referred to simply as the *Q-function.* It is tabulated in Appendix C, where a rational approximation is given that can easily be programmed on a programmable calculator or computer. Another integral that is often used is the error function, defined as[3]

$$\text{erf}(x) = \frac{2}{\sqrt{\pi}} \int_0^x e^{-u^2} du \tag{3-30}$$

The Q-function and the error function can be expressed in terms of each other, which is left to the problems. We will use the Q-function in this book. In terms of the Q-function, the Gaussian cdf is given by

$$F_X(x) = 1 - Q\left(\frac{x - m}{\sigma}\right) \tag{3-31}$$

The Gaussian pdf and cdf are plotted in Figure 3-7.

Example 3-9

Shims are manufactured to a nominal thickness of 5 mm. However, the stock from which they are cut is known to vary from this thickness, which from extensive measurements is modeled as Gaussian with $m = 5$ mm and $\sigma = 0.05$ mm. What is the probability that the thickness, T, of a shim is less than 4.9 mm or thicker than 5.1 mm?

Solution From (2-21), $P(T < 4.9 \text{ mm or } T > 5.1 \text{ mm}) = 1 - P(4.9 \text{ mm} \le T \le 5.1 \text{ mm})$, and from (3-9), $P(4.9 \text{ mm} \le T \le 5.1 \text{ mm}) = F_T(5.1) - F_T(4.9)$. Using (3-31) with appropriate parameter values, we have

$$\begin{aligned}
P(T < 4.9 \text{ mm or } T > 5.1 \text{ mm}) &= 1 - P(4.9 \text{ mm} \le T \le 5.1 \text{ mm}) \\
&= 1 - [F_T(5.1) - F_T(4.9)] \\
&= 1 - \left[1 - Q\left(\frac{5.1 - 5}{0.05}\right) - 1 + Q\left(\frac{4.9 - 5}{0.05}\right)\right] \\
&= 1 - [Q(-2) - Q(2)] = 1 - [1 - Q(|-2|) - Q(2)] \\
&= 2Q(2) = 0.02275
\end{aligned}$$

where the relationship $Q(x) = 1 - Q(|x|), x < 0$, has been used.

Exponential Random Variable

An exponential random variable with parameter α has the pdf

$$f_X(x) = \alpha e^{-\alpha x} u(x), \quad \alpha > 0 \tag{3-32}$$

[3]MATLAB and Mathcad both have built in functions for computing erf(x). MATLAB has the additional function erfc(x) $= 1 -$ erf(x).

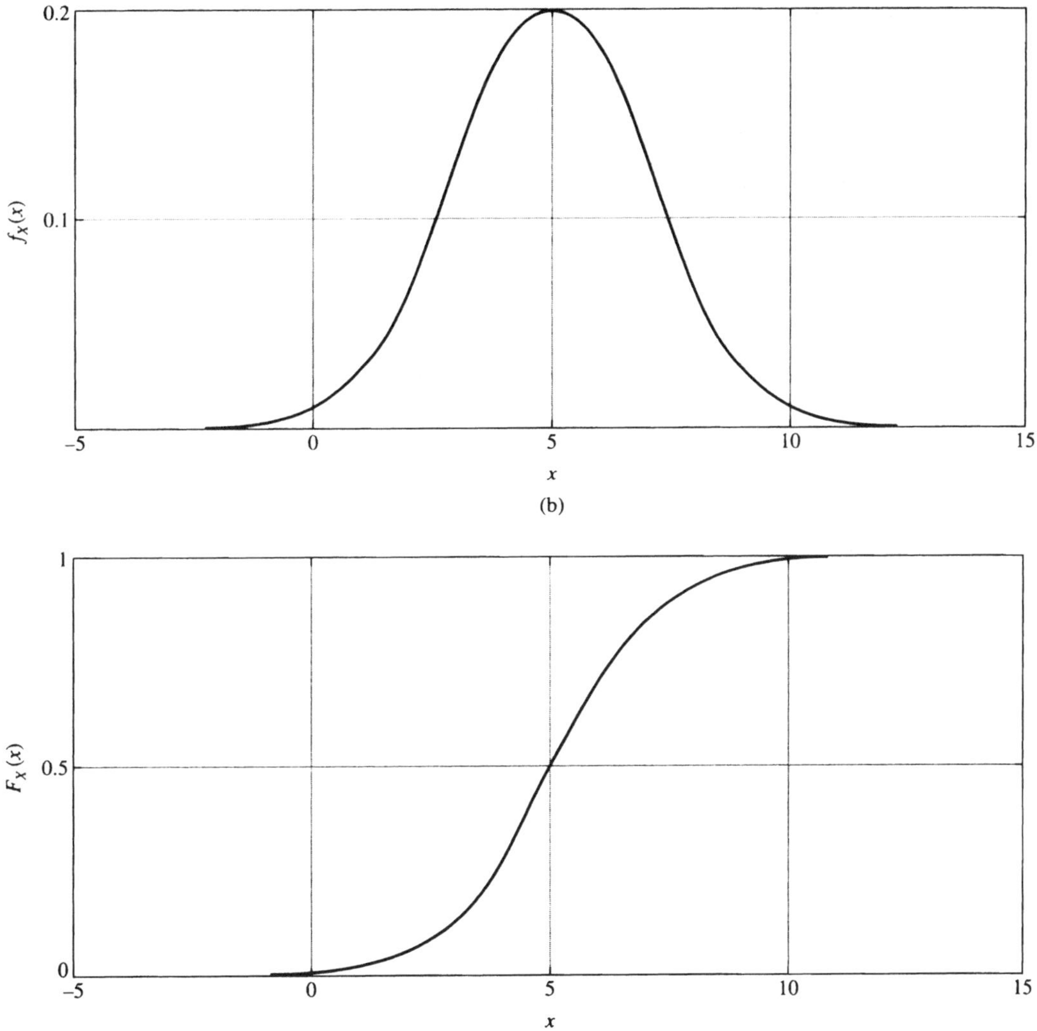

Figure 3-7 Gaussian distribution functions for $m = 5$ and $\sigma = 2$: (a) probability density function; (b) cumulative distribution function.

and cdf

$$F_X(x) = (1 - e^{-\alpha x})u(x) \tag{3-33}$$

where $u(x)$ is the unit step function. The exponential distribution is important in various applications, including intervals between message arrivals in telephone systems. The student should plot the exponential pdf and cdf and make sure that they have the appropriate properties.

Example 3-10

The interval of service (the duration from beginning to completion of service) for a customer in a line at a bank is exponential with parameter $\alpha = 0.033\ \text{s}^{-1}$. You are next to be served. What is the probability that 15 s or less will elapse until you are finished being served? One minute or greater?

Solution Let the waiting time be Δt. Then the cdf of the waiting time is

$$F_{\Delta t}(\zeta) = (1 - e^{-0.033\zeta})u(\zeta)$$

(note that it makes no difference what we call the independent variable). Thus

$$P(\Delta t \le 15\text{s}) = F_{\Delta t}(15) - F_{\Delta t}(0) = 0.39$$

and

$$P(\Delta t \ge 60\text{ s}) = 1 - P(\Delta t \le 60\text{ s})$$

$$= 1 - [F_{\Delta t}(60) - F_{\Delta t}(0)] = 1 - (1 - e^{0.033\times 60}) = e^{-1.98} = 0.138$$

Gamma Random Variable

This random variable has the pdf

$$f_X(x) = \frac{c^b}{\Gamma(b)} x^{b-1} e^{-cx} u(x), \qquad b, c > 0 \tag{3-34}$$

where $\Gamma(b)$ is the gamma function, which is given by the integral

$$\Gamma(b) = \int_0^\infty y^{b-1} e^{-y} dy \tag{3-35}$$

By evaluation of (3-35), we can show that

$$\Gamma(1) = 1 \quad \text{and} \quad \Gamma\left(\frac{1}{2}\right) = \sqrt{\pi} \tag{3-36}$$

Replacing b by $b + 1$ in (3-35), we can also show that

$$\Gamma(b + 1) = b\Gamma(b) \tag{3-37}$$

If b is an integer, it follows from (3-36) and (3-37) that $\Gamma(n + 1) = n!$.

Two special cases of the gamma pdf are the *chi-square* and the *Erlang* pdf's. These are given by

$$f_X(x) = \frac{1}{2^{n/2}\Gamma(n/2)} x^{(n/2)-1} e^{-x/2} u(x), \quad n \text{ integer (chi-square)} \tag{3-38}$$

and

$$f_X(x) = \frac{c^n}{(n-1)!} x^{n-1} e^{-cx} u(x), \quad n \text{ integer (Erlang)} \tag{3-39}$$

respectively. The chi-square random variable arises in statistics, and the Erlang random variable is used in queuing theory. Plots of the chi-square and Erlang pdf's are shown in Figure 3-8. We will make use of all of these probability models later.

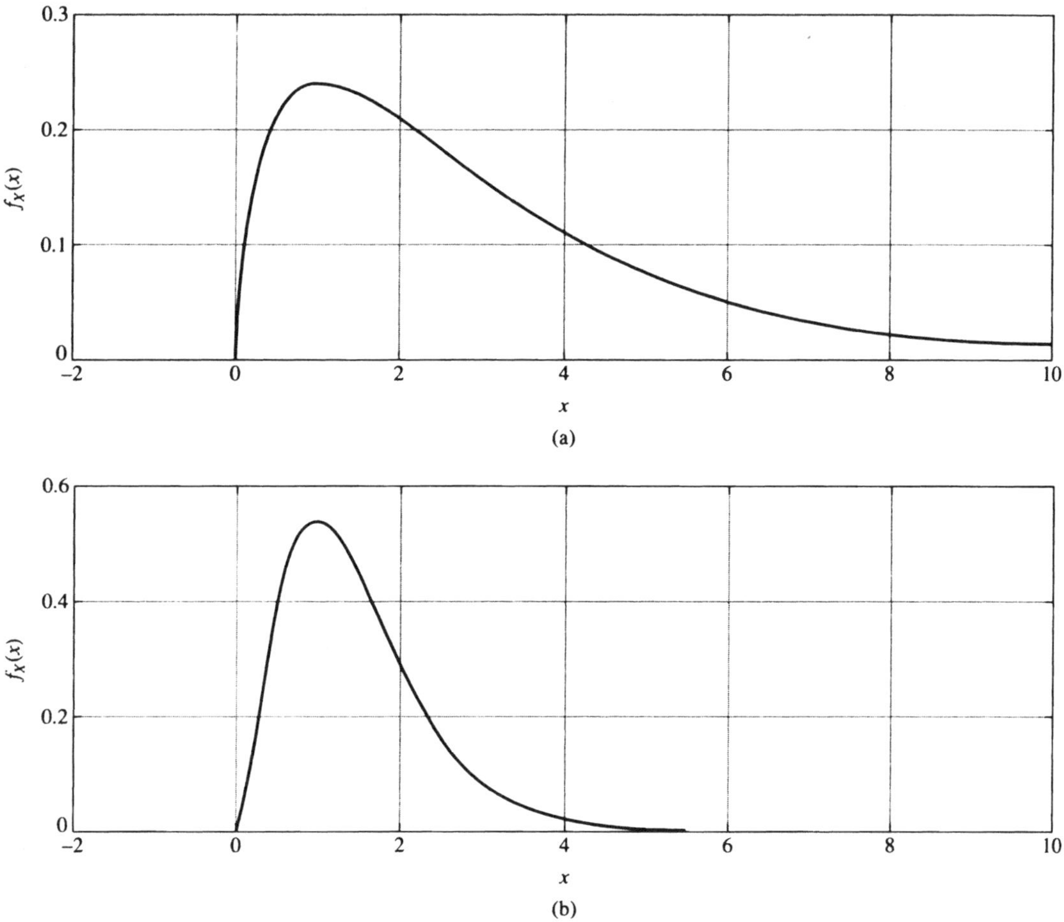

Figure 3-8 Plots of the chi-square and Erlang pdf's, which are special cases of the gamma pdf: (a) chi-square pdf for $n = 3$; (b) Erlang pdf for $c = 2$ and $n = 3$.

Cauchy Random Variable

This random variable has the pdf

$$f_X(x) = \frac{\alpha/\pi}{x^2 + \alpha^2} \tag{3-40}$$

The corresponding cdf is given by

$$F_X(x) = \frac{1}{2} + \frac{1}{\pi}\tan^{-1}\frac{x}{\alpha} \tag{3-41}$$

The student should note that this was the cdf considered in Example 3-5; it is plotted in Figure 3-3 for $\alpha = 1$.

Binomial Random Variable

The random variable X is binomially distributed if it takes on the values 0, 1, 2, . . . , n with probabilities

$$P(X = k) = \binom{n}{k} p^k q^{n-k}, \qquad k = 0,1, \ldots, n \tag{3-42}$$

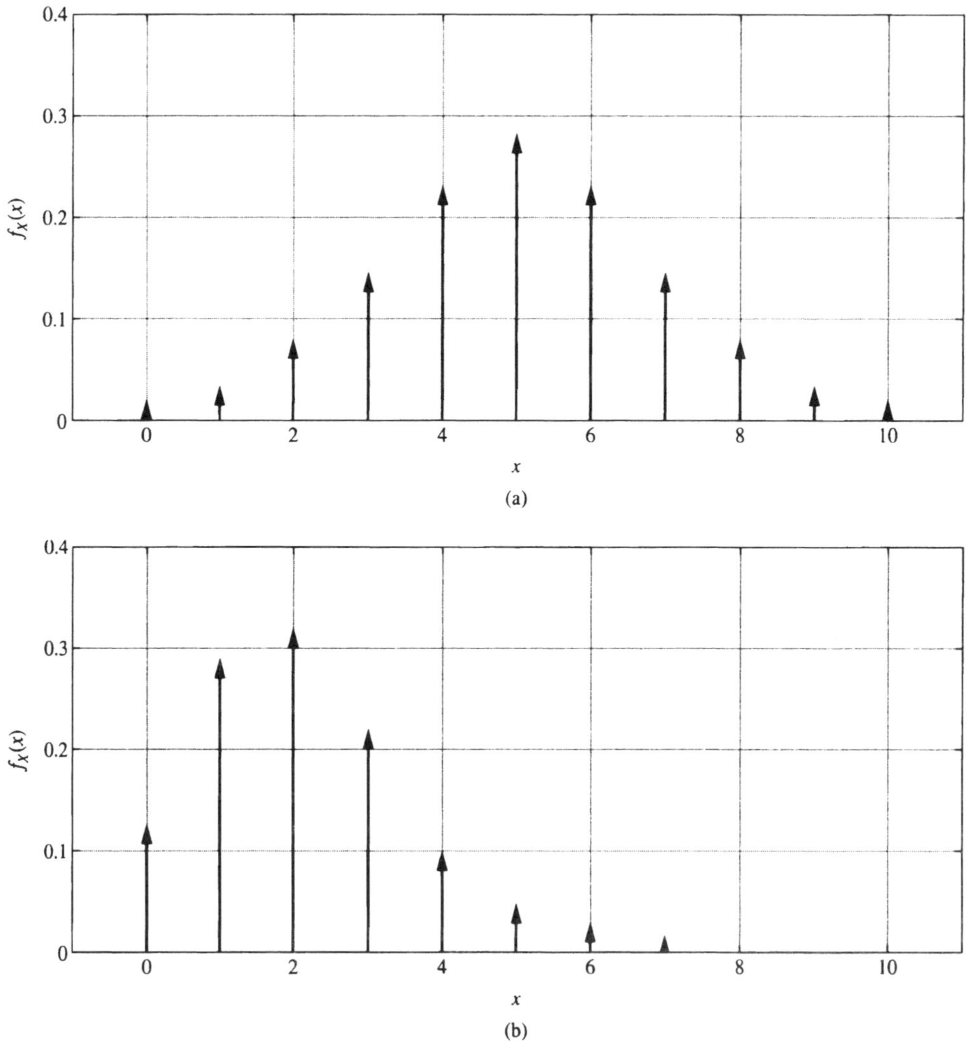

Figure 3-9 Binomial pdf for $n = 10$: (a) $p = q = \frac{1}{2}$; (b) $p = \frac{1}{5}, q = \frac{4}{5}$.

where $p + q = 1$. The corresponding pdf and cdf are given by

$$f_X(x) = \sum_{k=0}^{n} \binom{n}{k} p^k q^{n-k} \delta(x - k), \tag{3-43}$$

and

$$F_X(x) = \sum_{k \le x} \binom{n}{k} p^k q^{n-k} \tag{3-44}$$

respectively. This is the distribution describing the number of heads occurring in n tosses of a coin. If the coin is fair, $p = q = \frac{1}{2}$. Figure 3-9 gives some plots.

Geometric Random Variable

Suppose that we are interested in the probability of the first head occurring after 10 flips of a biased coin where the probability of a head is p and of a tail $1 - p$. Thus we must have 10 tails and then a head. The probability of 10 tails in a row is $(1 - p)^{10}$, and the probability of 10 tails followed by a head is $(1 - p)^{10}p$. Generalizing this to the probability of the first head at the kth toss, we have

$$P(\text{first success at trial } k) = P(X = k) = (1 - p)^{k-1}p, \quad k = 1, 2, \ldots \tag{3-45}$$

where p is the probability of success. This is called the *geometric distribution.* Note that for any value of $k \neq 1$, the probability value is the geometric mean of the $k - 1$ and $k + 1$ values.

Poisson Random Variable

A random variable X is Poisson with parameter a if it takes the values 0, 1, 2, . . . with probability

$$P(X = k) = \frac{a^k}{k!} e^{-a}, \qquad k = 0, 1, 2, \ldots \tag{3-46}$$

The cdf of a Poisson random variable is

$$F_X(x) = \sum_{k \le x} \frac{a^k}{k!} e^{-a} \tag{3-47}$$

Plots for the pdf of a Poisson random variable are provided in Figure 3-10.

Limiting Forms of the Binomial and Poisson Distributions

The *De Moivre–Laplace theorem* states that for n sufficiently large, the binomial distribution can be approximated by samples of a Gaussian curve properly scaled and shifted:

$$\binom{n}{k} p^k q^{n-k} \simeq \frac{e^{-(k-m)^2}}{\sqrt{2\pi\sigma^2}}, \qquad n \gg 1 \tag{3-48}$$

where

$$m = np \qquad \text{and} \qquad \sigma^2 = npq \tag{3-49}$$

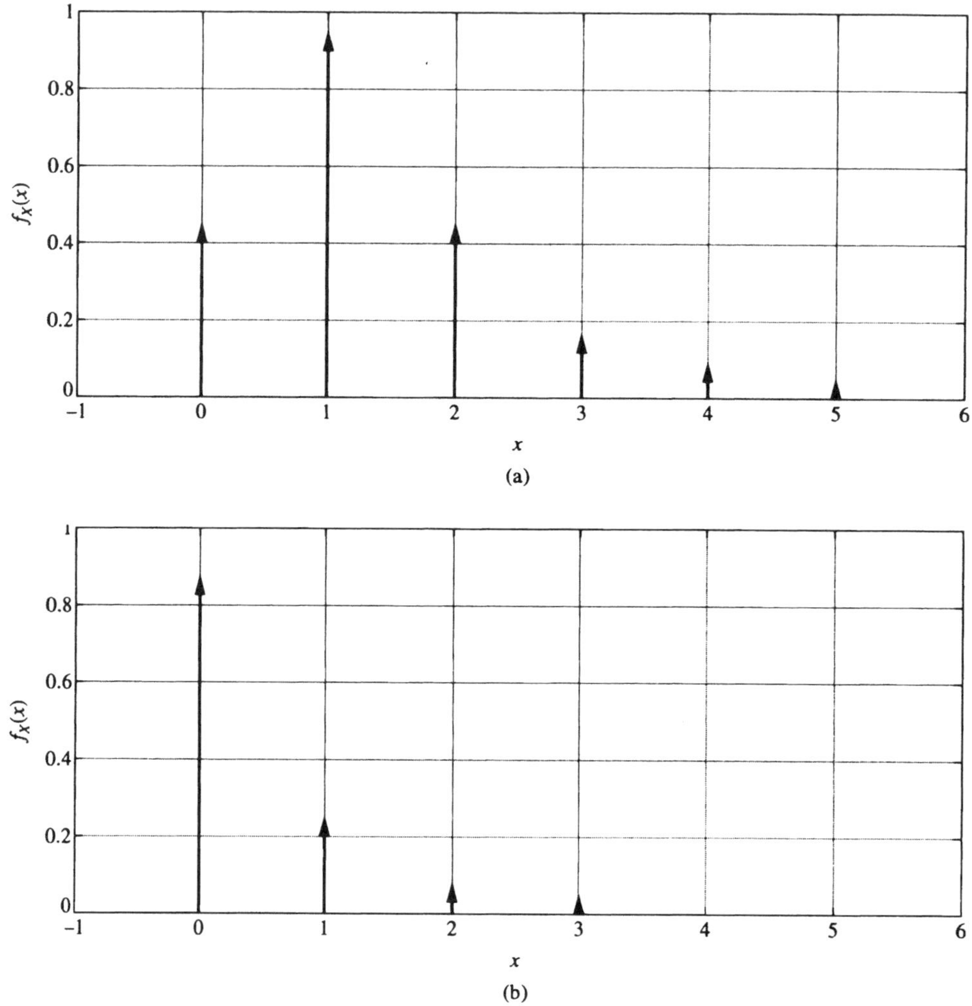

Figure 3-10 Poisson pdf; (a) $a = 0.9$; (b) $a = 0.2$.

This is a difficult relation to prove, and it is suggested that the student plot both sides of (3-48) for moderate values of n to be convinced of its applicability. Given the approximation (3-48), it follows that the cdf of a binomial random variable can be approximated by

$$F_X(x) \simeq 1 - Q\left(\frac{x - np}{\sqrt{npq}}\right) \tag{3-50}$$

where $Q(\,\cdot\,)$ is the Q-function defined by (3-29). Comparisons between the binomial cdf and De Moivre–Laplace approximation to it, given by (3-50), are shown in Figure 3-11.

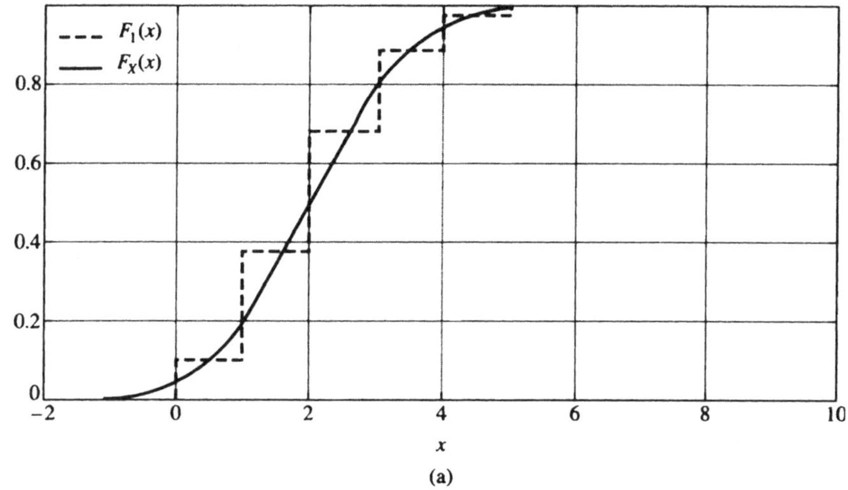

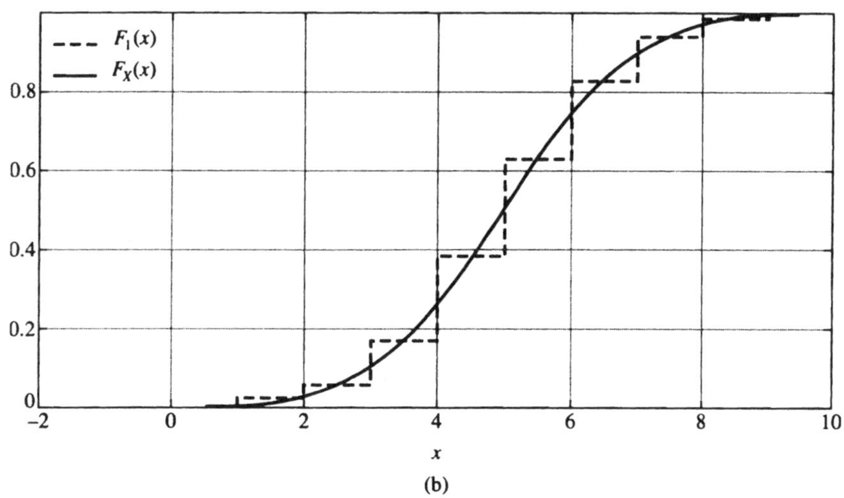

Figure 3-11 Binomial cdf and De Moivre—Laplace approximation for $n = 10$: (a) $p = 0.2$; (b) $p = 0.5$. $F_1(x)$ is binomial; $F_X(x)$ is Gaussian.

The Poisson distribution approaches a Gaussian distribution for $a \gg 1$, because it can be shown that a binomial distribution approaches the Poisson distribution with $a = np$ if $n \gg 1$ and $p \ll 1$. Thus, by the De Moivre–Laplace theorem, we have

$$\frac{a^k}{k!}e^{-a} \simeq \frac{e^{-(k-a)^2/2a}}{\sqrt{2\pi a}}, \qquad n \gg 1, \;\; p \ll 1, \;\; np = a \tag{3-51}$$

The binomial distribution approaches the Poisson distribution under the conditions stated above can be shown by writing the binomial distribution as (from above, $p = a/n$)

$$\binom{n}{k}p^k(1-p)^{n-k} = \frac{n!}{k!(n-k)!}p^k(1-p)^{n-k}$$
$$= \frac{n(n-1)\cdots(n-k+1)}{k(k-1)\cdots 1}\left(\frac{a}{n}\right)^k\left(1-\frac{a}{n}\right)^{n-k} \tag{3-52}$$
$$= \frac{\left(1-\frac{1}{n}\right)\left(1-\frac{2}{n}\right)\cdots\left(1-\frac{k-1}{n}\right)}{k!}a^k\left(1-\frac{a}{n}\right)^{n-k}$$

As $n\to\infty$, the numerator of the fraction in the last line approaches 1; we can write the last factor as $(1 - a/n)^{n-k} = [(1 - a/n)^n](1 - a/n)^{-k}\to e^{-a}, n\to\infty$ where $\lim_{n\to\infty} (1 - a/n)^n = e^{-a}$ has been used.[4] [Note that $(1 - a/n)^{-k}\to 1$ as $n \to \infty$ for any fixed k.] The limit of the binomial distribution under the conditions stated is

$$\lim_{n\to\infty,\ a=np,\ p\ll 1}\left[\binom{n}{k}p^k(1-p)^{n-k}\right] = \frac{a^k}{k!}e^{-a} \tag{3-53}$$

From (3-48), we may then approximate the binomial distribution with a Poisson distribution under the conditions above as

$$\binom{n}{k}p^k q^{n-k} \simeq \frac{(np)^k}{k!}e^{-np}, \quad n \gg 1,\ p \ll 1 \tag{3-54}$$

We will refer to this as the *Poisson approximation to the binomial distribution.*

Example 3-11

Consider a digital communications system in which 1's and 0's are transmitted. The probability of an error (i.e., a 1 received as a 0, or vice versa) is 10^{-3}. What is the probability of three errors in the transmission of 5000 bits (a bit is a 1 or a 0)?

Solution The Poisson approximation to the binomial distribution can be used here with $n = 5000, p = 10^{-3}$, and $k = 3$. Thus, the probability of exactly three errors is approximately

$$P(3\text{ errors}) \simeq \frac{5^3}{3!}e^{-5} = 0.14037$$

The exact result using the binomial distribution is

$$\binom{5000}{3}(10^{-3})^3(1-10^{-3})^{4997} = 0.14036$$

Poisson Points and the Exponential Probability Density Function

Suppose that we consider a finite interval T on the real line, as shown in Figure 3-12, with the probability that k occurrences of an event in this interval (e.g., arrivals of electrons at a certain point) obey a Poisson distribution:

[4]The student might check this limit out using MATLAB or Mathcad. Consider $(1 - 1/k)^k$ for various values of k: For $k = 100$, the result is 0.366; for $k = 1000$, computation gives 0.3677. The result for e^{-1} to four places is 0.3679.

$$P(X = k) = \frac{(\lambda T)^k}{k!}e^{-\lambda T}, \quad k = 0, 1, 2, \ldots \tag{3-55}$$

where λ is the number of events per unit of time. Then the pdf of the interval from an arbitrarily selected point and the next event is a random variable, W, that obeys an exponential distribution:

$$f_W(w) = \lambda e^{-\lambda w}u(w) \tag{3-56}$$

To show this, consider the probability that $W \leq w$, which is the cdf of W. Clearly, this is the case if there is at least one Poisson event in the interval $(0, w)$. The probability of this is

$$P(W \leq w) = F_W(w) = 1 - P(X = 0) = 1 - e^{-\lambda w}, \quad w \geq 0 \tag{3-57}$$

Differentiating, we obtain (3-56).

Example 3-12

Raindrops impinge on a corrugated metal roof at the rate of 100 s^{-1}. What is the probability that the interval between adjacent raindrops is greater than 1 millisecond (ms)? 10 ms?

Solution The raindrops falling on the roof can be approximated as a Poisson point process with $\lambda = 100$ drops per second. Thus

$$P(W \geq 10^{-3}\text{s}) = 1 - P(W \leq 10^{-3}\text{s})$$

$$= 1 - (1 - e^{-100 \times 0.001}) = e^{-0.1} = 0.905$$

and

$$P(W \geq 10^{-2}\text{s}) = 1 - P(W \leq 10^{-2}\text{s})$$

$$= 1 - (1 - e^{-100 \times 0.01}) = e^{-1} = 0.368$$

Pascal Random Variable

The random variable X has a Pascal distribution if it takes on the values 1, 2, 3, . . . with probability

$$P(X = k) = \binom{n-1}{k-1}p^k q^{n-k}, \quad k = 1, 2, \ldots, q = 1 - p \tag{3-58}$$

For example, the probability of obtaining the kth head at toss n in a sequence of coin tossings obeys a Pascal distribution (biased coin with the probability of a head $= p$). We can get the first $k - 1$ heads in any order in the first $n - 1$ tosses (this is binomial); on the nth toss we must get a head. Note that the geometric distribution is a special case of the Pascal distribution.

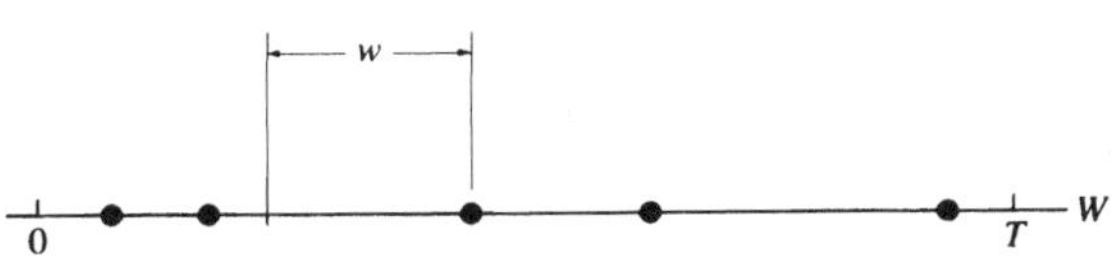

Figure 3-12 Poisson points on the real line.

Example 3-13

What is the probability that the third head in a sequence of tossings of a fair coin will turn up on the twentieth toss? On the fifth toss? On the third toss?

Solution From (3-58) with $k = 20$ and $p = q = \frac{1}{2}$ we obtain

$$P(X = 20) = \binom{19}{4}\left(\frac{1}{2}\right)^3\left(\frac{1}{2}\right)^{17} = 3876 \times \left(\frac{1}{2}\right)^{20} = 0.0037$$

For five tosses we obtain 0.1875. For three tosses we obtain $(\frac{1}{2})^3 = 0.125$.

Hypergeometric Random Variable

Consider the sampling without replacement of a lot of N items, K of which are different (say, defective). The probability of obtaining k defective items in a selection of n items without replacement obeys a hypergeometric distribution:

$$P(X = k) = \frac{\binom{K}{k}\binom{N-K}{n-k}}{\binom{N}{n}}, \qquad k = 0, 1, 2, \ldots, n \tag{3-59}$$

Example 3-14

A jar contains 90 beans and 10 jelly beans. Five objects are drawn from the jar without replacement. What is the probability that one of the items is a jelly bean? Two of the items? None of the items?

Solution We use a hypergeometric distribution with $N = 100$, $K = 10$, $n = 5$, and $k = 1$, 2, and 0 in turn. Thus the probabilities are

$$P(X = 1) = \frac{\binom{10}{1}\binom{90}{5-1}}{\binom{100}{5}} = 0.3394$$

$$P(X = 2) = \frac{\binom{10}{2}\binom{90}{5-2}}{\binom{100}{5}} = 0.07022$$

$$P(X = 0) = \frac{\binom{10}{0}\binom{90}{5}}{\binom{100}{5}} = 0.5838$$

The probability of obtaining no jelly beans is surprisingly the highest.

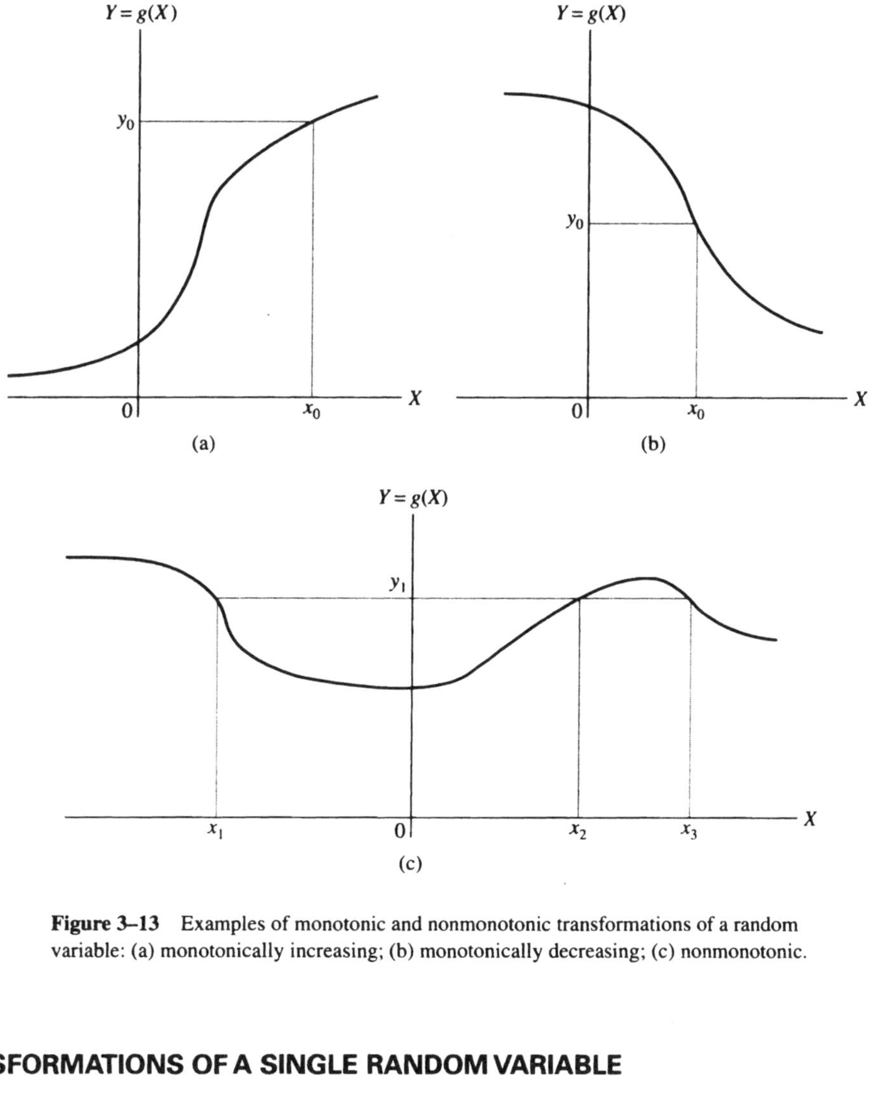

Figure 3–13 Examples of monotonic and nonmonotonic transformations of a random variable: (a) monotonically increasing; (b) monotonically decreasing; (c) nonmonotonic.

3–4 TRANSFORMATIONS OF A SINGLE RANDOM VARIABLE

It is often useful to consider functions of random variables. For the single-random-variable case considered in this chapter, several examples were provided by (3-4). In general, we consider

$$Y = g(X) \tag{3-60}$$

where $g(\cdot)$ is a single-valued function of its argument. Note that if X is a random variable, then so is Y. Given the cdf or pdf of X, we desire the cdf or pdf of Y. Initially, we consider the case where $g(\cdot)$ is monotonic. Examples are provided in Figure 3-13a and b. In this case, for an arbitrary value of the random variable X, say x_0, we have

$$y_0 = g(x_0) \tag{3-61}$$

and for $g(x)$ monotonically increasing (we remove this restriction shortly), the probability that $Y \le y_0$, which is the cdf of Y, is given by

$$F_Y(y_0) = P(Y \le y_0) = P[g(X) \le g(x_0)] = P(X \le x_0) = F_X(x_0) \tag{3-62}$$

as can be seen from Figure 3-13a. Changing to arbitrary values of x and y and differentiating with respect to y, we obtain the pdf of the random variable Y. The result is

$$f_Y(y) = \frac{dF_Y(y)}{dy} = \frac{dF_X(x)}{dx}\frac{dx}{dy}\bigg|_{x=g^{-1}(y)} = f_X(x)\frac{dx}{dy}\bigg|_{x=g^{-1}(y)} \tag{3-63a}$$

The substitution $x = g^{-1}(y)$ is used to obtain a function of y after we are done differentiating.

If $g(X)$ is monotonically decreasing as shown in Figure 3-13b, we then have $F_Y(y_0) = 1 - F_X(x_0)$. Differentiation to get the pdf after substitution of arbitrary values of x and y then results in a minus sign. Since pdf's must be nonnegative, we modify (3-63a) by putting an absolute value around the derivative to get

$$f_Y(y) = \frac{dF_Y(y)}{dy} = \frac{dF_X(x)}{dx}\left|\frac{dx}{dy}\right|_{x=g^{-1}(y)} = f_X(x)\left|\frac{dx}{dy}\right|_{x=g^{-1}(y)} \tag{3-63b}$$

An example will illustrate the procedure.

Example 3-15

Consider the transformation

$$Y = aX + b \tag{3-64}$$

where X is a random variable with exponential pdf (parameter α) given by (3-32). Find the pdf of Y.

Solution We apply (3-63), for which

$$x = \frac{1}{a}(y - b) \tag{3-65a}$$

and

$$\frac{dx}{dy} = \frac{1}{a} \tag{3-65b}$$

Assume that $a > 0$. Then

$$f_Y(y) = \frac{\alpha}{a}e^{-(\alpha/a)(y-b)}u(y - b) \tag{3-66a}$$

If $a < 0$, then $|a| = -a$, and

$$f_Y(y) = \frac{\alpha}{|a|}e^{-(\alpha/|a|)(y-b)}u(y - b) \tag{3-66b}$$

Plots for both cases are given in Figure 3-14.

We next consider the case where the transformation is not monotonic. In this case, there will be more than one solution to (3-63), as illustrated in Figure 3-13c. Suppose that these solutions are denoted by

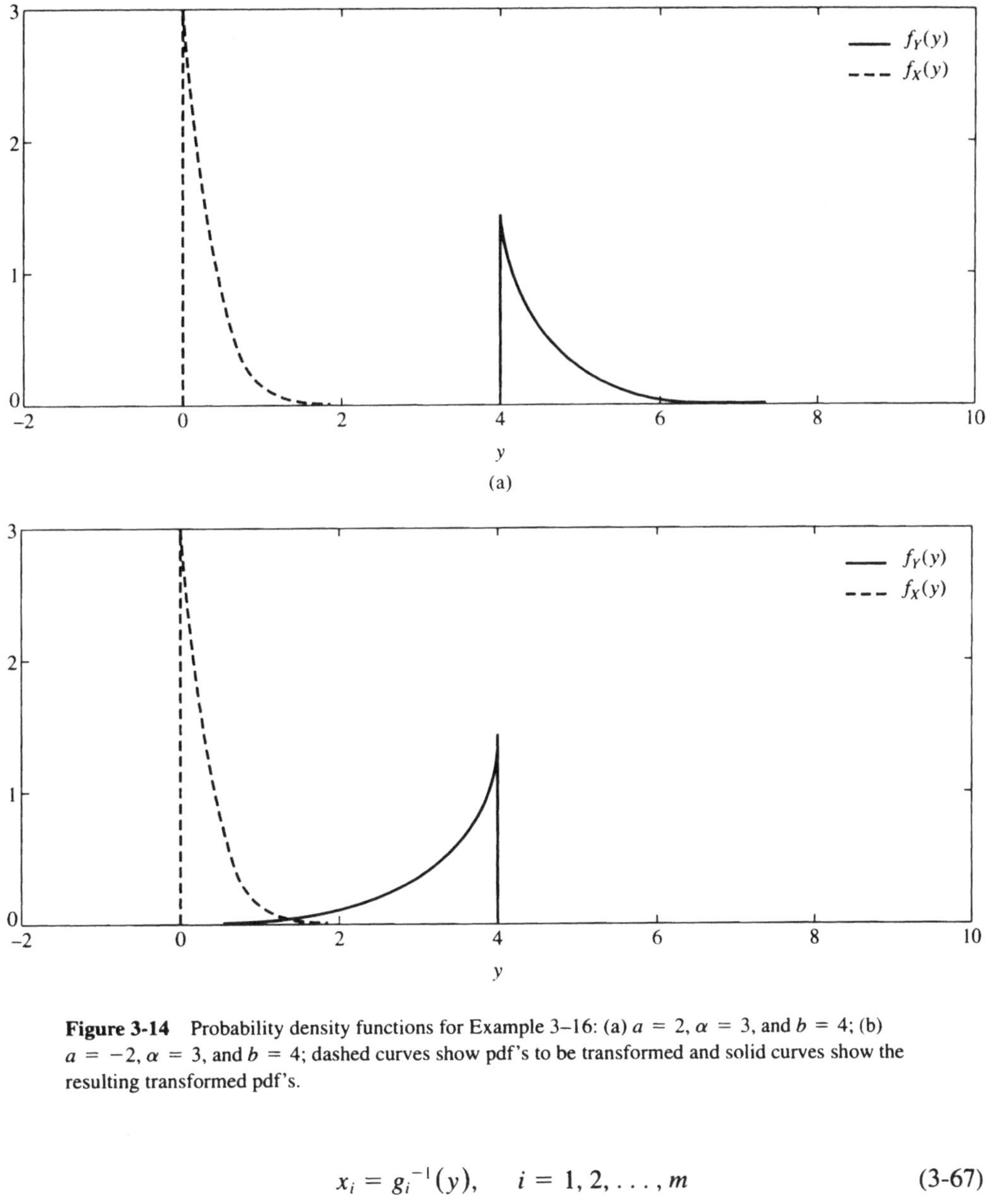

Figure 3-14 Probability density functions for Example 3–16: (a) $a = 2, \alpha = 3$, and $b = 4$; (b) $a = -2, \alpha = 3$, and $b = 4$; dashed curves show pdf's to be transformed and solid curves show the resulting transformed pdf's.

$$x_i = g_i^{-1}(y), \quad i = 1, 2, \ldots, m \tag{3-67}$$

Then (3-63) generalizes to

$$f_Y(y) = \sum_{i=1}^{m} f_X(x) \left|\frac{dx_i}{dy}\right|_{x_i = g_i^{-1}(y)} \tag{3-68}$$

An example will illustrate the procedure used to evaluate (3-68).

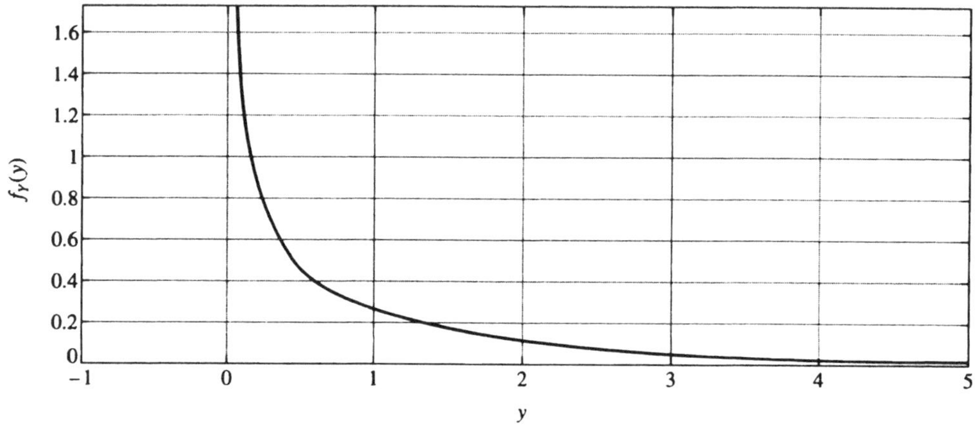

Figure 3-15 Probability density function for the result of Example 3-16 ($\sigma^2 = 1$).

Example 3-16

Suppose that the transformation to be considered is

$$y = x^2 \tag{3-69}$$

and the random variable X has a Gaussian pdf (3-28) with $m = 0$. Find $f_Y(y)$.

Solution The inverse solutions to (3-69) are

$$x_1 = \sqrt{y} \quad \text{and} \quad x_2 = -\sqrt{y} \tag{3-70}$$

Thus

$$\left|\frac{dx_i}{dy}\right| = \frac{1}{2\sqrt{y}}, \quad i = 1, 2 \tag{3-71}$$

Therefore, (3-68) becomes

$$\begin{aligned} f_Y(y) &= \frac{1}{2\sqrt{y}}\left[\frac{e^{-x^2/2\sigma^2}}{\sqrt{2\pi\sigma^2}}\right]_{x_1=\sqrt{y}} + \frac{1}{2\sqrt{y}}\left[\frac{e^{-x^2/2\sigma^2}}{\sqrt{2\pi\sigma^2}}\right]_{x_2=-\sqrt{y}} \\ &= \frac{e^{-y/2\sigma^2}}{\sqrt{2\pi\sigma^2 y}}, \quad y \geq 0 \end{aligned} \tag{3-72}$$

Because y is nonnegative, $f_Y(y)$ is zero for $y < 0$. The pdf is plotted in Figure 3-15.

In the next example we consider a case where values over a certain range of the transformation map into the same output value. Depending on the pdf of the independent variable, this may give a finite probability for the given output value (i.e. the output random variable is of the mixed type).

Example 3-17

Let the transformation be

$$y = \begin{cases} ax, & x \geq 0, \\ 0, & x < 0 \end{cases} \quad a > 0 \tag{3-73}$$

Assume that X has a Gaussian pdf, as given by (3-28), with $m = 0$. Find $f_Y(y)$. (See if you can deduce the answer before doing any actual calculations.)

Solution For $x \geq 0$, we have $x = y/a$ and $dx/dy = 1/a$. The pdf for Y is therefore

$$f_Y(y) = \frac{e^{-y^2/2a^2\sigma^2}}{\sqrt{2\pi a^2\sigma^2}}, \quad y \geq 0 \tag{3-74}$$

For $-\infty < X < 0$, we have $Y = 0$. Thus

$$P(Y = 0) = P(X < 0) = \frac{1}{2} \tag{3-75}$$

which follows due to the symmetry of the pdf for X about zero. We can represent the finite probability of Y taking on the value zero by a delta function in the pdf, and represent $f_Y(y)$ for all y as

$$f_Y(y) = \frac{e^{-y^2/2a^2\sigma^2}}{\sqrt{2\pi a^2\sigma^2}}u(y) + \frac{1}{2}\delta(y) \tag{3-76}$$

The fact that $X \leq 0$ maps into $Y = 0$ is reflected by the presence of the unit impulse function in the result for $f_Y(y)$.

3–5 AVERAGES OF RANDOM VARIABLES

While the pdf or cdf of a random variable describes it completely, we may not be interested in such a complete description in many cases. For example, a complete record of the daily prices of a certain stock throughout a year may provide too much detail, and we therefore compute the average price of the stock over a year's time or a month's time. The idea of the average of a random variable is similar to this example, except, rather than using an average over time, we compute the average using the random variable's pdf according to the equation

$$E(X) = \int_{-\infty}^{\infty} x f_X(x)\, dx \tag{3-77}$$

The notation $E(\cdot)$ stands for *expectation,* and $E(X)$ is read "the expectation of X." Note that if X is a discrete random variable, the integral becomes a sum. For example, representing the pdf of a discrete random variable that takes on the values $x_1, x_2, \ldots, x_n$ with probabilities $P(X = x_i)$, $i = 1, 2, \ldots, n$, (3-77) becomes

$$E(X) = \int_{-\infty}^{\infty} x \sum_{i=1}^{n} P(X = x_i)\delta(x - x_i)\, dx = \sum_{i=1}^{n} x_i P(X = x_i) \quad (X\text{ discrete}) \tag{3-78}$$

by using the sifting property of the unit impulse function.

Example 3-18

To show the reasonableness of the definitions above, consider a class of 100 students who have just taken a test. To simplify matters, the instructor rounds off the test scores to the nearest multiple of 5. This results in Table 3-2.

Applying (3-78) and using relative frequency to approximate probability, we obtain the average score in the class to be

Table 3-2 TEST SCORES BY A CLASS OF 100 STUDENTS

Score	Number of students	Relative frequency
100	2	0.02
95	5	0.05
90	10	0.10
85	20	0.20
80	33	0.33
75	15	0.15
70	7	0.07
65	4	0.04
60	3	0.03
55	1	0.01

$$\begin{aligned} E(X) &= 100 \times 0.02 + 95 \times 0.05 + 90 \times 0.1 + 85 \times 0.2 + 80 \times 0.33 \\ &\quad + 75 \times 0.15 + 70 \times 0.07 + 65 \times 0.04 + 60 \times 0.03 + 55 \times 0.01 \\ &= 80 \end{aligned} \tag{3-79}$$

We also could have calculated the average score as

$$\text{average} = \tag{3-80}$$

$$\frac{100 \times 2 + 95 \times 5 + 90 \times 10 + 85 \times 20 + 80 \times 33 + 75 \times 15 + 70 \times 7 + 65 \times 4 + 60 \times 3 + 55 \times 1}{100}$$

which would have resulted in the same answer. We are accustomed to using the second method from grammar school days (i.e. compute the total of all scores and divide by the total number of grades). Thus we see that the definition of an average given by (3-78) is reasonable.

We are also interested in averages, or expectations, of functions of random variables. For example, we are often interested in the average spread of the scores about the average score on a test. This is called the *standard deviation* and is given by

$$\sigma = \sqrt{E\{[X - E(X)]^2\}} \tag{3-81}$$

Note that not just any expectation will do for showing the spread of a random variable about its mean. If $E[X - E(X)]$ is used, for example, positive and negative values for the difference $X - E(X)$ will cancel and a small measure (actually 0) of the deviation may result even though the spread around the mean is very large. We might try to rectify this with the expectation $E[|X - E(X)|]$, which indeed would cure the problem of positive and negative values canceling. However, $E[|X - E(X)|]$ would not be very convenient from an analytic standpoint because the absolute value is hard to work with. Fortunately, using the squared error gives useful results.

In general, for any function of $g(X)$ of a random variable X, we define the expectation of this function to be[5]

[5]This has been referred to as the *law of the unconscious statistician* (Ross, 1994) because many books take it as a definition. That (3-82) holds follows by transformation of variables. By (3-77), the expectation of $y = g(X)$ is $E(Y) = \int_{-\infty}^{\infty} y f_Y(y)\, dy$. Assuming that $g(X)$ is monotonic for simplicity, application of (3-63) gives (3-82).

$$E[g(X)] = \int_{-\infty}^{\infty} g(x)f_X(x)\,dx \tag{3-82}$$

which holds for either continuous or discrete random variables if we use the unit impulse representation for the pdf of discrete random variables.

If $g(X) = X^m$, where m is an integer, we say that the resulting average is the mth *moment* of the random variable. The first moment ($m = 1$) of a random variable is called its *mean* and its second moment ($m = 2$) is called its *mean-square value.*

Example 3-19

Find the mean and mean-square value for the uniform random variable with pdf given by (3-24).

Solution The mean is given by

$$E(X) = \int_a^b \frac{x\,dx}{b-a} = \left.\frac{x^2}{2(b-a)}\right|_a^b = \frac{b^2-a^2}{2(b-a)} = \frac{a+b}{2} \tag{3-83}$$

The mean-square value is given by

$$E(X^2) = \int_a^b \frac{x^2dx}{b-a} = \left.\frac{x^3}{3(b-a)}\right|_a^b = \frac{b^3-a^3}{3(b-a)} = \frac{a^2+ab+b^2}{3} \tag{3-84}$$

If we subtract the mean from a random variable and find the expectation of integer powers of this difference, we obtain the *central moments* of the random variable. Thus, if $\mu_x = E(X)$ is the mean of the random variable X, its central moments are given by

$$m_n = E[(X-\mu_X)^n] = \int_{-\infty}^{\infty} (x-\mu_X)^n f_X(x)\,dx, \qquad n = 1, 2, \ldots \tag{3-85}$$

The second central moment of a random variable is called its *variance;* it is usually denoted by the symbol σ^2, and for the random variable X is given by

$$\sigma_X^2 = m_2 = E[(X-\mu_X)^2] = \int_{-\infty}^{\infty} (x-\mu_X)^2 f_X(x)\,dx \tag{3-86}$$

The variance of the uniform random variable in (a, b), whose mean and mean-square values were computed in Example 3-19, can be shown to be $\sigma^2 = (b-a)^2/12$.

Example 3-20

(a) Show that the mean of a Gaussian random variable, with pdf given by (3-28), is $\mu_X = m$.

(b) Find the central moments of a Gaussian random variable.

Solution (a) The mean is given by

$$\begin{aligned}\mu_X &= \int_{-\infty}^{\infty} x\frac{e^{-(x-m)^2/2\sigma^2}}{\sqrt{2\pi\sigma^2}}\,dx \qquad (\text{let } u = x - m)\\ &= \int_{-\infty}^{\infty}(u+m)\frac{e^{-u^2/2\sigma^2}}{\sqrt{2\pi\sigma^2}}\,du = \int_{-\infty}^{\infty} u\frac{e^{-u^2/2\sigma^2}}{\sqrt{2\pi\sigma^2}}\,du + m\int_{-\infty}^{\infty}\frac{e^{-u^2/2\sigma^2}}{\sqrt{2\pi\sigma^2}}\,du\end{aligned} \tag{3-87}$$

The second-to-last integral on the right is zero because the integrand is odd and the range of integration is symmetric about $u = 0$. The last integral evaluates to unity because the integrand is a pdf. Hence we have shown that the mean is $\mu_X = m$.

(b) Employing the definition of the central moments of a random variable given by (3-85) along with (3-28), we obtain

$$\begin{aligned} m_n = E\left[(X - \mu_X)^n\right] &= \int_{-\infty}^{\infty} (x - \mu_X)^n \frac{e^{-(x-\mu_X)^2/2\sigma^2}}{\sqrt{2\pi\sigma^2}} dx \\ &= \int_{-\infty}^{\infty} u^n \frac{e^{-u^2/2\sigma^2}}{\sqrt{2\pi\sigma^2}} du, \qquad n = 1, 2, \ldots \end{aligned} \tag{3-88}$$

For n odd, the integrand is odd and the integral evaluates to zero. For n even, we can double the integral and employ a table of integrals to get the result

$$m_{2k} = 2\int_0^{\infty} u^{2k} \frac{e^{-u^2/2\sigma^2}}{\sqrt{2\pi\sigma^2}} du = 1 \cdot 3 \cdot \ldots \cdot (2k-1)\sigma^{2k} \qquad k = 1, 2, \ldots \tag{3-89}$$

The special case $2k = n = 2$ shows that the variance of a Gaussian random variable is σ^2.

Several properties of the expectation computation are handy:

1. The expectation of a constant is the constant itself:
$$E[a] = a, \qquad a = \text{constant} \tag{3-90}$$
2. The expectation of a constant times a function of a random variable is the constant times the expectation of the function of the random variable:
$$E[ag(X)] = aE[g(X)], \qquad a = \text{constant} \tag{3-91}$$
3. The expectation of the sum of two functions of a random variable is the sum of their separate expectations:
$$E[g_1(X) + g_2(X)] = E[g_1(X)] + E[g_2(X)] \tag{3-92}$$

These can be proved by writing down the various expectations in terms of the integral definition of the expectation and using the properties of integrals.

Example 3-21

Show that the variance of a random variable can be computed according to

$$\sigma_X^2 = E[(X - \mu_X)^2] = E(X^2) - [E(X)]^2 \tag{3-93}$$

Solution We use the foregoing properties:

$$\begin{aligned} \sigma_X^2 &= E[(X - \mu_X)^2] \\ &= E(X^2 - 2\mu_X X + \mu_X^2) \\ &= E(X^2) - 2\mu_X E(X) + E(\mu_X^2) \\ &= E(X^2) - 2\mu_X^2 + \mu_X^2 = E(X^2) - \mu_X^2 \end{aligned} \tag{3-94}$$

Example 3-22

Let the random variables X and Y be linearly related by (3-64); show that

$$\mu_Y = a\mu_X + b \quad \text{and} \quad \sigma_Y^2 = a^2\sigma_X^2 \tag{3-95}$$

Solution Again, we use the properties given above. For the relationship between the means, we have

$$\mu_Y = E[aX + b] = aE(X) + E(b) = a\mu_X + b$$

For the relationship between the variances, we write

$$\sigma_Y^2 = E[(Y - \mu_Y)^2] = E\{[(aX + b) - (a\mu_X + b)]^2\} = E[a^2(X - \mu_X)^2] = a^2\sigma_X^2$$

Example 3-23

Find the mean and variance of a binomial random variable with probability distribution given by (3-42).

Solution The mean is given by

$$E(X) = \sum_{k=0}^{n} k\binom{n}{k}p^k q^{n-k} = \sum_{k=1}^{n} k\frac{n!}{k!\,(n-k)!}p^k q^{n-k} = \sum_{k=1}^{n} \frac{n!}{(k-1)!\,(n-k)!}p^k q^{n-k}$$

In the last sum, let $j = k - 1$:

$$E(X) = \sum_{j=0}^{n-1} np\frac{(n-1)!}{j!\,(n-j-1)!}p^j q^{n-j} = np(p+q)^{n-1} = np \quad \text{(binomial r.v.)} \tag{3-96}$$

The sum is evaluated by means of the binomial theorem from algebra and by noting that $p + q = 1$.

We obtain the variance by first computing the mean-square value:

$$E(X^2) = \sum_{k=0}^{n} k^2\binom{n}{k}p^k q^{n-k} = \sum_{k=1}^{n} k^2\frac{n!}{k!\,(n-k)!}p^k q^{n-k} = \sum_{k=1}^{n} k\frac{n!}{(k-1)!\,(n-k)!}p^k q^{n-k}$$

where the lower sum limit can be set to 1 because $k = 0$ in the sum. We add and subtract the sum

$$\sum_{k=1}^{n} \frac{n!}{(k-1)!\,(n-k)!}p^k q^{n-k}$$

from the right-hand side to get (the first sum can start at $k = 2$ because the $k = 1$ term is zero)

$$\begin{aligned} E(X^2) &= \sum_{k=2}^{n}(k-1)\frac{n!}{(k-1)!\,(n-k)!}p^k q^{n-k} + \sum_{k=1}^{n}\frac{n!}{(k-1)!\,(n-k)!}p^k q^{n-k} \\ &= \sum_{k=2}^{n}\frac{n!}{(k-2)!\,(n-k)}p^k q^{n-k} + \sum_{k=1}^{n}\frac{n!}{(k-1)!\,(n-k)!}p^k q^{n-k} \end{aligned}$$

The last sum on the right-hand side is the sum that evaluated to np in (3-96). Now make the change of summation index to $j = k - 2$ in the first sum on the right-hand side to get

$$\begin{aligned} E(X^2) &= \sum_{j=0}^{n-2} n(n-1)p^2\frac{(n-1)!}{j!(n-2-j)!}p^j q^{n-2-j} + np \\ &= n(n-1)p^2(p+q)^{n-2} + np = n(n-1)p^2 + np \end{aligned}$$

Using this result in (3-93) along with (3-96), we obtain

$$\sigma_X^2 = n(n-1)p^2 + np - (np)^2 = np(1-p) = npq \quad \text{(binomial r.v.)} \tag{3-97}$$

3-6 CHARACTERISTIC FUNCTION

The *characteristic function* is the special average

$$M_X(jv) = E(e^{jvX}) = \int_{-\infty}^{\infty} f_X(x)e^{jvx}\,dx \tag{3-98}$$

It can be shown to be absolutely convergent for any pdf. It is useful for a number of reasons:

1. The *m*th moment of a random variable can be obtained by differentiating the characteristic functions *m* times with respect to its argument.
2. It is sometimes possible to obtain the characteristic function of a random variable when the pdf is very difficult or impossible to obtain.
3. The characteristic function and pdf are Fourier transform pairs.
4. The Maclaurin series expansion of the characteristic function can be expressed in terms of the moments of the random variable.

To show the first statement, we differentiate the characteristic function with respect to v to obtain

$$\frac{dM_X(jv)}{dv} = \int_{-\infty}^{\infty} f_X(x)\frac{d}{dv}e^{jvx}\,dx = j\int_{-\infty}^{\infty} xf_X(x)e^{jvx}\,dx \tag{3-99a}$$

Now set $v = 0$ and divide through by j to obtain

$$-j\left.\frac{dM_X(jv)}{dv}\right|_{v=0} = \int_{-\infty}^{\infty} xf_X(x)\,dx = E(X) \tag{3-99b}$$

Thus we have obtained the first moment by differentiating. Generalizing, we see that

$$E(X^n) = (-j)^n\left.\frac{d^nM_X(jv)}{dv^n}\right|_{v=0} \tag{3-100}$$

Example 3-24

Obtain the characteristic function for the Cauchy random variable with pdf given by (3-40).

Solution From (3-98) with (3-40) substituted, we obtain

$$M_X(jv) = \int_{-\infty}^{\infty}\frac{\alpha/\pi}{x^2+\alpha^2}e^{jvx}\,dx = \int_{-\infty}^{\infty}\frac{\alpha/\pi}{x^2+\alpha^2}[\cos(vx) + j\sin(vs)]\,dx = \frac{\alpha}{\pi}\int_{-\infty}^{\infty}\frac{\cos(vx)}{x^2+\alpha^2}\,dx$$

where the integral over the sine is zero due to the oddness of the integrand. A table of definite integrals allows us to evaluate the remaining integral over the cosine as

$$M_X(jv) = e^{-\alpha|v|} \tag{3-101}$$

Because of the discontinuity of slope at the origin for this characteristic function, it is not differentiable. Thus we cannot use (3-100) to evaluate the moments in this case. Its moments, in fact, do not exist.

Example 3-25

Find the characteristic function corresponding to the double-sided exponential pdf (also called the *Laplace* pdf), given by

$$f_X(x) = \frac{\alpha}{2} e^{-\alpha|x|}, \quad \alpha > 0 \tag{3-102}$$

Solution By the definition of the characteristic function (3-98), we have

$$M_X(jv) = \int_{-\infty}^{\infty} \frac{\alpha}{2} e^{-\alpha|x|} e^{jvx}\, dx = \int_{-\infty}^{\infty} \frac{\alpha}{2} e^{-\alpha|x|}[\cos(vx) + j\sin(vx)]\, dx = \frac{\alpha}{2}\int_{-\infty}^{\infty} \cos(vx)e^{-\alpha|x|}\, dx$$

The integral over the sine is zero because of the odd symmetry of the integrand. The integral over the cosine can be doubled and integrated from 0 to ∞, and a table of definite integrals can be used to obtain

$$M_X(jv) = \alpha\int_0^{\infty} \cos(vx)e^{-\alpha x}\, dx = \frac{\alpha^2}{\alpha^2 + v^2} \tag{3-103}$$

Differentiation shows that the first moment is zero and the second moment is $2/\alpha^2$.

3–7 CHEBYSHEV'S INEQUALITY

As discussed in connection with (3-81), a random variable's standard deviation gives a measure of the spread of realizations of that random variable about its mean. *Chebyshev's inequality* provides a bound on the probability that a random variable deviates more than k standard deviations from its mean. Although this bound can be loose, very little need be known about the random variable. Chebyshev's inequality is given by

$$P(|X - \mu_X| \geq k\sigma_X) \leq \frac{1}{k^2} \tag{3-104}$$

Since the events

$$|X - \mu_X| \geq k\sigma_X \quad \text{and} \quad |X - \mu_X| < k\sigma_X$$

are mutually exclusive, it follows that

$$P(|X - \mu_X| < k\sigma_X) > 1 - \frac{1}{k^2} \tag{3-105}$$

To prove (3-105), consider the random variable

$$Y = X - \mu_X \tag{3-106}$$

and let $a = k\sigma_X$. Consider the left side of (3-104), which in terms of the notation just defined can be written as

$$P(|Y| \geq a) = P(Y \leq -a) + P(Y \geq a) \tag{3-107}$$

which follows because the event $|Y| \geq a$ corresponds to the mutually exclusive events $Y \geq a$ and $Y \leq -a$. Now consider the second moment of Y, which is

$$E(Y^2) = \int_{-\infty}^{\infty} y^2 f_Y(y)\, dy \geq \int_{-\infty}^{-a} y^2 f_Y(y)\, dy + \int_a^{\infty} y^2 f_Y(y)\, dy, \quad a > 0 \tag{3-108}$$

The latter inequality follows because the integrand for the second-moment integral is nonnegative, and we have left part of the integral out. Now note that $y^2 > a^2$ in both integrals on the right-hand side of the inequality in (3-108). Thus we can write (3-108) as

$$E(Y^2) \geq a^2\left[\int_{-\infty}^{-a} f_Y(y)\, dy + \int_a^{\infty} f_Y(y)\, dy\right] = a^2[P(Y \leq -a) + P(Y \geq a)], \quad a > 0 \tag{3-109}$$

Solving, we obtain

$$P(Y \le -a) + P(Y \ge a) \le \frac{E(Y^2)}{a^2} \tag{3-110}$$

Substituting into (3-107), we have

$$P(|Y| \ge a) \le \frac{E(Y^2)}{a^2} \tag{3-111}$$

Chebyshev's inequality given by (3-104) results by letting $Y = X - \mu_X$ so that $E(Y^2) = \sigma_X^2$ and $a = k\sigma_X$.

Example 3-26

(a) Find a bound on the probability that a random variable is within three standard deviations of its mean.

(b) If the random variable has a Gaussian pdf given by (3-28), find the exact probability of this event and compare with the bound.

Solution

(a) We use (3-105) to obtain

$$P(|X - \mu_X| < 3\sigma_X) > 1 - \frac{1}{3^2} = 0.889 \tag{3-112}$$

(b) For the Gaussian random variable with pdf given by (3-28), we have

$$\begin{aligned} P(|X - \mu_X| < 3\sigma_X) &= \int_{\mu_X - 3\sigma_X}^{\mu_X + 3\sigma_X} \frac{e^{-(x-\mu_X)^2/2\sigma_X^2}}{\sqrt{2\pi\sigma_X^2}}\,dx \\ &= \int_{-3}^{3} \frac{e^{-u^2/2}}{\sqrt{2\pi}}\,du = 2\int_{0}^{3} \frac{e^{-u^2/2}}{\sqrt{2\pi}}\,du \qquad (3\text{-}113) \\ &= 1 - 2Q(3) = 1 - 2 \times 0.00135 \\ &= 0.9973 \end{aligned}$$

That is, for a Gaussian random variable, 99.73% of its realizations are within three standard deviations of the mean. In this case, Chebyshev's inequality does not provide a very tight bound. For pdf's less concentrated about their means, the bound would be closer to the actual probability. This is more fully examined in the problems.

3-8 COMPUTER GENERATION OF RANDOM VARIABLES

As discussed in Section 1-3, the simulation of random phenomena on computers is often useful. There are, in fact, simulation languages that allow the simulation of complex systems subject to randomness, such as communication networks as well as communications, control, and radar systems. The generation of uniformly distributed random numbers was discussed in Section 1-3. In this section we extend this discussion to generation of random variables with other distributions and discuss simple simulation ideas. Some computer exercises involving simulation of probabilistic phenomena are given at the end of the chapter. Their solution can be carried out with a high-level programming language such as

Pascal, Fortran, or C. However, as discussed in Chapter 1, use of one of the many mathematics packages now available is recommended (these include Mathematica, Mathcad, and MATLAB).

As discussed in Section 1-3, most computer programming languages include a subprogram that will generate a pseudorandom variable uniformly distributed over a given range, usually [0, 1]. We assume the existence of such a random number generator in the following. In addition, a generator for Gaussian pseudorandom numbers is often available. In MATLAB, the statement $X = \text{randn}(p, q)$ will generate an array of Gaussian pseudorandom numbers with p rows and q columns, each with zero mean and unity variance. This array can, of course, be transformed to an array of Gaussian pseudorandom numbers with means m and standard deviations σ (both assumed to be scalars) by the transformation $Y = \sigma X + m$.

We often desire random numbers distributed in some other fashion than uniformly or Gaussian, however. We accomplish this by performing a transformation of variables. Let U and V be random variables related by the monotonic transformation

$$V = g(U) \tag{3-114}$$

where U is uniformly distributed in [0, 1]. Then, by (3-63b), we have

$$\begin{aligned} f_V(v) &= f_U(u)\left|\frac{du}{dv}\right|_{u=g^{-1}(v)} \\ &= \begin{cases} \left|\dfrac{du}{dv}\right| = \left|\dfrac{dg^{-1}(v)}{dv}\right|, & 0 \le u \le 1 \\ 0, & \text{otherwise} \end{cases} \end{aligned} \tag{3-115}$$

where the last equation follows because $f_U(u)$ is unity in the interval [0, 1] and zero otherwise. Now the absolute value of a quantity is equal to that quantity when it is greater than zero, and, equal to minus that quantity. Thus, (3-115) can be rewritten as

$$f_V(v) = \begin{cases} \dfrac{dg^{-1}(v)}{dv}, & \dfrac{dg^{-1}(v)}{dv} > 0 \\ -\dfrac{dg^{-1}(v)}{dv}, & \dfrac{dg^{-1}(v)}{dv} < 0 \end{cases} \tag{3-116}$$

Integrating and solving for $g^{-1}(v)$, we obtain

$$g^{-1}(v) = \begin{cases} \displaystyle\int_{-\infty}^{v} f_v(\lambda)\, d\lambda = F_V(v), & \dfrac{dg^{-1}(v)}{dv} > 0 \\ \displaystyle -\int_{-\infty}^{v} f_v(\lambda)\, d\lambda = -F_V(v), & \dfrac{dg^{-1}(v)}{dv} < 0 \end{cases} \tag{3-117}$$

where $F_V(v)$ is the cdf of V. The next example will illustrate the application of this result.

Example 3-27

Find the transformation that will generate an exponential pdf, given by

$$f_V(v) = 2e^{-2v}u(v) \tag{3-118}$$

from a random variable uniform in [0, 1].

Solution To find the desired transformation, we require that

$$F_V(v) = \int_{-\infty}^{v} f_V(\lambda)\, d\lambda = \begin{cases} 0, & v < 0 \\ 1 - e^{-2v}, & v > 0 \end{cases} \tag{3-119}$$

Therefore, from (3-117), we obtain

$$u = g^{-1}(v) = 1 - e^{-2v}, \qquad v > 0 \tag{3-120}$$

(Note that this inverse transformation always has positive slope.) Solving for v and recalling that it is the relationship between the random variables that we want, we have

$$\begin{aligned} V &= -0.5 \ln (1 - U) \\ &= -0.5 \ln U, \qquad U \text{ uniform in } [0, 1] \end{aligned} \tag{3-121}$$

where the last equation results because if U is uniform in [0, 1], so is $1 - U$.

3-9 SUMMARY

In this chapter, the concept of a random variable was introduced. Random variables are convenient in probability theory because we can numerically describe the outcomes of chance experiments through their use. Various probability distributions for describing random variables were introduced and their properties given. Transformations of random variables were then considered, followed by averages of functions of random variables. Finally, Chebyshev's inequality was derived and the generation of random numbers by computer was discussed.

1. A *random variable* is a function that assigns a real number to each outcome of a chance experiment: that is, it is a function whose domain of definition is the sample space of a chance experiment and whose range of values is the real line. Uppercase letters near the end of the English alphabet are usually used to denote a random variable, and the values that it takes on is denoted by the corresponding lowercase letter.
2. If X is a random variable, any function of X is also a random variable; for example, $Y = \cos X$ is a random variable if X is.
3. Random variables can be *discrete, continuous,* or *mixed.* A discrete random variable takes on a countable number of values with finite probability. A continuous random variable takes on a continuum of values; the probability of taking on a specific value is infinitesimal. Mixed random variables take on both discrete values as well as a continuum of values.
4. A discrete random variable can be described by its probability mass distribution, which is simply a tabulation or plot of the probability that the random variable takes on a given value versus that value [i.e., $P(X = x_i)$, $i = 1, 2, 3, \ldots$, is plotted versus x_i, where X is the random variable and the x_i's are the values it takes on with probabilities $P(X = x_i)$].

5. Two other possible descriptions of a random variable are in terms of its *cumulative probability distribution function* (cdf) and its *probability density function* (pdf). For a random variable X, the cdf $F_X(x)$ is defined as

$$F_X(x) = P(X \leq x)$$

where the independent variable x is a running value. This description works for discrete, continuous, or mixed random variables. The cdf of a discrete or mixed random variable includes jump discontinuities. The pdf $f_X(x)$ of a continuous random variable is defined as

$$f_X(x) = \frac{F_X(x)}{dx}$$

Since the cdf of a discrete or mixed random variable includes jump discontinuities, we have to agree that the derivative of such a discontinuity gives an impulse or delta function in order to apply the pdf to all three types of random variables.

6. Properties of a cdf include the following:
   (a) Limit of 0 as its argument goes to $-\infty$, and limit of 1 as its argument goes to ∞.
   (b) Monotonically nondecreasing function of its argument.
   (c) Continuous from the right at a point of discontinuity.
   (d) The probability that the random variable lies in a given range is

$$P(x_1 < X \leq x_2) = F_X(x_2) - F_X(x_1)$$

7. Properties of a pdf include the following:
   (a) It is nonnegative everywhere.
   (b) The area under the pdf curve is unity.
   (c) Its integral is the cdf

$$F_X(x) = \int_{-\infty}^{x} f_X(u)\, du$$

   (d) As a generalization of the integral above,

$$P(x_1 < X \leq x_2) = \int_{x_1}^{x_2} f_X(u)\, du$$

8. Examples of useful types of random variables include the following:
   (a) *Continuous:* uniform, Gaussian, exponential, Laplace (double-sided exponential), gamma, chi-square, Erlang, and Cauchy
   (b) *Discrete:* binomial, Poisson, geometric, Pascal, and hypergeometric

9. An important theorem is the De Moivre–Laplace theorem, which states that a binomial distribution can be approximated by samples of a Gaussian distribution if the number of trials is large. If, in addition, the probability of success is small, the binomial distribution can be approximated by the Poisson distribution. This is very handy for problems where the number of trials is large,

making the computation of the binomial coefficient in the binomial distribution tedious.

10. If a number of random points in a finite interval on the real line (quite often, time) obeys a Poisson distribution, we say that we have a *Poisson point process.* Picking an arbitrary point, it can be shown that the interval to the next Poisson point for such a process has an exponential pdf.

11. If $Y = g(X)$ is a monotonic transformation, the pdf of Y given in terms of the pdf of X is

$$f_Y(y) = \left[f_X(x) \left| \frac{dx}{dy} \right| \right]_{x=g^{-1}(x)} \qquad \text{monotonic}$$

For nonmonotonic transformations, one must consider the various solutions to the inverse transformation and include as many terms in the equation above as there are solutions; that is,

$$f_Y(y) = \sum_{i=1}^{m} \left[f_X(x) \left| \frac{dx}{dy} \right| \right]_{x=g_i^{-1}(x)} \qquad \text{nonmonotonic}$$

12. The expectation, or average, of a function $g(\cdot)$ of a random variable X is defined as

$$E[g(X)] = \sum_{i=1}^{m} g(x_i)P(X = x_i), \qquad X \text{ discrete}$$

$$E[g(X)] = \int_{-\infty}^{\infty} g(x)f_X(x)\, dx, \qquad X \text{ continuous}$$

13. Important expectations include the following:

(a) The mean: $\mu_X = E(X)$

(b) The moments: $E(X^n)$

(c) The central moments: $m_n = E[(X - \mu_X)^n]$

(d) The variance: $\sigma_X^2 = E\left[(X - \mu_X)^2\right]$

(e) The characteristic function: $M_X(jv) = E[e^{jvX}]$

14. The moments of a random variable can be obtained by repeatedly differentiating its characteristic function and setting its argument to zero:

$$E(X^n) = (-j)^n \left. \frac{d^n M_X(jv)}{dv^n} \right|_{v=0}$$

15. Chebyshev's inequality is

$$P(|X - \mu_X| \ge k\sigma_X) \le \frac{1}{k^2}$$

Depending on the pdf of the random variable, it can provide reasonably tight bounds on the probability that the random variable deviates from its mean by at most a certain number of standard deviations.

16. Random variables with arbitrary pdf's can be generated from uniformly distributed random variables using transformations found by (3-117).

3-10 FURTHER READING

The references given in Chapters 1 and 2 provide further information and alternative treatments of the topics covered in this chapter. In addition, Ross (1994) provides a very readable introduction to this material.

3-11 PROBLEMS

Section 3-1

3-1. A chance experiment consists of rolling a pair of dice. A random variable is numerically equal to the difference between the number of spots up on die 1 and die 2. Make up a table describing this random variable similar to the one given in Table 3-1. Assuming fair dice, give the probability mass distribution for this experiment and random variable assignment.

3-2. A chance experiment consists of tossing three coins simultaneously. Make up a table describing the random variable that equals the number of heads up. For fair coins, give the probability mass distribution.

3-3. An urn contains 10 red balls, 20 blue balls, and 20 white balls. Three balls are drawn out of the urn without replacement. If a red ball is drawn, the number 1 is assigned. If a blue ball is drawn, the number 3 is assigned. If a white ball is drawn, the number 5 is assigned. Make up a table describing the random variable that is equal to the numbers assigned to each ball drawn.

Section 3-2

3-4. Are the following functions proper cdf's? Justify your answers.
(a) $F_1(x) = (1 - e^{-2x})u(x)$.
(b) $F_2(x) = 0.5\,u(x) + 0.1\,u(x - 1) + 0.3\,u(x - 3)$.
(c) $F_3(x) = 0.5\,u(x) + 0.1\,u(x - 1) + 0.4\,u(x - 3)$.

3-5. Obtain the cdf for Problem 3-1.

3-6. Obtain the cdf for Problem 3-2.

3-7. Given the cdf

$$F_X(x) = (1 - e^{-x/5})u(x)$$

Obtain the following probabilities: **(a)** $P(5 \leq X \leq 7)$; **(b)** $P(X \geq 3)$; **(c)** $P(X < 3)$.

3-8. Obtain the pdf corresponding to the cdf of Problem 3-7.

3-9. Given the Gaussian pdf

$$f_X(x) = \frac{e^{-(x-2)^2/4}}{\sqrt{4\pi}}$$

Express the following probabilities in terms of the Q-function. Look up numerical answers in a table of Q-functions or by computer evaluation (see Appendix C): **(a)** $P(2 \leq X \leq 4)$; **(b)** $P(X > 4)$; **(c)** $P(X < 2)$. **(d)** How are the results for parts (a) to (c) related?

3-10. Plot the cdf of a binomial random variable with **(a)** $n = 6$ and $p = 0.5$; **(b)** $n = 6$ and $p = 0.1$; **(c)** $n = 7$ and $p = 0.5$; **(d)** $n = 7$ and $p = 0.1$. **(e)** Deduce expressions for the maximum of the binomial distribution in terms of n and p for the two cases of n even and n odd. Check your formulas with the results computed in parts (a) to (d).

3-11. A digital transmission system has a probability of error of 10^{-2}. Find the probability of more than three errors occurring in 1000 transmissions. Compare the exact calculation using the binomial distribution with the approximation resulting from application of the Poisson distribution. (*Hint:* This is equal to 1 minus the probability of 0, 1, 2, or 3 errors.)

3-12. What is the probability of 3 girls in a 4-child family, assuming the probability of the birth of either gender is the same?

3-13. Relate the positive constants A and α in the functions of x given below so that a proper pdf results in each case.

(a) $f_1(x) = \dfrac{A}{(\alpha + x)^2} u(x)$

(b) $f_2(x) = \begin{cases} A(a - |x|), & |x| \le a \\ 0, & \text{otherwise} \end{cases}$

3-14. Buckshot is poured onto a tin plate at the rate of 50 shots per second. What is the probability that the interval between adjacent hits of the shot is greater than **(a)** 10 ms? **(b)** 5 ms?

3-15. At a certain telephone exchange, calls arrive at the exchange at the rate of 100 per hour. What is the probability that the interval between adjacent calls is **(a)** 1 min or greater? **(b)** 30 s or greater?

3-16. In a certain child's game, the object is to catch a ball tethered by a string to a small bucket in the bucket. The probability of catching the ball on a single try is 0.1. What is the probability that the first catch occurs on **(a)** trial 5? **(b)** trial 10?

3-17. In a certain digital communications system, 1s and 0s are transmitted with probability 0.01 that either will be received in error. What is the probability that the third error will occur at **(a)** transmission number 25? **(b)** transmission number 100? **(c)** transmission number 500?

3-18. A manufacturing line turns out transistors, 0.1% of which are defective. A lot of 1000 is to be shipped. To determine whether it should be shipped or not, a sample of 5 transistors is selected from the lot without replacement. If one or more of the selected transistors is found to be defective, the lot is scrapped. What is the probability that the lot will be shipped?

Section 3-4

3-19. A random variable X is uniformly distributed in the interval $[1, 4]$. Find and sketch the pdf of the random variable $Y = -3X + 5$.

3-20. The random variable X has an exponential pdf given by

$$f_X(x) = 3e^{-3x}u(x)$$

Find and sketch the pdf of the random variable Y where **(a)** $Y = X^2$; **(b)** $Y = |X|$; **(c)** $Y = X\,u(X)$, where $u(X)$ is the unit step function.

3-21. If Θ is a random variable uniformly distributed in the interval $(-\pi, \pi)$, show that the pdf of $Y = \cos \Theta$ is given by

$$f_Y(y) = \begin{cases} \dfrac{1}{\pi\sqrt{1 - y^2}}, & |y| < 1 \\ 0, & \text{otherwise} \end{cases}$$

3-22. A random variable X is uniformly distributed in the interval $[0, 1]$. Find the pdf of the random variable $Y = -\ln X$.

Section 3-5

3-23. Show that the mean and variance of a Poison random variable with pdf (3-46) are equal and are given by a.

3-24. A random variable has pdf

$$f_X(x) = 0.1\delta(x - 1) + 0.3\delta(x - 3) + 0.5\delta(x - 4) + 0.1\delta(x - 7)$$

(a) Find an expression for the nth moment of X.
(b) Find the variance of X.
(c) Find an expression for the nth central moment of X.

3-25. The *mode* of a random variable is the value for which its pdf is maximum. Its *median* is the value for which $P(X > \text{median}) = P(X < \text{median})$. Find the mean, mode, and median for the triangularly distributed random variable with pdf

$$f_X(x) = \begin{cases} 0, & x < 0 \\ 2x, & 0 \le x \le 1, \\ 0, & x > 1 \end{cases}$$

3-26. A fair die is rolled. Find the expected number of spots up and the variance of the number of spots up.

3-27. Let X denote the number of rolls of a pair of fair dice until a 7 shows (i.e. the sum of the number of spots up is seven). What is $E(X)$?

3-28. Find the constants a and b such that if $Y = aX + b$ with the mean and standard deviation of X are 2 and 3, respectively, the mean of Y is 0 and its standard deviation is 1.

3-29. (a) The random variable X has a uniform pdf in the interval $[2, 5]$ and $Y = X^3$. Find the mean and variance of Y.
(b) Repeat if the random variable X is exponentially distributed with parameter $\alpha = 3$.

Section 3-6

3-30. Find the characteristic function of the random variable of Problem 3-24. Check the mean and second moment found in Problem 3-24 by using the characteristic function approach.

3-31. Obtain the characteristic function corresponding to the pdf of Problem 3-20. Find the mean and variance of this random variable using the characteristic function.

3-32. Show that a Maclaurin series expansion of the characteristic function in terms of the moments of the random variable is given by

$$M_X(jv) = \sum_{m=0}^{\infty} \frac{E(X^m)}{m!}(jv)^m$$

3-33. (a) Find the characteristic function of a Poisson random variable.
(b) Obtain the moments of a Poisson random variable using the expansion of Problem 3-32.
(c) Check the results given in Problem 3-23 for the mean and variance of a Poisson random variable using the results obtained in part (b).

Section 3-7

3-34. **(a)** Find an upper bound on the probability that a random variable is more than two standard deviations from its mean.

(b) Compare the bound found in part (a) with the exact result for a uniformly distributed random variable.

(c) Compare the bound found in part (a) with the exact result for an exponentially distributed random variable.

3-35. Repeat Problem 3-34 for the probability that a random variable departs from its mean by more than three standard deviations.

Section 3-8

3-36. A Rayleigh random variable has pdf

$$f_V(v) = \begin{cases} \dfrac{v}{\sigma^2} e^{-v^2/2\sigma^2}, & v > 0 \\ 0, & v < 0 \end{cases}$$

Find the transformation that will generate this random variable from a random variable, U, that is uniformly distributed in $[0, 1]$.

3-37. Repeat Problem 3-36 for a Cauchy random variable with cdf given by (3-41).

3-12 COMPUTER EXERCISES

3-1. Implement a Rayleigh pseudorandom number generator. Generate 1000 Rayleigh pseudorandom numbers for $\sigma = 1$, plot a histogram, and compare with the theoretical Rayleigh pdf. To do so, you will have to normalize the histogram by the total number of samples generated (1000) and the bin width. Alternatively, you can multiply the theoretical Rayleigh pdf by the product of the number of samples and the bin width used for the histogram. Experiment with the bin width to get a reasonable number of samples per bin, but not so large that the histogram plot is too coarse for comparison with the theoretical pdf. If you have a fairly fast computer, generate 10,000 Rayleigh random variables and compare with the theoretical pdf. (See Problem 3-36.)

3-2. Redo Computer Exercise 3-1 for a Cauchy random variable. (See Problem 3-37.)

3-3. **(a)** Generate 1000 binomially distributed pseudorandom numbers by comparing 1000 uniform pseudorandom vectors of length n with $p < 1$ (the probability of success on each Bernoulli trial) to get vectors of 1s and 0s and doing a sum on each vector (resulting in a binomial random variable for each vector). Check the sample mean and variance of your generated pseudorandom binomial numbers with (3-96) and (3-97).

(b) Check out Chebyshev's inequality

$$P(|X - \mu| \geq k\sigma) \leq \frac{1}{k^2}$$

by determining the number of binomial pseudorandom numbers that you generated falling within $\pm k\sigma$ of their mean, taking the ratio of this number to the total number (1000) to get the relative frequency of success, subtract from 1 and comparing the difference with the bound for various k.

4

Probability Distributions for More Than One Random Variable

4–1 WHAT ARE BIVARIATE RANDOM VARIABLES?

It often happens in measurement problems and the analysis of systems with random inputs or random components that we need to consider probability distributions for pairs of random variables, which will be referred to as *bivariate*. In Section 4-14, we also consider how the probability distributions for bivariate random variables generalize to more than two. For example, in the case of a system we may wish to consider the joint probability that the input and output are in certain ranges at a given time instant. Another example is provided in the case of shooting a projectile at a target, where two variables are needed to describe the impact point—say either Cartesian or polar coordinates. There are many more such examples, and we will consider some of them in this chapter.

Of course, there are also problems where more than two random variables are required. Because the techniques used to describe pairs of random variables are easily extended to more than two random variables, we will stay with the case of two random variables for the most part. In Section 4-14 we point out briefly how this extension to more than two random variables can be carried out.

To summarize, in this chapter we wish to consider the probability that a pair of random variables, say X and Y, are in a certain region in the plane. We consider a basic way of doing this in the next section, where the *bivariate,* or *joint, cumulative distribution function* (joint cdf) is defined.

4–2 BIVARIATE CDF

Given two random variables, X and Y, we define their bivariate, or joint, cdf as

$$F_{XY}(x, y) = P(X \le x, Y \le y) \tag{4-1}$$

That is, we consider the joint probability that the random variables are in the lower left-hand portion of the x–y plane, which is shown shaded in Figure 4-1a.

From the defining relation (4-1), several properties of the joint cdf can be inferred. First, if either x or y or both is minus infinity, the joint cdf is zero because no outcomes of the chance experiment are included. That is,

$$F_{XY}(-\infty, y) = F_{XY}(x, -\infty) = F_{XY}(-\infty, -\infty) = 0 \tag{4-2}$$

It also follows that if both x and y are infinity, the joint cdf is unity because all outcomes of the chance experiment are included. That is,

$$F_{XY}(\infty, \infty) = 1 \tag{4-3}$$

Second, if we set one of the independent variables equal to infinity, we get the cdf of the other random variable:

$$\begin{aligned} F_{XY}(x, \infty) &= F_X(x) \\ F_{XY}(\infty, y) &= F_Y(y) \end{aligned} \tag{4-4}$$

To explain why, consider the first of these relationships. In accordance with (4-1), we are asking for the probability that $X \le x$ and that Y takes on *any* value. This is simply the cdf of X alone, called the *marginal cdf* of X. A similar argument holds for the second relationship.

Third, as either x or y is increased, we include no fewer outcomes (and possibly more) of our chance experiment. Thus the joint cdf is a nondecreasing function of its arguments [see (4-1)].

Finally, the joint cdf is continuous from the right on either independent variable. That is,

$$\begin{aligned} \lim_{x \to x_0^+} F_{XY}(x, y) &= F_{XY}(x_0, y) \\ \lim_{y \to y_0^+} F_{XY}(x, y) &= F_{XY}(x, y_0) \end{aligned} \tag{4-5}$$

Joint cdf's can be defined for both bivariate continuous random variables and bivariate discrete random variables. We discuss the discrete random variable case in Section 4-4.

The probability of the random variables lying in a certain range is found from the relationship

$$\begin{aligned} P(x_1 \le X \le x_2, y_1 \le Y \le y_2) &= P(x_1 \le X \le x_2, Y \le y_2) - P(x_1 \le X \le x_2, Y \le y_1) \\ &= P(X \le x_2, Y \le y_2) - P(X \le x_1, Y \le y_2) - P(X \le x_2, Y \le y_1) + P(X \le x_1, Y \le y_1) \\ &= F_{XY}(x_2, y_2) - F_{XY}(x_1, y_2) - F_{XY}(x_2, y_1) + F_{XY}(x_1, y_1) \end{aligned} \tag{4-6}$$

which can be found by applying the axioms of probability after suitably defining rectangular regions in the X–Y plane as shown in Figure 4-1b [the student should verify (4-6) from Figure 4-1b].

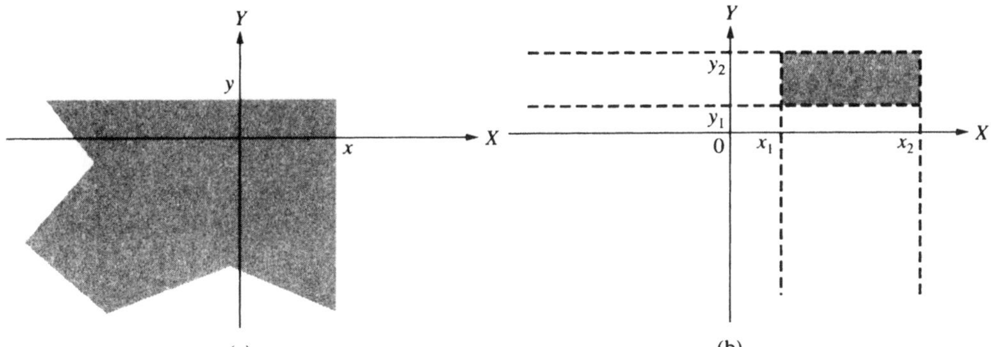

Figure 4-1 Definitions pertinent to the joint cdf: (a) region of the X–Y plane defining the joint cdf; (b) definitions pertinent to computing probability of being in a rectangle in X–Y plane.

Example 4-1

Consider the function

$$\tilde{F}(x, y) = [1 - \exp(-x)]\,[1 - \exp(-y)]u(x)u(y) \tag{4-7}$$

(a) Is it suitable for a joint cdf? Tell why.

(b) Find the probability that the random variables lie in the rectangle

$$-10 \le X \le 2 \quad \text{and} \quad 3 \le Y \le 5$$

Solution (a) Yes, it is. It is zero for either x or y or both minus infinity. It approaches unity as x and y approach infinity. If y (x) is infinity, it becomes a function only of x (y), which is an appropriate marginal cdf. It is also a nondecreasing function of its arguments and is continuous everywhere and therefore continuous from the right. A plot of this cdf is shown in Figure 4-2.

(b) According to (4-7),

$$P(-10 \le X \le 2, 3 \le Y \le 5) = F_{XY}(2, 5) - F_{XY}(-10, 5) - F_{XY}(2, 3) + F_{XY}(-10, 3)$$

From (4-7) we have

$$F_{XY}(2, 5) = (1 - e^{-2})(1 - e^{-5}) = 0.859$$

$$F_{XY}(-10, 5) = 0$$

$$F_{XY}(2, 3) = (1 - e^{-2})(1 - e^{-3}) = 0.822$$

$$F_{XY}(-10, 3) = 0$$

Thus the desired probability is

$$P(-10 \le X \le 2, 3 \le Y \le 5) = 0.037$$

4-3 BIVARIATE PDF

The bivariate, or joint, probability density function (joint pdf) of two random variables is defined in terms of their joint cdf as

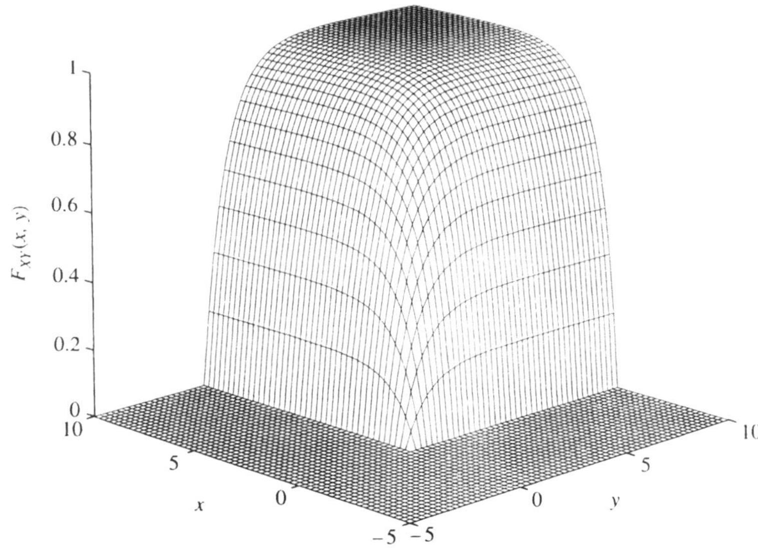

Figure 4-2 Plot of the bivariate cdf of Example 4-1.

$$f_{XY}(x, y) = \frac{\partial^2 F_{XY}(x, y)}{\partial x \, \partial y} \tag{4-8}$$

If $F_{XY}(x, y)$ is the joint cdf of bivariate continuous random variables, the differentiation in (4-8) presents no problems. If the random variables are mixed or discrete, the joint cdf possesses jump discontinuities and is not differentiable at these locations. In that case, the joint pdf is not defined unless we represent its derivative at jump discontinuities by delta functions. Since we can have line discontinuities, this is a somewhat more complex problem than that for the single random variable case; in the next section, we discuss joint probability mass functions. In this section we assume that the joint cdf is everywhere differentiable.

Example 4-2

Find the joint pdf corresponding to the cdf of Example 4-1.

Solution Carrying out the differentiation, we obtain

$$f_{XY}(x, y) = e^{-x}e^{-y}u(x)u(y) = e^{-(x+y)}u(x)u(y) \tag{4-9}$$

This joint pdf is plotted in Figure 4-3.

The joint pdf has several properties which follow from those of the joint cdf. First, by integrating (4-8) on both variables and using (4-2) to evaluate the constant of integration, we obtain

$$F_{XY}(x, y) = \int_{-\infty}^{y} \int_{-\infty}^{x} f_{XY}(u, v) \, du \, dv \tag{4-10}$$

From this relationship and property (4-3), we deduce that the volume under the joint pdf surface and above the x–y plane is unity. Second, since the joint cdf is a nondecreasing

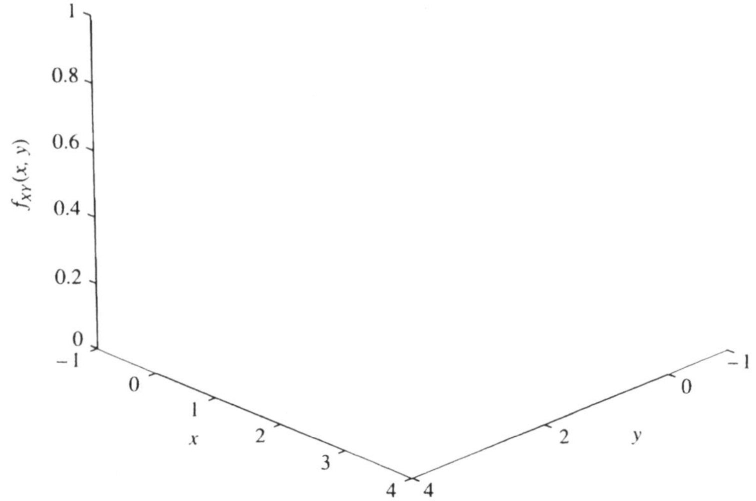

Figure 4-3 Joint pdf of Example 4-2.

function of its arguments and (4-8) gives the slope of the surface, we conclude that the joint pdf is nonnegative. Next, from (4-4), we can set $y = \infty$ to obtain

$$F_{XY}(x, \infty) = F_X(x) = \int_{-\infty}^{\infty}\int_{-\infty}^{x} f_{XY}(u, v)\, du\, dv = \int_{-\infty}^{x}\int_{-\infty}^{\infty} f_{XY}(u, v)\, dv\, du = \int_{-\infty}^{x} f_X(u)\, du \quad \text{(4-11a)}$$

where the right-hand integral follows from (3-16). Similarly,

$$F_Y(y) = \int_{-\infty}^{y}\int_{-\infty}^{\infty} f_{XY}(u, v)\, du\, dv = \int_{-\infty}^{y} f_Y(v)\, dv \quad \text{(4-11b)}$$

By differentiating the double integral in (4-11a) with respect to x, we obtain the relationship

$$f_X(x) = \int_{-\infty}^{\infty} f_{XY}(x, y)\, dy \quad \text{(4-12a)}$$

Performing a similar operation on (4-11b), we obtain

$$f_Y(y) = \int_{-\infty}^{\infty} f_{XY}(x, y)\, dx \quad \text{(4-12b)}$$

Because they are obtained by integrating the joint pdf, $f_X(x)$ and $f_Y(y)$ are called *marginal pdf's*.

Finally, we can use the axioms of probability as we did to get (4-7) to obtain

$$P(x_1 \le X \le x_2, y_1 \le Y \le y_2) = \int_{x_1}^{x_2}\int_{y_1}^{y_2} f_{XY}(x, y)\, dx\, dy \quad \text{(4-13)}$$

Example 4-3

Consider the function of two variables

$$f_{XY}(x, y) = \begin{cases} Axy, & 0 < x < y, \quad 0 < y < 1 \\ 0, & \text{otherwise} \end{cases}$$

(a) Find A such that this is a proper pdf.

(b) Find the probability that $0 < X < 0.5$ and $0.5 < Y < 1$.

(c) Obtain the marginal pdf's for X and Y.

Solution (a) Since the volume under the joint pdf must be 1 [see the remark after (4-10)], we compute

$$\int_{-\infty}^{\infty}\int_{-\infty}^{\infty} f_{XY}(x, y)\, dx\, dy = \int_0^1\int_0^y Axy\, dx\, dy = \frac{A}{2}\int_0^1 y^3\, dy = \frac{A}{2}\frac{y^4}{4}\bigg|_0^1 = \frac{A}{8} = 1$$

from which we deduce that $A = 8$. In the integration of the $f_{XY}(x, y)$, we have made use of the fact that the function is nonzero over a triangle defined by x between the y-axis and the line $x = y$ and on y from 0 to 1. The student should sketch this area and deduce that the limits of integration are proper.

(b) From (4-13), the desired probability may be computed as

$$P(0 < X < 0.5, 0.5 < Y < 1) = 8\int_{0.5}^{1}\int_0^{0.5} xy\, dx\, dy = 0.375$$

(c) The marginal pdf for X is obtained by integrating the joint pdf over all y as given by (4-12a):

$$f_X(x) = \int_x^1 8xy\, dy = 8x\frac{y^2}{2}\bigg|_x^1 = 4x(1 - x^2),\ 0 < x < 1$$

and zero elsewhere. The limits of integration are deduced by noting the region in the x–y plane, where $f_{XY}(x, y)$ is nonzero. The marginal pdf for Y is obtained by integrating the joint pdf over all x as given by (4-12b):

$$f_Y(y) = \int_0^y 8xy\, dx = 8y\frac{x^2}{2}\bigg|_0^y = 4y^3,\ 0 < y < 1$$

and zero elsewhere. Note that both $f_X(x)$ and $f_Y(y)$ integrate to 1, as they should.

Example 4-4

(**a**) Use (4-12) to obtain the probability that X is between -10 and 2 and Y is between 3 and 5 for the pdf of Example 4-2.

(**b**) What are the marginal pdf's for X and Y?

Solution (**a**) Putting the given values into (4-13), we obtain

$$\begin{aligned} P(-10 \le X \le 2, 3 \le Y \le 5) &= \int_{-10}^{2}\int_3^5 e^{-(x+y)}u(x)u(y)\, dx\, dy \\ &= \int_0^2\int_3^5 e^{-(x+y)}\, dx\, dy \end{aligned}$$

$$= [-e^{-x}]_0^2 [-e^{-y}]_3^5 = (1 - e^{-2})(e^{-3} - e^{-5}) = 0.037$$

as obtained in Example 4-1.

(b) Integration of the joint pdf over all y shows that the marginal pdf for X is $\exp(-x)u(x)$, and integration over all x gives the marginal pdf for Y as $\exp(-y)u(y)$.

4-4 DISCRETE RANDOM VARIABLE PAIRS

Suppose that a pair of random variables (X, Y) take on the exhaustive set of values $\{(x_i, y_j), i = 1, 2, \ldots, m, j = 1, 2, \ldots, n\}$ with probabilities $\{P_{ij}, i = 1, 2, \ldots, m, j = 1, 2, \ldots, n\}$. Since the set of values taken on is exhaustive, we have

$$\sum_{i=1}^{m} \sum_{j=1}^{n} P_{ij} = 1 \tag{4-14}$$

The probability mass function consists of a set of points above the x–y plane as shown in the example in Figure 4-4 for m = n = 3.

Example 4-5

A pair of fair dice is thrown. Let X and Y denote the number of spots up on the first and the second, respectively. Find the joint probability mass function for this pair of random variables.

Solution Since it is implicit that each die rolls independently of each other, the joint probability mass function is

$$P(X = i, Y = j) = P(X = i)P(Y = j) = \tfrac{1}{6} \times \tfrac{1}{6} = \tfrac{1}{36}, \quad i, j = 1, 2, \ldots, 6$$

A plot of this would consist of lines $\frac{1}{36}$ high at the integer values in the plane at (1, 1), (1, 2), . . . , (6, 6). The joint cdf for this random variable can be found from the double sum

$$F_{XY}(x, y) = \begin{cases} \sum_{i \le x} \sum_{j \le y} \frac{1}{36} & i, j \le 6 \\ 1, & i \text{ and } j > 6 \end{cases} \tag{4-15}$$

The student should sketch traces of this joint cdf in vertical planes parallel to the x and y axes.

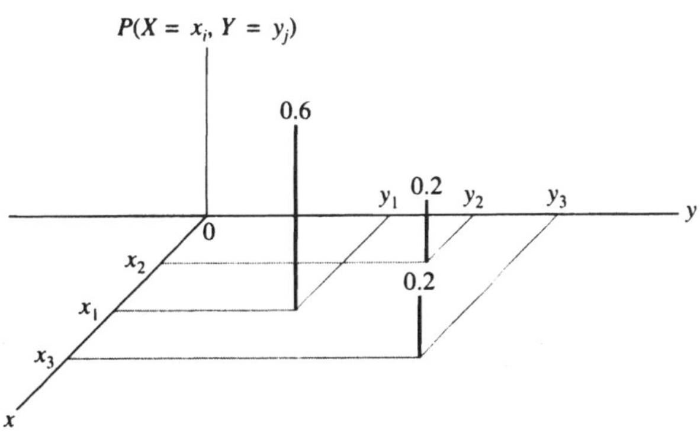

Figure 4-4 Probability mass function.

Example 4-6

Referring to the rolling of a pair of dice, find the probability mass function for the random variables X and Y defined as

$$X = \text{sum of number of spots up}, \quad Y = \text{difference of number of spots up}$$

Solution: Construct a table of sums and differences (Table 4-1). From the table we deduce the set of joint probabilities listed in Table 4-2, which define the joint probability mass function.

The marginal probability mass functions for X and Y may be obtained by summing over the columns and the rows, respectively, of Table 4-2. For example, the probability mass function for X is

$$P(X = 2) = \tfrac{1}{36}, \quad P(X = 3) = \tfrac{2}{36}, \quad P(X = 4) = \tfrac{3}{36}, \quad P(X = 5) = \tfrac{4}{36},$$
$$P(X = 6) = \tfrac{5}{36}, \quad P(X = 7) = \tfrac{6}{36}, \quad P(X = 8) = \tfrac{5}{36}, \quad P(X = 9) = \tfrac{4}{36},$$
$$P(X = 10) = \tfrac{3}{36}, \quad P(X = 11) = \tfrac{2}{36}, \quad P(X = 12) = \tfrac{1}{36}$$

When plotted, this would look similar to Figure 3-5 except that we have lines in place of impulses.

4-5 CONDITIONAL CUMULATIVE DISTRIBUTION AND CONDITIONAL PROBABILITY DENSITY FUNCTIONS

It is often convenient to condition the outcomes of random variables on the occurrence or nonoccurrence of a hypothesis such as rain or no rain, noise alone present or signal plus noise present, random variable Z in a certain range, and so on. Recall the definition of the conditional probability of one event given another from Chapter 2, given by (2–31) or (2–32). We can similarly consider the cdf of a random variable is conditioned on the event that a second, related random variable is in a given range. In (2-31), for example, we can let $A = \{X \leq x\}$ and $B = \{Y \in R_y\}$ to obtain the conditional probability

$$P(X \leq x \mid Y \in R_y) = \frac{P(X \leq x, Y \in R_y)}{P(Y \in R_y)} = F_X(x \mid Y \in R_y) \tag{4-16}$$

The right-hand side is called the conditional cdf of X given the event $\{Y \in R_y\}$. Extending this idea somewhat, we take the events A and B to be

Table 4-1 SUMS AND DIFFERENCES (IN PARENTHESES) WHEN A PAIR OF DICE ARETHROWN)

	Die 2					
Die 1;	1	2	3	4	5	6
1	2 (0)	3 (1)	4 (2)	5 (3)	6 (4)	7 (5)
2	3 (−1)	4 (0)	5 (1)	6 (2)	7 (3)	8 (4)
3	4 (−2)	5 (−1)	6 (0)	7 (1)	8 (2)	9 (3)
4	5 (−3)	6 (−2)	7 (−1)	8 (0)	9 (1)	10 (2)
5	6 (−4)	7 (−3)	8 (−2)	9 (−1)	10 (0)	11 (1)
6	7 (−5)	8 (−4)	9 (−3)	10 (−2)	11 (−1)	12 (0)

Table 4-2 JOINT PROBABILITY MASS FUNCTION FOR RANDOM VARIABLES DEFINED AS THE SUM AND DIFFERENCE OF THE SPOTS UP ON TWO FAIR DIES

Y \ X	2	3	4	5	6	7	8	9	10	11	12
−5	0	0	0	0	0	1/36	0	0	0	0	0
−4	0	0	0	0	1/36	0	1/36	0	0	0	0
−3	0	0	0	1/36	0	1/36	0	1/36	0	0	0
−2	0	0	1/36	0	1/36	0	1/36	0	1/36	0	0
−1	0	1/36	0	1/36	0	1/36	0	1/36	0	1/36	0
0	1/36	0	1/36	0	1/36	0	1/36	0	1/36	0	1/36
1	0	1/36	0	1/36	0	1/36	0	1/36	0	1/36	0
2	0	0	1/36	0	1/36	0	1/36	0	1/36	0	0
3	0	0	0	1/36	0	1/36	0	1/36	0	0	0
4	0	0	0	0	1/36	0	1/36	0	0	0	0
5	0	0	0	0	0	1/36	0	0	0	0	0

$$\begin{aligned} A &= \{x < X \leq x + \Delta x\} \\ B &= \{y < X \leq y + \Delta y\}, \quad |\Delta x|, \ |\Delta y| \ll 1 \end{aligned} \tag{4-17}$$

and the conditional probability of A given B, expressed by (2-31), becomes

$$\begin{aligned} P(x < X \leq x + \Delta x \,|\, y < Y \leq y + \Delta y) &= \frac{P(x < X \leq x + \Delta x, y < Y \leq y + \Delta y)}{P(y < Y \leq y + \Delta y)} \\ &\simeq \frac{f_{XY}(x, y)\, \Delta x\, \Delta y}{f_Y(y)\, \Delta y} \\ &= f_{X|Y}(x \,|\, y)\, \Delta x \end{aligned} \tag{4-18}$$

where the approximation (3-14) and its extension to two dimensions has been used. From (4-18), we define the conditional pdf of X given $Y = y$ to be

$$f_{X|Y}(x \,|\, y) = \frac{f_{XY}(x, y)}{f_Y(y)} \tag{4-19}$$

Equation (4-18) provides an interpretation of this conditional pdf: in particular, that the conditional pdf of x given y multiplied by a small increment in x; Δx, is the probability that random variable X is in a Δx interval around x given that Y is in a small Δy interval around y. Similarly, we define the conditional pdf of the random variable Y given that the random variable $X = x$ to be

$$f_{Y|X}(y \,|\, x) = \frac{f_{XY}(x, y)}{f_X(x)} \tag{4-20}$$

with an interpretation similar to that used to derive (4-19).

Conditional cdf's and pdf's have the same properties that regular cdf's and pdf's have. For example, if we integrate (4-19) over x from $-\infty$ to ∞ and use (4-12b), we obtain

$$\int_{-\infty}^{\infty} f_{X|Y}(x \mid y)\, dx = \int_{-\infty}^{\infty} \frac{f_{XY}(x, y)}{f_Y(y)}\, dx = \frac{f_Y(y)}{f_Y(y)} = 1 \tag{4-21}$$

Example 4-7

Obtain the conditional pdf of X given Y for the joint pdf of Example 4-3.

Solution From Example 4-3c, we have the required marginal pdf. Substituting the joint pdf and marginal pdf for Y into (4-19), we obtain

$$f_{X|Y}(x \mid y) = \begin{cases} \dfrac{2x}{y^2}, & 0 < x < y,\ 0 < y < 1 \\ 0 & \text{otherwise} \end{cases}$$

The conditional pdf for Y given X may be found similarly. Note that the integral of this conditional density function over all x gives 1.

Example 4-8

A certain signal may or may not be present in noise. If the signal is present, the voltage at the output of the detector has the pdf

$$f_{X|S}(x|S) = \frac{e^{-(x-2)^2/8}}{\sqrt{8\pi}} \tag{4-22a}$$

If the signal is absent, the pdf at the output of the detector is

$$f_{X|\bar{S}}(x \mid \bar{S}) = \frac{e^{-x^2/8}}{\sqrt{8\pi}} \tag{4-22b}$$

A threshold of 1 V is set at the output of the detector. If the detector output is greater than the threshold, the decision is made that the signal was present.

(a) What is the probability that, if present, the signal is not detected (called the *probability of a miss*)?

(b) What is the probability that if the signal was not present, the decision is that it was present (called the *probability of a false alarm*)?

Solution (a) The probability that the signal is not detected even though present is the probability that the random variable X does not cross the threshold given the signal was present. In terms of the first conditional pdf, this is

$$P_{\text{miss}} = \int_{-\infty}^{1} \frac{e^{-(x-2)^2/8}}{\sqrt{8\pi}}\, dx = \int_{-\infty}^{-0.5} \frac{e^{-u^2/2}}{\sqrt{2\pi}}\, du = \int_{0.5}^{\infty} \frac{e^{-u^2/2}}{\sqrt{2\pi}}\, du = Q(0.5) = 0.309 \tag{4-23a}$$

where the change of variables $u = (x - 2)/2$ was used to go from the first integral to the second one, and the evenness of the integrand was used to go from the second to the third. Finally, once P_{miss} is expressed in terms of the Q-function, tables are used to find the answer numerically.

(b) The probability of a false alarm is the probability that noise alone crosses the threshold. In terms of the second pdf given in the problem statement, it is

$$P_{FA} = \int_{1}^{\infty} \frac{e^{-x^2/8}}{\sqrt{8\pi}} dx = \int_{0.5}^{\infty} \frac{e^{-u^2/2}}{\sqrt{2\pi}} du = 0.309 \tag{4-23b}$$

These are high probabilities of making an error. (Typical values for a communication system are 10^{-2} to 10^{-8}.) They are equal because the threshold was set halfway between the signal present and signal not present means (i.e., 2 and 0), and because the Gaussian pdf is even. To make these probabilities smaller, we can either increase the mean of the signal present pdf (i.e. the 2 in the exponent), or decrease the variance of both pdf's (4 in this case). The latter corresponds to decreasing the noise.

Example 4-9

Metal meter sticks are manufactured on a production line where the true length can be modeled as Gaussian with mean 1 m and standard deviation 2 mm. The meter sticks are checked as they come off the production line and, if more than 1 mm short or long, they are rejected.

(**a**) What is the pdf of the meter sticks after rejecting those too long or too short?

(**b**) After culling, what is the probability that a meter stick will be within ± 0.5 mm of 1 m?

Solution (**a**) Let X be a random variable denoting the actual length of a meter stick. With all quantities expressed in terms of millimeters, this pdf is given by

$$f_X(x) = \frac{e^{-(x-1000)^2/8}}{\sqrt{8\pi}} \tag{4-24}$$

Define the event B as

$$B = (999 < X \leq 1001) \tag{4-25}$$

The probability of B is

$$\begin{aligned} P(B) &= \int_{999}^{1001} \frac{e^{-(x-1000)^2/8}}{\sqrt{8\pi}} dx = \int_{-0.5}^{0.5} \frac{e^{-u^2/2}}{\sqrt{2\pi}} du \\ &= 1 - 2\int_{0.5}^{\infty} \frac{e^{-u^2/2}}{\sqrt{2\pi}} du = 1 - 2Q(0.5) \qquad (4\text{-}26) \\ &= 1 - 2(0.3085) = 0.383 \end{aligned}$$

We obtain the conditional pdf of X by first getting its conditional cdf. Thus, we consider the events $\{X \leq x\}$ and B. Using (4-25), we consider different values for the running variable x and the joint event $\{X \leq x\} \cap B$:

$$\{X \leq x\} \cap B = \begin{cases} \varnothing, & x \leq 999 \\ 999 < X \leq x, & 999 < x \leq 1001 \\ 999 < X \leq 1001, & x > 1001 \end{cases} \tag{4-27}$$

The conditional cdf of X given B is

$$F_{X|B}(x|B) = \frac{P(X \leq x \cap B)}{P(B)} \tag{4-28}$$

From (4-26), the denominator of (4-28) is 0.383. From (4-27) and properties of the cdf, it follows that the numerator of (4-28) is

$$P(X \le x \cap B) = \begin{cases} 0, & x \le 999 \\ F_X(x) - F_X(999), & 999 < X \le 1001 \\ P(B), & x > 1001 \end{cases} \quad (4\text{-}29)$$

Writing down the integral for F_X (999) and doing a change of variables, we find it to be equal to Q(0.5) = 0.309. Using this along with (4-29) and (4-26) in (4-28), we obtain

$$F_X(x \mid B) = \begin{cases} 0, & x \le 999 \\ \dfrac{F_X(x)\text{–}0.309}{0.383}, & 999 < x \le 1001 \\ 1, & x > 1001 \end{cases} \quad (4\text{-}30)$$

Differentiating, we find the conditional pdf of the lengths, given they are in the range 999 to 1001 mm, to be

$$f_X(x \mid B) = \frac{1}{0.383} f_X(x) = 2.611 f_X(x), \qquad 999 < x \le 1001 \quad (4\text{-}31)$$

and zero otherwise.

(**b**) The desired probability is

$$\begin{aligned} P(999.5 < X \le 1000.5 \mid B) &= \int_{999.5}^{1000.5} f_X(x \mid B)\, dx \\ &= \int_{999.5}^{1000.5} 2.611 \frac{e^{-(x-1000)^2/8}}{\sqrt{8\pi}}\, dx \\ &= 2.611 \int_{-0.25}^{0.25} \frac{e^{-u^2/2}}{\sqrt{2\pi}}\, du \\ &= 2.611[1 - 2Q(0.25)] = 0.516 \end{aligned} \quad (4\text{-}32)$$

If we are not given the event B (that is, the meter sticks are not culled after coming off the production line), then $P(999.5 < X \le 1000.5) = 1 - 2Q(0.25) = 0.197$—a considerably smaller probability than $P(999.5 < X \le 1000.5|\text{B})$. The student is encouraged to derive the probability $\text{P}(999.5 < X \le 1000.5)$.

4–6 STATISTICALLY INDEPENDENT RANDOM VARIABLES

In Chapter 2, statistically independent events were defined as events whose joint probability could be factored into the product of their marginal probabilities [see (2-37)]. We can apply this directly to two random variables via the definition of the joint cdf given by (4-1) and conclude that

$$F_{XY}(x, y) = P(X \le x, Y \le y) = P(X \le x)P(Y \le y) = F_X(x)F_Y(y) \quad (4\text{-}33a)$$

(statistically independent random variables)

From the definition of the joint pdf (4-8) and by differentiating (4-33a) partially with respect to x and then y, it follows that

$$f_{XY}(x, y) = f_X(x)f_Y(y) \tag{4-33b}$$

(statistically independent random variables)

Finally, from the definition of the conditional pdf (4-19) and (4-20), it follows that

$$\begin{aligned} f_{X|Y}(x \mid y) &= f_X(x) \\ f_{Y|X}(y \mid x) &= f_Y(y) \end{aligned} \tag{4-33c}$$

(statistically independent random variables)

Any one of these statements defines statistically independent random variables; all are equivalent.

Example 4-10

The random variables described by the joint cdf of Example 4-1 are statistically independent since their joint cdf factors. From Example 4-2, we see that their joint pdf also factors, as it should.

Example 4-11

The random variables defined by the joint pdf of Example 4-3 are not statistically independent, as may be checked by noting that their joint pdf is not equal to the product of their respective marginal pdf's.

4–7 AVERAGES OF FUNCTIONS OF TWO RANDOM VARIABLES

Given a function of two random variables, say $g(X, Y)$, it is a random variable also. The expectation or average of this function of the two random variables X and Y is obtained by an extension of the definition of the expectation of a function of expectations of a single random variable given by (3-82). For continuous random variables, we have

$$E[g(X, Y)] = \int_{-\infty}^{\infty}\int_{-\infty}^{\infty} g(x, y)f_{XY}(x, y)\, dx\, dy \tag{4-34a}$$

where $f_{XY}(x, y)$ is the joint pdf of X and Y. If X and Y are discrete random variables, then

$$E[g(X, Y)] = \sum_{i=1}^{m}\sum_{j=1}^{n} g(x_i, y_j)P(X = x_i, Y = y_j) \tag{4-34b}$$

where $P(X = x_i, Y = y_j)$ is the joint probability mass function of X and Y. Some important joint expectations are considered below.

Before doing so, however, some useful properties of expectations of functions of two random variables will be derived:

1. If $g(X, Y)$ is a product of two functions, one of which is a function of X alone and the other a function of Y alone, and if X and Y are statistically independent, then

$$E[g_1(X)g_2(Y)] = \int_{-\infty}^{\infty}\int_{-\infty}^{\infty} g_1(X)g_2(X)f_X(x)f_Y(y)\,dx\,dy$$
$$= \int_{-\infty}^{\infty} g_1(X)f_X(x)\,dx \int_{-\infty}^{\infty} g_2(Y)f_Y(y)\,dy = E[g_1(X)]E[g_2(Y)] \tag{4-35}$$

2. If $g(X, Y) = a_1g_1(X, Y) + a_2g_2(X, Y)$, where $g_1(X, Y)$ and $g_2(X, Y)$ are separate functions of X and Y and a_1 and a_2 are constants, then

$$E[a_1g_1(X, Y) + a_2g_2(X, Y)] = \int_{-\infty}^{\infty}\int_{-\infty}^{\infty} [a_1g_1(x,y) + a_2g_2(x,y)]f_{XY}(x, y)\,dx\,dy$$
$$= \int_{-\infty}^{\infty}\int_{-\infty}^{\infty} a_1g_1(x,y)f_{XY}(x, y)\,dx\,dy + \int_{-\infty}^{\infty}\int_{-\infty}^{\infty} a_2g_2(x,y)f_{XY}(x, y)\,dx\,dy$$
$$= a_1E[g_1(X,Y)] + a_2E[g_2(X,Y)] \tag{4-36}$$

This holds whether or not X and Y are statistically independent.

Example 4-12

Given the joint pdf

$$f(x, y) = \frac{\alpha\beta}{4}\exp(-\alpha|x| - \beta|y|)$$

(**a**) Find the expectation of $g(X, Y) = XY$.

(**b**) Obtain the expectation of $h(X, Y) = X^2 + Y^2$.

Solution The joint pdf of X and Y factors into a function of x alone and a function of y alone. Thus X and Y are statistically independent and (4-35) holds. We have that

$$E(XY) = E(X)E(Y) = 0$$

which follows because the pdf's of X and Y are even functions.

(b) We may apply (4-36), which results in

$$E(X^2 + Y^2) = E(X^2) + E(Y^2)$$

But

$$E(X^2) = \int_{-\infty}^{\infty} x^2\frac{\alpha}{2}\exp(-\alpha|x|)\,dx$$
$$= \alpha\int_0^{\infty} x^2\exp(-\alpha x)\,dx = \frac{1}{\alpha^2}\int_0^{\infty} u^2\exp(-u)\,du = \frac{2}{\alpha^2}$$

where the evenness of the integrand has been used to integrate from 0 to ∞ and multiply by 2 rather than from $-\infty$ to ∞ (note that $|x| = x$ for $x > 0$). A similar integration shows that $E(Y^2) = 2/\beta^2$. Thus

$$E(X^2 + Y^2) = \frac{2}{\alpha^2} + \frac{2}{\beta^2}$$

4-8 SOME SPECIAL AVERAGES OF FUNCTIONS OF TWO RANDOM VARIABLES

Joint Moments

The *mn*th joint moments of two random variables *X* and *Y* are defined as

$$m_{mn} = E[X^m Y^n], \; m, n = 1, 2, \ldots \tag{4-37}$$

Special cases of (4-37) include the means of *X* and *Y* obtained, respectively, by setting $m = 1$ and $n = 0$ and $m = 0$ and $n = 1$. Note that the joint moments of statistically independent random variables factor for all *m* and *n*.

Example 4-13

Find the joint moments of the random variables with joint probability mass function shown in Figure 4-4 if

$$x_1 = 1, \; x_2 = 2, \; x_3 = 3, \; y_1 = 3, \; y_2 = 3, \; y_3 = 4$$

Solution Substituting into (4-37) and using (4-34b) for the expectation, we obtain

$$E(X^m Y^n) = 1^m 3^n \times 0.2 + 2^m 3^n \times 0.6 + 3^m 4^n \times 0.2$$

Several special cases are given in Table 4-3.

Joint Central Moments

These are obtained by first subtracting from *X* and *Y* their respective means and then finding the joint moments of these new random variables:

Table 4-3 JOINT MOMENTS FOR THE RANDOM VARIABLES WITH PROBABILITY MASS FUNCTION SHOWN IN FIGURE 4-4

m	n	$E(X^m Y^n)$
0	0	1
1	0	2.0
0	1	3.2
1	1	6.6

$$\mu_{mn} = E[(X - \mu_X)^m (Y - \mu_Y)^n] \quad m, n = 1, 2, \ldots \tag{4-38}$$

where μ_X and μ_Y are the means of X and Y, respectively. The special cases $m = 2, n = 0$ and $m = 0, n = 2$ give the variances of X and Y, respectively.

Covariance

This is a special case of the joint central moments with $m = n = 1$:

$$C_{XY} = E[(X - \mu_X)(Y - \mu_Y)] \tag{4-39a}$$

By expanding the expectation, this can be put into the form

$$C_{XY} = E[X\,Y] - \mu_X\mu_Y = R_{XY} - \mu_X\mu_Y \tag{4-39b}$$

where R_{XY} is called the *correlation*; it is a special case of (4-37) for $m = n = 1$.

Correlation Coefficient

The *correlation coefficient* is defined as

$$\rho_{XY} = \frac{C_{XY}}{\sigma_X\sigma_Y} \tag{4-40}$$

where C_{XY} denotes the covariance (4-39a). It has several properties of interest, which are listed below.

1. Its absolute value is bounded by unity:

   $$-1 \le \rho_{XY} \le 1 \tag{4-41}$$

   We can show this by considering the quantity

   $$E\left[\left(\frac{X - \mu_X}{\sigma_X} \pm \frac{Y - \mu_Y}{\sigma_Y}\right)^2\right] \ge 0 \tag{4-42a}$$

   By expanding the square and taking the expectation term by term, we have that

   $$1 \pm 2\rho_{XY} + 1 \ge 0 \tag{4-42b}$$

   which is equivalent to (4-41).

2. If X and Y are statistically independent, then

   $$\rho_{XY} = 0 \tag{4-43}$$

   The condition of statistical independence is a sufficient condition for this to be the case. There are other instances where the random variables are not statistically independent and the correlation coefficient is still zero. Any pair of random variables whose correlation coefficient is zero are said to be *uncorrelated*. Thus, *statistically independent random variables are always uncorrelated; it is not generally true that uncorrelated random variables are statistically independent,* although there are special cases for which this statement is true. To show (4-43), we apply (4-35) to find the expectation of (4-39a) when X and Y are statistically independent. Let

$$g_1(X) = X - \mu_X \quad \text{and} \quad g_2(Y) = Y - \mu_Y \tag{4-44}$$

Thus (4-39a) becomes

$$\begin{aligned} E[(X - \mu_X)(Y - \mu_Y)] & \\ &= [E(X) - \mu_X][E(Y) - \mu_Y] \\ &= (\mu_X - \mu_X)(\mu_Y - \mu_Y) = 0 \end{aligned} \tag{4–45}$$

where (3-90) and (3-92) have been used to go from the first to the second equation.

3. If $Y = aX + b$, where a and b are constants, then $\rho_{XY} = \pm 1$. To show this, note that $\mu_Y = a\mu_X + \text{b}$ and substitute into (4-39a) to obtain

$$C_{XY} = E\{(X - \mu_X)[aX + b - (a\mu_X + b)]\} = aE[(X - \mu_X)^2] = a\sigma_X^2 \tag{4-46}$$

We also note that $\sigma_Y = \pm\, a\sigma_X$ which, when substituted into (4-40) with the result for C_{XY} above gives the desired result of $\rho_{XY} = \pm 1$.

Example 4-14

Find the correlation, covariance, and correlation coefficient for the random variables with joint pdf given in Example 4-3.

Solution By definition, the correlation is

$$R_{XY} = E[X\,Y] = \int_0^1 \int_0^y xy(8xy)\,dx\,dy = \frac{4}{9}$$

The means are

$$\mu_X = \int_0^1 x[4x(1 - x^2)]\,dx = \frac{8}{15} \quad \text{and} \quad \mu_Y = \int_0^1 y(4y^3)\,dy = \frac{4}{5}$$

where the marginal pdf's obtained in Example 4-3 were used. The covariance is

$$C_{XY} = \frac{4}{9} - \frac{8}{15}\frac{4}{5} = \frac{4}{225}$$

To get the correlation coefficient, we need the variances. We compute the mean-square values first. They are

$$E[X^2] = \int_0^1 x^2[4x(1 - x^2)]\,dx = \frac{1}{3} \quad \text{and} \quad E[Y^2] = \int_0^1 y^2(4y^3)\,dy = \frac{2}{3}$$

Thus the variances are

$$\sigma_X^2 = \frac{11}{225} \quad \text{and} \quad \sigma_Y^2 = \frac{2}{75}$$

Substituting into (4-40), we find the correlation coefficient to be

$$\rho_{XY} = \frac{4/225}{(\sqrt{11}/15)(\sqrt{2}/5\sqrt{3})} = 2\sqrt{2/33} = 0.4924$$

Joint Characteristic Function

When generalized to the bivariate case, (3-98) becomes

$$M_{XY}(ju, jv) = E[e^{j(uX+vY)}] = E(e^{juX}e^{jvY}) \tag{4-47}$$

Some special cases are of interest:

1. If X and Y are statistically independent, it follows by applying (4-35) that

$$M_{XY}(ju, jv) = E(e^{juX})E(e^{jvY}) = M_X(ju)M_Y(jv) \tag{4-48}$$

   That is, the joint characteristic function of two statistically independent random variables is the product of their separate characteristic functions.

2. Another useful special case occurs for the characteristic function of the sum of statistically independent random variables, $Z = X + Y$. In this case

$$\begin{aligned} M_Z(jv) &= E[e^{jv(X+Y)}] = E(e^{jvX}e^{jvY}) \\ &= E(e^{jvX})E(e^{jvY}) = M_X(jv)M_Y(jv) \end{aligned} \tag{4-49}$$

   This is similar to (4-48) except that the arguments of the two right-hand-side characteristic functions are the same in (4-49); they are different in (4-48). Also, the left-hand side is a marginal, or single-variable, characteristic function in (4-49).[1]

Example 4-15

Find the joint characteristic function for the pdf of Example 4-2.

Solution As shown in Example 4-10, these are statistically independent random variables. According to (4-48), their characteristic function is the product of the marginal characteristic functions. Thus

$$M_X(ju) = \int_{-\infty}^{\infty} e^{jux}e^{-x}u(x)\,dx = \int_0^{\infty} e^{-(1-ju)x}\,dx = -\frac{e^{-(1-ju)x}}{1-ju}\bigg|_0^{\infty} = \frac{1}{1-ju} \tag{4-50}$$

with a similar result for the characteristic function of Y, but with jv in place of ju. Therefore, their joint characteristic function is

$$M_{XY}(ju, jv) = \frac{1}{(1-ju)(j-jv)} \tag{4-51}$$

4-9 JOINT GAUSSIAN PDF

An important joint pdf is the Gaussian pdf, given by

[1]The characteristic function is the Fourier transform of the pdf; thus, from (4-49), the pdf of a sum of two independent random variables is the inverse Fourier transform of the product of their separate characteristic functions. From Fourier theory, it follows that the *pdf of the sum random variable is the convolution of the separate pdf's.*

$$f_{XY}(x, y) = \frac{\exp\left[-\dfrac{(x-\mu_X)^2/\sigma_X^2 - 2r(x-\mu_X)(y-\mu_Y)/\sigma_X\sigma_Y + (y-\mu_Y)^2/\sigma_Y^2}{2(1-r^2)}\right]}{2\pi\sigma_X\sigma_Y\sqrt{1-r^2}} \tag{4-52}$$

By tedious but straightforward integrations, it can be shown that

$$\mu_X = E(X) = \text{mean of } X, \qquad \mu_Y = E(Y) = \text{mean of } Y$$

$$\sigma_X^2 = E[(X-\mu_X)^2] = \text{variance of } X, \qquad \sigma_Y^2 = E[(Y-\mu_Y)^2] = \text{variance of } Y$$

$$r = \frac{E[(X-\mu_X)(Y-\mu_Y)]}{\sigma_X\sigma_Y} = \text{correlation coefficient of } X \text{ and } Y \tag{4-53}$$

The joint Gaussian pdf is a bell-shaped volume above the x–y plane (in general, with cuts through it parallel to the x–y plane being ellipses) with its maximum (the top of the bell) centered at $x = \mu_X$ and $y = \mu_Y$. The fatness of the ellipse cuts of the bell is determined by σ_X and σ_Y. We postpone consideration of the influence of σ_X and σ_Y until later; to simplify our initial discussion of this pdf, we introduce the new variables

$$x_n = \frac{x-\mu_X}{\sigma_X} \qquad \text{and} \qquad y_n = \frac{y-\mu_Y}{\sigma_Y} \tag{4-54}$$

or dropping the subscripts, we consider the simplified pdf

$$f(x, y) = \frac{\exp[-(x^2-2rxy+y^2)/2(1-r^2)]}{2\pi\sqrt{1-r^2}} \tag{4-55}$$

That is, we are considering a Gaussian pdf for which the random variables have zero means and unity variances. The lines of constant probability density are obtained by setting the expression in the exponent equal to a constant or, equivalently,

$$x^2 - 2rxy + y^2 = K^2 \tag{4-56}$$

This is the equation of a family of ellipses in the x–y plane, although not with their major and minor axes along the x and y coordinate axes, in general.[2]

Now consider a rotation of coordinates about the origin. It is readily shown that a counterclockwise rotation to new coordinates, x' and y' (Kreyzig, 1988), by an angle θ is given by

$$\begin{aligned} x' &= x\cos\theta + y\sin\theta \\ y' &= -x\sin\theta + y\cos\theta \end{aligned} \tag{4-57a}$$

or

$$\begin{aligned} x &= x'\cos\theta - y'\sin\theta \\ y &= x'\sin\theta + y'\cos\theta \end{aligned} \tag{4-57b}$$

[2]Recall that the equation of an ellipse of semimajor and semiminor axes of a and b along the x–y coordinate axes is

$$\frac{x^2}{a^2} + \frac{y^2}{b^2} = 1$$

Substituting these into (4-56), we obtain

$$(x'\cos\theta - y'\sin\theta)^2 - 2r(x'\cos\theta - y'\sin\theta)(x'\sin\theta + y'\cos\theta) + (x'\sin\theta + y'\cos\theta)^2 = K \tag{4-58}$$

or

$$x'^2 + y'^2 - r(x'^2 - y'^2)\sin 2\theta - 2rx'y'\cos 2\theta = K$$

We now find the angle θ such that the rotated coordinate axes are along the axes of the ellipse. For this to be the case, the $x'y'$ term must be zero, or $\cos 2\theta = 0$ or $\theta = \pm\pi/4$. If this is the case, (4-58) can be put into the form

$$\frac{x'^2}{K/(1-r)} + \frac{y'^2}{K/(1+r)} = 1 \tag{4-59}$$

This is the form of an ellipse with its axes along the x'–y' coordinate axes. The semimajor axes lengths are $a = [K/(1-r)]^{1/2}$ and $b = [K/(1+r)]^{1/2}$ (see footnote 2). If σ_X and σ_Y are not 1, as assumed here, the fatness of the ellipse is changed accordingly. Plots of the contours of (4-55) parallel to the x–y plane are shown in Figure 4-5 for several choices of r.

To get an idea of the effect of values for the standard deviations of X, and Y, consider the contours shown in Figure 4-6, which are for $\sigma_X = 1.25$ and $\sigma_Y = 1$. Three-dimensional representations of the joint Gaussian pdf with zero means and unit variances are shown in Figure 4-7 for $r = 0.9$ and -0.9. If $r = 0$ in (4-52), the middle term in the exponent vanishes, and we obtain

$$f_{XY}(x, y) = \frac{\exp\left[-(x-\mu_X)^2/2\sigma_X^2 + (y-\mu_Y)^2/2\sigma_Y^2\right]}{2\pi\sigma_X\sigma_Y}$$

$$= \frac{\exp[-(x-\mu_X)^2/2\sigma_X^2]\ \exp[-(y-\mu_Y)^2/2\sigma_Y^2]}{\sqrt{2\pi\sigma_X^2}\ \sqrt{2\pi\sigma_Y^2}} \tag{4-60}$$

Thus the joint pdf in this case is the product of a function of x alone and one of y alone. These functions, given by

$$f_X(x) = \frac{\exp\left[-(x-\mu_X)^2/2\sigma_X^2\right]}{\sqrt{2\pi\sigma_X^2}} \quad \text{and} \quad f_Y(y) = \frac{\exp\left[-(y-\mu_Y)^2/2\sigma_Y^2\right]}{\sqrt{2\pi\sigma_Y^2}} \tag{4-61}$$

are seen from (3-28) to be Gaussian pdf's. Recalling (4-33b), which states that if the joint pdf of two random variables factors into the product of the marginal pdf's the random variables are statistically independent we see from (4-52), (4-60), and (4-61) that if $r = 0$, jointly Gaussian random variables are statistically independent. It was stated in (4-53) that r is the correlation coefficient of X and Y. Random variables whose correlation coefficient is zero are said to be uncorrelated. Thus we have shown that *uncorrelated Gaussian random variables are also statistically independent,* which is not generally true (recall that statistically independent random variables *are* uncorrelated).

Example 4-16

Measurements are made on a random voltage at two time instants. The mean is found to be zero, and the variance (average power) is found to be 4 watts (volts^2 per ohm). Further mea-

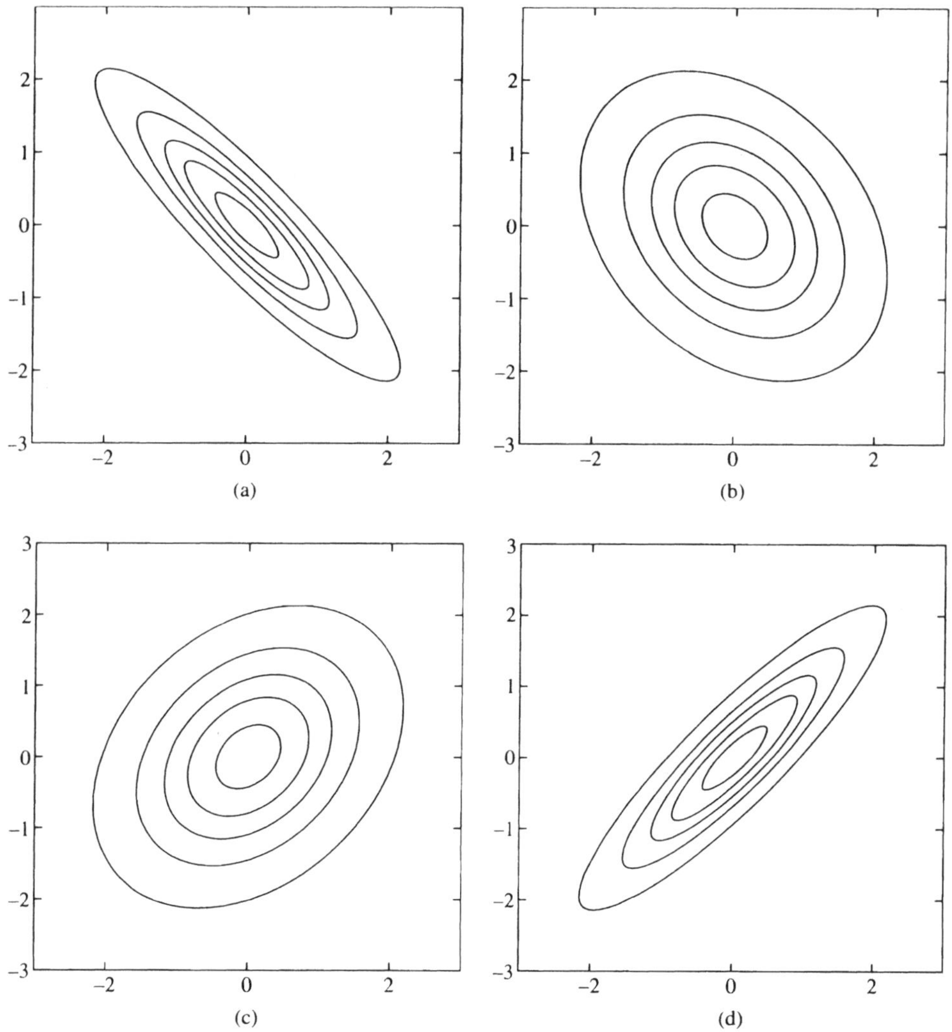

Figure 4-5 Equal probability density contours for a joint Gaussian pdf (zero means and unity variances with contour cuts at 0.9, 0.7, 0.5, 0.3, and 0.1 times the maximum): (a) $r = -0.9$; (b) $r = -0.3$; (c) $r = 0.3$; (d) $r = 0.9$.

surements also show that the two measurements are uncorrelated. Write down the joint pdf of the two measurements.

Solution Uncorrelated Gaussian random variables are statistically independent, so we know that the joint pdf is of the form (4-60). Let the random variables X and Y refer to the measurements at the two time instants. We assume that the statistics of the random voltage do not vary with time (called *statistical stationarity*), so

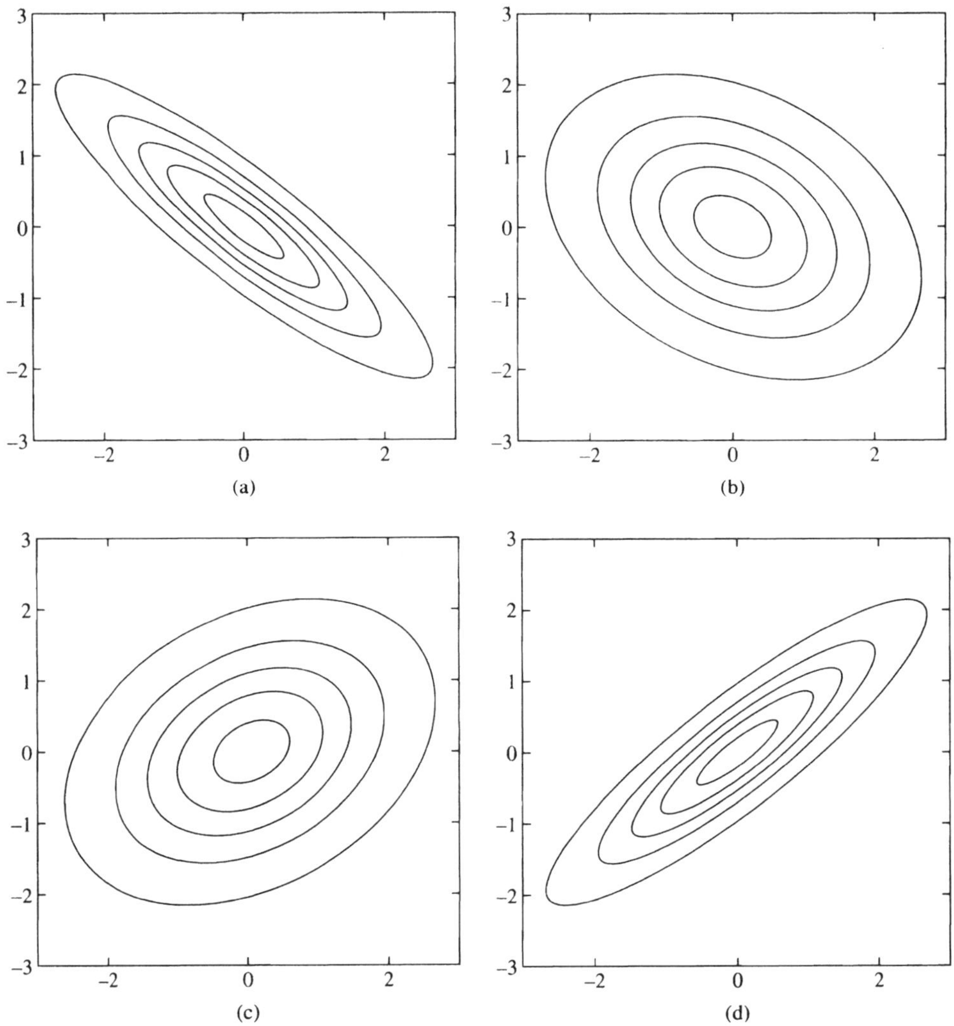

Figure 4-6 Equal probability density contours for a joint Gaussian pdf (zero means; $\sigma_X = 1.25$ and $\sigma_Y = 1$; contour cuts at 0.9, 0.7, 0.5, 0.3, and 0.1 times the maximum): (a) $r = -0.9$; (b) $r = -0.3$; (c) $r = 0.3$; (d) $r = 0.9$.

$$\mu_X = \mu_Y = 0$$

and

$$\sigma_X^2 = \sigma_Y^2 = 4$$

Thus, from (4-60), the joint pdf is

$$f_{XY}(x, y) = \frac{e^{-(x^2+y^2)/2(4)}}{2\pi(4)} = \frac{e^{-(x^2+y^2)/8}}{8\pi}$$

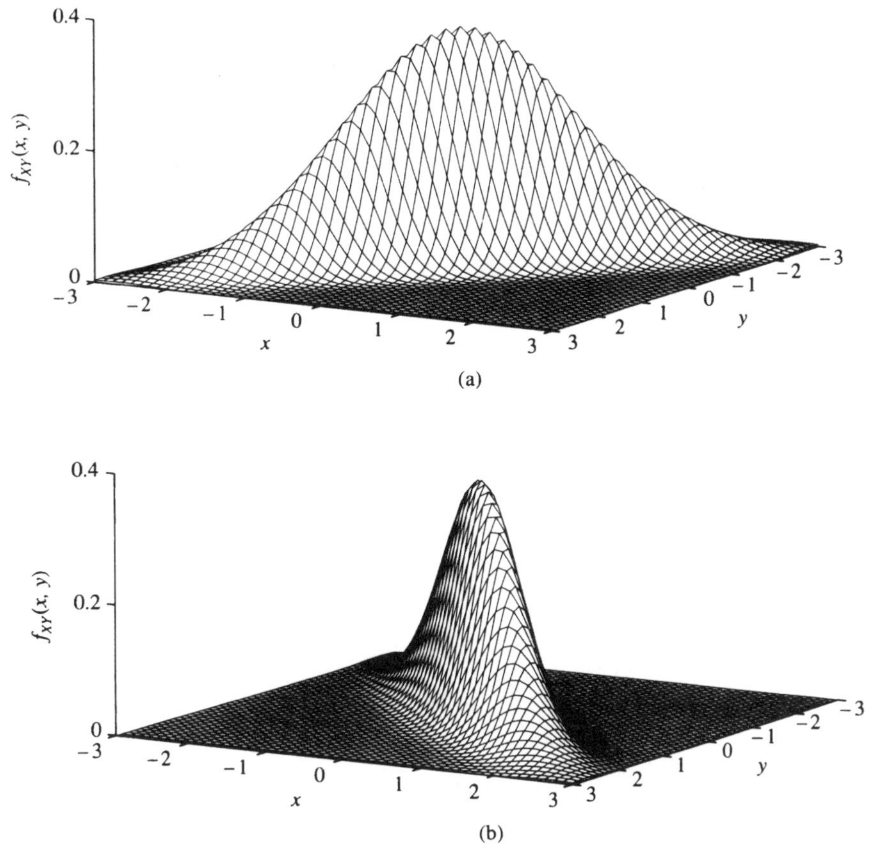

Figure 4-7 Three-dimensional representations of normalized bivariate Gaussian pdf's (zero means and unit variances: (a) $r = -0.9$; (b) $r = 0.9$.

4-10 FUNCTIONS OF TWO RANDOM VARIABLES

It often happens that we have a function of two random variables, say

$$Z = g(X, Y) \tag{4-62}$$

with a given joint pdf, and we wish to obtain the pdf of Z. One way to do this is to first find the cdf of Z, which is

$$F_Z(z) = P(Z \leq z) = P[g(X, Y) \leq z] \tag{4-63}$$

and then differentiate this cdf to get the pdf of Z. The condition $g(X, Y) \leq z$ defines some region in the x–y plane, so finding the cdf by evaluating (4-63) involves integration of the joint pdf of X and Y over this region. Some examples will illustrate the technique.

Example 4-17

Let the transformation be

$$Z = \sqrt{X^2 + Y^2} \tag{4-64}$$

and suppose that X and Y are independent Gaussian random variables with zero means and variances σ^2. Find the pdf of Z.

Solution We evaluate

$$P(\sqrt{X^2 + Y^2} \leq z) = F_Z(z) = \iint\limits_{R(z)} \frac{e^{-(x^2+y^2)/2\sigma^2}}{2\pi\sigma^2}\, dx\, dy \tag{4-65}$$

where the region $R(z)$ is the area inside the circle

$$x^2 + y^2 = z^2 \tag{4-66}$$

shown in Figure 4-8. The integral in (4-65) is easily evaluated by changing to polar coordinates, defined by

$$y = r\cos\theta \qquad \text{and} \qquad x = r\sin\theta \tag{4-67}$$

or

$$r = \sqrt{x^2 + y^2} \qquad \text{and} \qquad \theta = \tan^{-1}\frac{y}{x} \tag{4-68}$$

The differential area $dx\, dy$ transforms to

$$dx\, dy \rightarrow r dr\, d\theta \tag{4-69}$$

The region $R(z)$ is defined in terms of polar coordinates by

$$0 \leq r \leq z \qquad \text{and} \qquad 0 \leq \theta \leq 2\pi \tag{4-70}$$

Thus the integral (4-65) becomes

$$\begin{aligned} F_Z(z) &= \int_{\theta=0}^{2\pi}\int_0^z \frac{e^{-r^2/2\sigma^2}}{2\pi\sigma^2} r\, dr\, d\theta = \int_0^{2\pi}\left[\frac{e^{-r^2/2\sigma^2}}{2\pi}\right]_0^z d\theta \\ &= \int_0^{2\pi} (1 - e^{-z^2/2\sigma^2})\frac{d\theta}{2\pi} = 1 - e^{-z^2/2\sigma^2}, \qquad z \geq 0 \end{aligned} \tag{4-71}$$

Differentiation of (4-71) with respect to z gives the pdf:

$$f_Z(z) = \frac{z}{\sigma^2} e^{-z^2/2\sigma^2}, \qquad z \geq 0 \tag{4-72}$$

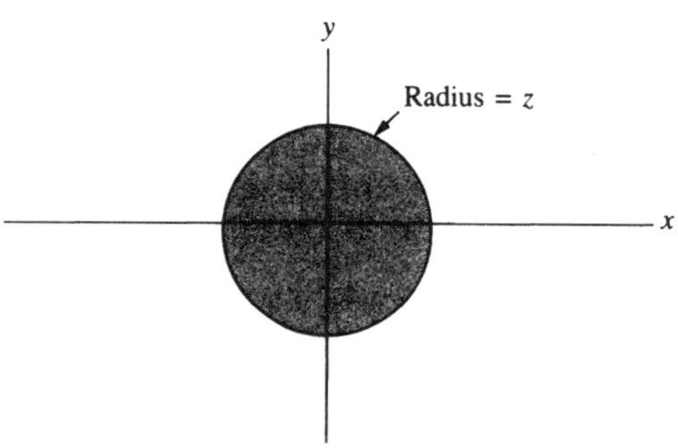

Figure 4-8 Area of integration.

This is called a *Rayleigh pdf.* It is plotted in Figure 4-9. By straightforward differentiation, the maximum can be shown to be at $z = \sigma$ with a value of $e^{-1/2}/\sigma = 0.6065/\sigma$. Also, the mean of a Rayleigh random variable is $E(Z) = (\pi/2)^{1/2}\sigma$; its mean-square value is $E(Z^2) = 2\sigma^2$ and its variance is $\sigma_Z^2 = (2 - \pi/2)\sigma^2$. Note that σ^2 is a parameter of the distribution, *not* the variance.

Example 4-18

Given the transformation

$$Z = X + Y \tag{4-73}$$

where X and Y are independent random variables. Find the pdf of Z.

Solution The area of integration is shown in Figure 4-10. Applying (4-13), we have

$$F_Z(z) = P(X + Y \le z) = \int_{-\infty}^{\infty}\int_{-\infty}^{z-y} f_{XY}(u, v)\, du\, dv \tag{4-74}$$

Differentiating and using statistical independence, we obtain the pdf of the sum random variable:

$$f_Z(z) = \int_{-\infty}^{\infty} f_X(z - y) f_Y(y)\, dy \tag{4-75}$$

Thus the pdf of the sum of two statistically independent random variables is the convolution of their separate pdf's as mentioned in connection with (4-49). For example, if the pdf's of X and Y are uniform in $(-0.5, 0.5)$, the pdf of Z is a triangular function centered at $z = 0$:

$$F_Z(z) = \begin{cases} 0, & z < -1 \\ 1 + z, & -1 \le z \le 0 \\ 1 - z, & 0 < z \le 1 \\ 0, & z > 1 \end{cases}$$

Note that there are many ways to get $Z = 0$, but only one way to get $Z = \pm 1$. Thus the triangular shape is not surprising. The derivation of this result and other examples is left to the problems.

Example 4-19

Show that the sum of two zero-mean Gaussian random variables is Gaussian whether or not they are independent.

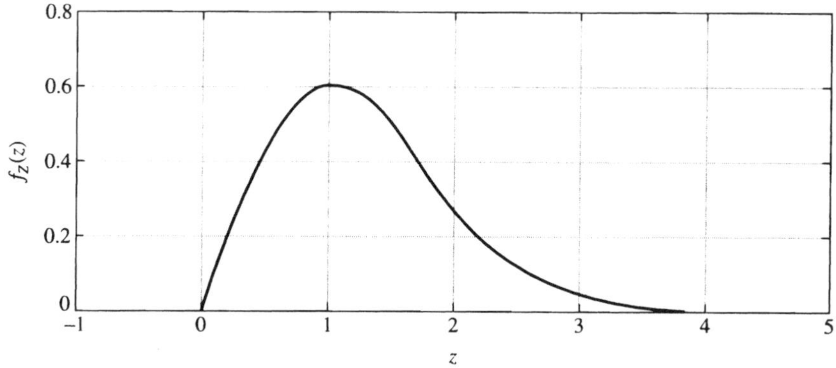

Figure 4-9 Plot of the Rayleigh pdf for $\sigma = 1$.

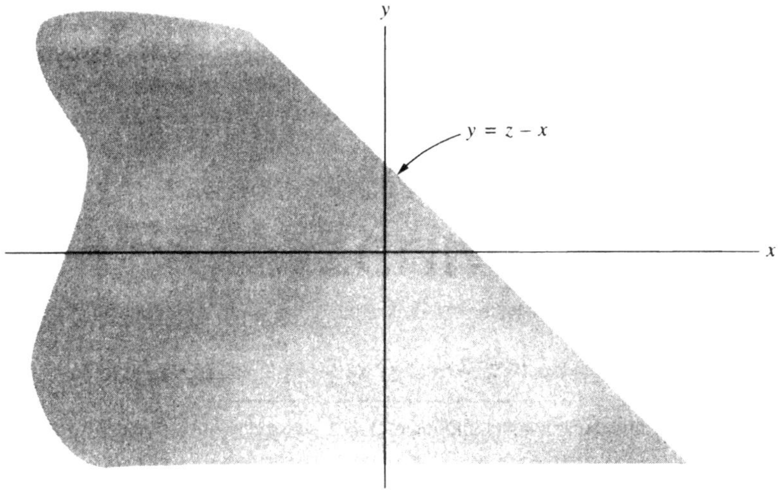

Figure 4-10 Transformation for finding the pdf of the sum of two random variables.

Solution Differentiation of (4-74) with respect to z gives

$$f_Z(z) = \int_{-\infty}^{\infty} f_{XY}(z - y, y)\, dy$$

Substitution of (4-52) with $\mu_X = \mu_Y = 0$ gives

$$f_Z(z) = \int_{-\infty}^{\infty} \frac{\exp\{-[(z-y)^2/\sigma_X^2 - 2r(z-y)y/\sigma_X\sigma_Y + y^2/\sigma_Y^2]/2(1-r^2)\}}{2\pi\sigma_X\sigma_Y\sqrt{1-r^2}}\, dy$$

Expansion of the exponent yields

$$\text{exponent} = -\frac{1}{2(1-r^2)}\left[\frac{z^2}{\sigma_X^2} + \left(\frac{1}{\sigma_X^2} + \frac{2r}{\sigma_X\sigma_Y} + \frac{1}{\sigma_Y^2}\right)y^2 - 2\left(\frac{1}{\sigma_X^2} + \frac{r}{\sigma_X\sigma_Y}\right)zy\right]$$

$$= -\frac{z^2}{2(1-r^2)\sigma_X^2} - \frac{\sigma_X^2 + 2r\sigma_X\sigma_Y + \sigma_Y^2}{2(1-r^2)\sigma_X^2\sigma_Y^2}\left[y - \frac{\sigma_Y(\sigma_Y + r\sigma_X)}{\sigma_X^2 + 2r\sigma_X\sigma_Y + \sigma_Y^2}\right]^2$$

$$+ \frac{1}{2(1-r^2)\sigma_X^2}\,\frac{(\sigma_Y + r\sigma_X)^2 z^2}{(\sigma_X^2 + 2r\sigma_X\sigma_Y + \sigma_Y^2)}$$

The last equation results by completing the square in y. Now define

$$\sigma_u^2 = \frac{(1-r^2)\sigma_X^2\sigma_Y^2}{\sigma_X^2 + 2r\sigma_X\sigma_Y + \sigma_Y^2}$$

and

$$u = y - \frac{\sigma_Y(\sigma_Y + r\sigma_X)}{\sigma_X^2 + 2r\sigma_X\sigma_Y + \sigma_Y^2}$$

The integral for $f_Z(z)$ then takes the form

$$
\begin{aligned}
f_Z(z) &= \frac{\exp[-z^2/2(\sigma_X^2 + 2r\sigma_X\sigma_Y + \sigma_Y^2)]}{\sqrt{2\pi(\sigma_X^2 + 2r\sigma_X\sigma_Y + \sigma_Y^2)}} \int_{-\infty}^{\infty} \frac{\exp(-u^2/2\sigma_u^2)}{\sqrt{2\pi\sigma_u^2}} \\
&= \frac{\exp[-z^2/2(\sigma_X^2 + 2r\sigma_X\sigma_Y + \sigma_Y^2)]}{\sqrt{2\pi(\sigma_X^2 + 2r\sigma_X\sigma_Y + \sigma_Y^2)}}
\end{aligned}
\tag{4-76}
$$

This is a Gaussian pdf with zero mean and variance

$$\sigma_z^2 = \sigma_X^2 + 2r\sigma_X\sigma_Y + \sigma_Y^2$$

We can easily change the pdf of (4-76) to the case for nonzero means by a translation of variables.

4–11 TRANSFORMATION OF A PAIR OF RANDOM VARIABLES

Suppose that we have a transformation of a pair of random variables, whose joint pdf is known, to another pair of random variables, U and V:

$$
\begin{aligned}
U &= g(X, Y) \\
V &= h(X, Y)
\end{aligned}
\tag{4-77}
$$

The probability of X and Y being in some region in the x–y plane is given in terms of x and y's joint pdf as the integral

$$P(X, Y \in R) = \iint_R f_{XY}(x, y)\, dx\, dy \tag{4-78}$$

We want to change variables in the integral of (4-78) from x and y to u and v. The region R in the x–y plane transforms to some region R^* in the u–v plane under the transformation $u = g(x, y)$ and $v = h(x, y)$. Assume that this transformation is one-to-one and that the inverse transformation equations $x = g^{-1}(u, v)$ and $y = h^{-1}(u, v)$ have continuous first partial derivatives in R^*. Then it is known from the calculus of transformation of variables in double integrals (Kreyzig, 1988) that

$$\iint_R f_{XY}(x, y)\, dx\, dy = \iint_{R^*} f_{XY}\left[g^{-1}(u, v), h^{-1}(u, v)\right] \left|\frac{\partial(x, y)}{\partial(u, v)}\right| du\, dv \tag{4-79}$$

where

$$J = \frac{\partial(x, y)}{\partial(u, v)} = \begin{vmatrix} \dfrac{\partial x}{\partial u} & \dfrac{\partial x}{\partial v} \\ \dfrac{\partial y}{\partial u} & \dfrac{\partial y}{\partial v} \end{vmatrix} \tag{4-80}$$

is the Jacobian. From (4-79), we deduce that the pdf of U and V in terms of the known pdf of X and Y is

$$F_{UV}(u, v) = f_{XY}[g^{-1}(u, v), h^{-1}(u, v)] \left|\frac{\partial(x, y)}{\partial(u, v)}\right| \tag{4-81}$$

The partial derivatives in (4-80) are defined as

$$\frac{\partial x}{\partial u} = \frac{\partial}{\partial u} g^{-1}(u, v), \quad \frac{\partial x}{\partial v} = \frac{\partial}{\partial v} g^{-1}(u, v), \tag{4-82}$$

$$\frac{\partial y}{\partial u} = \frac{\partial}{\partial u} h^{-1}(u, v), \quad \frac{\partial y}{\partial v} = \frac{\partial}{\partial v} h^{-1}(u, v)$$

Example 4-20

Consider the transformation of two Gaussian random variables, X and Y, of mean zero, variance 2, and correlation coefficient 0.7 by the transformation

$$U = X - Y \qquad \text{and} \qquad V = X + 2Y \tag{4-83}$$

Find the pdf of U and V.

Solution Solve the transformation for x and y to obtain

$$x = \frac{2}{3}u + \frac{1}{3}v \qquad \text{and} \qquad y = -\frac{1}{3}u + \frac{1}{3}v$$

The Jacobian is

$$J\begin{pmatrix} x & y \\ u & v \end{pmatrix} = \begin{vmatrix} \dfrac{\partial x}{\partial u} & \dfrac{\partial x}{\partial v} \\ \dfrac{\partial y}{\partial u} & \dfrac{\partial y}{\partial v} \end{vmatrix} = \begin{vmatrix} \dfrac{2}{3} & \dfrac{1}{3} \\ -\dfrac{1}{3} & \dfrac{1}{3} \end{vmatrix} = \frac{1}{3}$$

From (4-52), the pdf of X and Y is

$$f_{XY}(x, y) = \frac{\exp\left\{-\dfrac{(x^2/2) - \left[2(0.7)xy/2\right] + (y^2/2)}{2(1 - (0.7)^2)}\right\}}{2\pi(2)\sqrt{1 - (0.7)^2}}$$

$$= 0.111e^{-(0.495x^2 - 0.686xy + 0.495y^2)}$$

Multiplying this by the Jacobian and substituting for x and y, we obtain

$$f_{UV}(u, v) = \frac{0.111}{3} \exp\{-[0.495(0.667u + 0.333v)^2 - 0.686(0.667u + 0.333v)$$

$$(-0.333u + 0.333v) + 0.495(-0.333u + 0.333v)^2)\}$$

$$= 0.037 \exp\{-(0.423u^2 + 0.033uv + 0.0325v^2)\}$$

Example 4-21

Given two independent, zero-mean Gaussian random variables with variances σ^2, find the joint pdf of the random variables

$$R = \sqrt{X^2 + Y^2} \qquad \text{and} \qquad \Theta = \tan^{-1}\frac{Y}{X} \tag{4-84}$$

That is, find the joint pdf in terms of polar coordinates. Also find the pdf's of R and Θ alone.

Solution The transformation from polar to Cartesian coordinates is

$$x = r\cos\theta \qquad \text{and} \qquad y = r\sin\theta,\ r \geq 0, \quad 0 \leq \theta < 2\pi \tag{4-85}$$

The Jacobian is

$$J = \begin{vmatrix} \cos\theta & -r\sin\theta \\ \sin\theta & r\cos\theta \end{vmatrix} = r$$

We let $\mu_X = \mu_Y = 0$ and $\sigma_X^2 = \sigma_Y^2 = \sigma^2$ in (4-52) and use (4-85) to transform to polar coordinates to obtain

$$f_{R\Theta}(r, \theta) = \frac{r}{2\pi\sigma^2} e^{-r^2/2\sigma^2}, \quad r \geq 0, \quad 0 \leq \theta < 2\pi \tag{4-86}$$

To obtain the pdf of R alone, we integrate over θ and to get the pdf of Θ alone we integrate over r [see (4-12)]. The results are

$$f_R(r) = \frac{r}{\sigma^2} e^{-r^2/2\sigma^2}, r \geq 0 \quad \text{and} \quad f_\Theta(\theta) = \frac{1}{2\pi}, 0 \leq \theta < 2\pi \tag{4-87}$$

These are the Rayleigh and uniform pdf's, respectively [see Example 4-17 and (3-24) with $a = 0$ and $b = 2\pi$]. Since (4-87) is the product of $f_R(r)$ and $f_\Theta(\theta)$, R and Θ are independent.

4–12 SUM OF MANY INDEPENDENT RANDOM VARIABLES: THE CENTRAL LIMIT THEOREM

In Example 4-18 we saw that the pdf of the sum of two independent random variables is the convolution[3] of their respective pdf's. Clearly, we can carry this on to four random variables by defining

$$Z_1 = X_1 + X_2 \quad \text{and} \quad Z_2 = X_3 + X_4 \tag{4-88}$$

If X_1, X_2, X_3, and X_4 are statistically independent random variables, we can first find the pdf's of Z_1 and Z_2 by convolving the pdf's of the constituent random variables and, finally, find the pdf of the sum

$$Z = Z_1 + Z_2 = X_1 + X_2 + X_3 + X_4 \tag{4-89}$$

by convolving the pdf's of Z_1 and Z_2. This procedure can, of course, be continued to include more than four random variables. We will stop at four, however, where surprisingly at first, the pdf of the sum resembles a Gaussian pdf relatively closely.

Example 4-22

Four statistically independent random variables all have the same pdf, which is given by

$$f_{X_i}(x_i) = \begin{cases} 1, & |x_i| \leq \frac{1}{2} \\ 0, & \text{otherwise} \end{cases} \tag{4-90}$$

Find the pdf of their sum.

Solution The convolution by pairs gives the pdf's

[3]Some students may not be familiar with convolution at this point in their studies. If this is the case, we advise that they focus on the results of this section and skip over the mechanics of getting those results.

$$f_{Z_i}(z_i) = \begin{cases} 1 - |z_i|, & |z_i| \le 1, \\ 0, & \text{otherwise} \end{cases} \quad i = 1, 2 \tag{4-91}$$

Convolution of Z_1 and Z_2 gives

$$f_Z(z) = \begin{cases} (1 - z) - \frac{1}{3}(1 - z)^3 + \frac{1}{6}z^3, & 0 \le z \le 1 \\ \frac{1}{6}(2 - z)^3, & 1 \le z \le 2 \end{cases} \tag{4-92}$$

for $z > 0$. The pdf is even, so we may use

$$f_z(-z) = f_Z(z) \tag{4-93}$$

to get the result for z negative. From (4-93), it follows that the mean of Z is zero (the pdf is symmetrical about 0), and its variance is $\frac{4}{12} = \frac{1}{3}$ (each constituent random variable has variance $\frac{1}{12}$ and statistical independence is used to note that the variance of Z is four times this).

We compare the pdf (4-92) with a Gaussian pdf with mean zero and variance $\frac{1}{3}$ in Figure 4-11. Note that the two pdf's look remarkably similar, with the largest deviation being at the tails and the peak. Had we found the pdf of, say, the sum of eight independent, uniformly distributed random variables and compared it with a Gaussian pdf with the same mean and variance, the resemblance would have been closer yet. The reason for this is summarized by a remarkable theorem, called the *central limit theorem,* stated below.

Central Limit Theorem

Let $X_1, X_2, X_3, \ldots, X_N$ be independent random variables with means $m_1, m_2, m_3, \ldots, m_N$ and standard deviations $\sigma_1, \sigma_2, \sigma_3, \ldots, \sigma_N$, respectively. Then

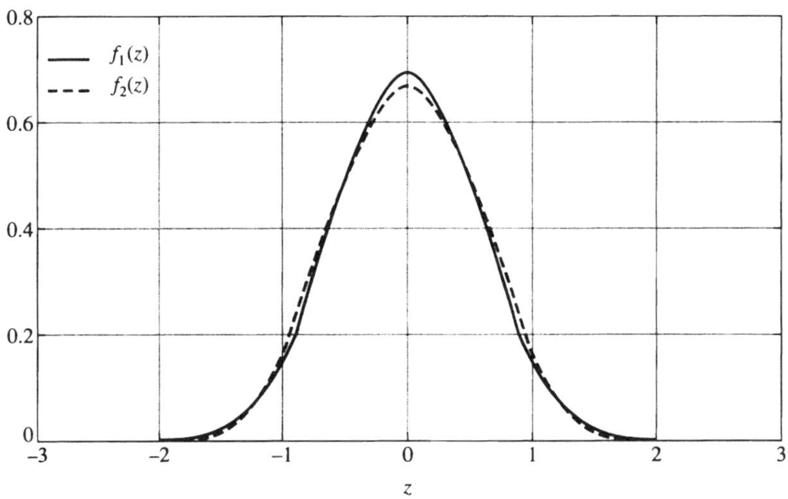

Figure 4-11 Comparison of the pdf of the sum of four uniformly distributed random variables [$f_2(z)$] and a Gaussian pdf with the same mean and variance [$f_1(z)$].

$$Z = \frac{1}{\sqrt{N}} \sum_{i=1}^{N} \frac{X_i - m_i}{\sigma_i} \tag{4-94}$$

approaches a Gaussian random variable with zero mean and unit variance as N becomes large, provided that

$$\lim_{N \to \infty} \frac{\sigma_i}{\sigma} = 0 \tag{4-95}$$

The pdf's of the component random variables need not be of any specified type as long as (4-95) is satisfied. Indeed, they need not all be identical. The condition (4-95) simply ensures that no one random variable in the sum (4-94) dominates. Note that if one random variable of the sum had a much larger variance, the width of its pdf would influence the shape of the final pdf of the sum through the convolution operation.

Example 4-23

Gear key pins are manufactured to have a nominal length of $L = 10$ mm. Extensive measurements have shown that the standard deviations of their lengths is 0.1 mm. What is the probability that the sample mean [see (1-1)] of the lengths of a random selection of 10 such pins differs by more than 0.1 standard deviation of their lengths from their nominal lengths? How large a sample should be taken to make this probability 0.05?

Solution We find the solution for an arbitrary number of pins, N. Let L_i denote the N separate measurements, each of mean μ and standard deviation σ. According to the central limit theorem, the following quantity is approximately Gaussian with zero mean and unit variance for N sufficiently large:

$$Z = \frac{1}{\sqrt{N}} \sum_{i=1}^{N} \frac{L_i - \mu}{\sigma} = \frac{1}{\sqrt{N}\sigma}\left(\sum_{i=1}^{N} L_i - N\mu\right) = \frac{\sqrt{N}}{\sigma}\left(\frac{1}{N}\sum_{i=1}^{N} L_i - \mu\right) = \frac{\sqrt{N}}{\sigma}(\bar{L} - \mu) \tag{4-96}$$

In (4-96),

$$\bar{L} = \frac{1}{N} \sum_{i=1}^{N} L_i \tag{4-97}$$

is the sample mean for N measurements. Using the fact that the right-hand quantity in (4-96) is closely Gaussian with zero mean and unit variance, we have that

$$\begin{aligned} P_{\text{exceed}} &= P[\,|\bar{L} - \mu| \geq 0.1\sigma] = P\left[Z = \frac{\sqrt{N}}{\sigma}|\bar{L} - \mu| \geq 0.1\sqrt{N}\right] \\ &= \int_{0.1\sqrt{N}}^{\infty} \frac{e^{-z^2/2}}{\sqrt{2\pi}} dz + \int_{-\infty}^{-0.1\sqrt{N}} \frac{e^{-z^2/2}}{\sqrt{2\pi}} dz \\ &= 2\int_{0.1\sqrt{N}}^{\infty} \frac{e^{-z^2/2}}{\sqrt{2\pi}} dz \\ &= 2Q(0.1\sqrt{N}) \end{aligned} \tag{4-98}$$

Tables (see Appendix C) or a rational approximation for the Q-function give the values for P_{exceed} versus N that are listed in Table 4-4. For $N = 10$, the probability that the sample mean of the measurements will exceed the boundaries of $\mu \pm 0.1\sigma$ is fairly high. It takes 300 measurements to make this probability less than 0.1 and almost 400 measurements to make it 0.05. This is because of the stringent condition put on the accuracy of the measurement (0.1σ). Surprising as it may seem, neither the nominal length nor the standard deviation of the pin lengths were needed to solve the problem as stated, as long as the central limit theorem can be applied.

4–13 WEAK LAW OF LARGE NUMBERS

The *weak law of large numbers* states that the probability of the average of n independent values of a random variable X differing from their means $\mu_X = E(X)$ by more than an arbitrary $\epsilon > 0$ goes to zero as $n \to \infty$. To show this, consider Chebyshev's inequality rewritten as

$$P(|X - \mu_X| \geq \epsilon) \leq \frac{\sigma_X^2}{\epsilon^2} \tag{4-99}$$

which follows from (3-104) by letting $\epsilon = k\sigma_X$. The average of n independent values of a random variable is $S_n = (1/n)\sum_{i=1}^{n} X_i$. Now rewrite (4-99) with S_n replacing X to get

$$P(|S_n - E(S_n)| \geq \epsilon) \leq \frac{\sigma_{S_n}^2}{\epsilon^2} \tag{4-100}$$

The expectation of S_n is just $E(S_n) = \mu_X$ which can be shown by taking the expectation of the sum term by term. The variance of S_n can be shown to be σ_X^2/n by writing S_n^2 as a double sum and taking the expectation term by term. These details are left to the problems. Thus (4-100) becomes

$$P(|S_n - \mu_X| \geq \epsilon) \leq \frac{\sigma_X^2}{n\epsilon^2} \tag{4-101}$$

As $n \to \infty$ with σ_X and ϵ fixed, the right-hand side of (4-101) goes to zero, which results in

$$\lim_{n\to\infty} P\left[\left|\frac{1}{n}\sum_{i=1}^{n} X_i - E(X)\right| \geq \epsilon\right] = 0, \qquad \epsilon > 0 \tag{4-102}$$

which is the weak law of large numbers.

Example 4-24

Show the probability that the relative frequency of an event differs from the probability of that event by more than an arbitrarily selected positive amount goes to zero as the number of repetitions of the underlying chance experiment goes to infinity.

Solution Denote the event by A and the number of times it occurs in n repetitions of the

Table 4-4 P_{EXCEED} VERSUS N

N	P_{exceed}
10	0.752
50	0.480
100	0.317
200	0.157
300	0.083
384	0.050
400	0.046

chance experiment by n_A. Its relative frequency of occurrence is n_A/n. The occurrence of A can be modeled as a series of Bernoulli trials, with each success occurring with probability p. Thus the mean of n_A is np [recall (3-96) for the mean of a binomial random variable], or $E(n_A/n) = p$. Selecting an arbitrary value for ϵ, from (4-102) we have that

$$\lim_{n\to\infty} P\left[\left|\frac{n_A}{n} - p\right| \geq \epsilon\right] = 0, \epsilon > 0 \tag{4-103}$$

which says that the probability of the relative frequency of an event differing by more than an arbitrarily chosen number ϵ is zero in the limit of an arbitrarily large number of trials of the underlying chance experiment.

4–14 EXTENSION OF JOINT DISTRIBUTIONS TO MORE THAN TWO RANDOM VARIABLES

All the definitions given in this chapter regarding joint random variables can be extended directly from two to $N > 2$ random variables. For example, if $X_1, X_2, X_3, \ldots, X_N$ are random variables, their joint cdf is defined as

$$F_N(x_1, x_2, \ldots, x_N) = P(X_1 \leq x_1, X_2 \leq x_2, \ldots, X_N \leq x_N) \tag{4-104}$$

and their joint pdf is defined as

$$f_N(x_1, x_2, \ldots, x_N) = \frac{\partial^N}{\partial x_1\, \partial x_2 \ldots \partial x_N} F_N(x_1, x_2, \ldots, x_N) \tag{4-105}$$

Both have properties that are direct extensions of the two-variable case. For example, (4-105), when integrated over all its variables from $-\infty$ to ∞, gives unity. When integrated over a single variable, we get the next-lowest-order pdf:

$$f_{N-1}(x_1, \ldots, x_{m-1}, x_{m+1}, \ldots, x_N) = \int_{x_m=-\infty}^{\infty} f_N(x_1, x_2, \ldots, x_N)\, dx_m \tag{4-106}$$

If all but one of the variables is integrated out, we get a first-order pdf of the remaining variable. Conditional pdf's can also be defined. For example,

$$f_N(x_1, x_2, \ldots, x_j \mid x_{j+1}, \ldots, x_N) = \frac{f_N(x_1, x_2, \ldots, x_N)}{f_{N-j}(x_{j+1}, \ldots, x_N)} \tag{4-107}$$

It is generally difficult to find explicit expressions for joint cdf's or pdf's. One exception is the N-fold jointly Gaussian pdf, which is expressed compactly in matrix form as

$$f_{\mathbf{x}}(\mathbf{x}) = |\mathbf{C}|^{-1/2}(2\pi)^{-N/2} \exp\left[-\frac{1}{2}(\mathbf{x} - \mathbf{m})^t\mathbf{C}^{-1}(\mathbf{x} - \mathbf{m})\right] \tag{4-108}$$

where the superscript t denotes transpose, and $\mathbf{x}$ and $\mathbf{m}$ are column matrices defined as

$$\mathbf{x} = \begin{bmatrix} x_1 \\ x_2 \\ \vdots \\ x_N \end{bmatrix} \quad \text{and} \quad \mathbf{m} = \begin{bmatrix} m_1 \\ m_2 \\ \vdots \\ m_N \end{bmatrix} \tag{4-109}$$

The matrix **C** is the matrix of covariances with elements

$$C_{ij} = E[(X_i - m_i)(X_j - m_j)], \quad i, j = 1, 2, \ldots, N \tag{4-110}$$

The joint characteristic function corresponding to (4-108) is

$$M_{\mathbf{X}}(j\mathbf{v}) = \exp\left(j\mathbf{m}'\mathbf{v} - \frac{1}{2}\mathbf{v}'\mathbf{C}\mathbf{v}\right) \tag{4-111}$$

where **v** is a column matrix with N components like **x**.

The expectation of a function of several random variables, say $g(X_1, X_2, X_3, \ldots, X_N)$, is a direct extension of (3-82) or (4-34), and is expressed in terms of their joint pdf as

$$E[g(X_1, X_2, \ldots, X_N)] = \int_{-\infty}^{\infty} \cdots \int_{-\infty}^{\infty} g(x_1, x_2, \ldots, x_N) f_N(x_1, x_2, \ldots, x_N)\, dx_1\, dx_2 \cdots dx_N \tag{4-112}$$

Properties of the expectation of a function of several random variables are directly analogous to those stated for two random variables, such as (4-35) and (4-36). For example, the mean of a sum of random variables is the sum of their means whether they are independent or not. It is not true that the variance of a sum of random variables is the sum of their separate variances, however, *unless they are statistically independent* (sufficient condition). A general result for the variance of a sum of random variables where the constituent random variables are not statistically independent may be derived as follows. Consider

$$\begin{aligned}
\mathrm{Var}\left(\sum_{i=1}^{n} X_i\right) &= E\left\{\left[\sum_{i=1}^{n} X_i - \sum_{i=1}^{n} \mu_i\right]^2\right\} = E\left\{\left[\sum_{i=1}^{n} (X_i - \mu_i)\right]^2\right\} \\
&= E\left\{\left[\sum_{i=1}^{n} (X_i - \mu_i)\right]\left[\sum_{j=1}^{n} (X_j - \mu_j)\right]\right\} \\
&= E\left[\sum_{i=1}^{n}\sum_{j=1}^{n} (X_i - \mu_i)(X_j - \mu_j)\right] \\
&= \sum_{i=1}^{n}\sum_{j=1}^{n} E[(X_i - \mu_i)(X_j - \mu_j)]
\end{aligned} \tag{4-113}$$

where the fact that the mean of a sum is the sum of the separate means, and manipulation of summations has led to being able to take the expectation inside the double sum at the last step. Focusing on this last sum, we see that for $i = j$ the expectations of the individual terms are of the form

$$E[(X_i - \mu_i)^2] = \mathrm{Var}(X_i) = \sigma_i^2 \tag{4-114}$$

There are n terms like this [imagine the double sum of (4-109) written out as an array—these terms are the ones along the diagonal]. The remaining $n(n - 1)$ terms are of the form

$$E[(X_i - \mu_i)(X_j - \mu_j)] = C_{X_i X_j} = \sigma_i \sigma_j \rho_{ij} \tag{4-115}$$

which follows from (4-39a) and (4-40). In (4-115), ρ_{ij} is the correlation coefficient of X_i and X_j. Thus

$$\text{Var}\left(\sum_{i=1}^{n} X_i\right) = \sum_{i=1}^{n} \sigma_i^2 + \sum_{\substack{i=1 \\ i \neq j}}^{n} \sum_{j=1}^{n} \sigma_i \sigma_j \rho_{ij} \tag{4-116}$$

Note that by this formula, the variance of the sum random variable found in Example 4-19 is correct.

By a simple extension, it follows that, with the a_i's constants,

$$\text{Var}\left(\sum_{i=1}^{n} a_i X_i\right) = \sum_{i=1}^{n} a_i^2 \sigma_i^2 + \sum_{\substack{i=1 \\ i \neq j}}^{n} \sum_{j=1}^{n} a_i a_j \sigma_i \sigma_j \rho_{ij} \tag{4-117}$$

Example 4-25

Find the variance of a sum of 10 random variables, all with variances of 1 and with the correlation coefficient between different random variables being 0.5.

Solution From (4-116), the result is

$$\text{Var}\left(\sum_{i=1}^{10} X_i\right) = \sum_{i=1}^{10} 1 + \sum_{\substack{i=1 \\ i \neq j}}^{10} \sum_{j=1}^{10} (1)(1)(0.5) = 10 + (10)(9)(0.5) = 55 \tag{4-118}$$

4-15 SUMMARY

In this chapter, distributions (cdf's and pdf's) for more than one random variable were defined and their properties discussed. Such distributions are referred to as joint, as opposed to marginal which describe a single random variable. Several examples involving averages and transformations of multiple random variables were also given.

1. The joint cdf of two random variables, X and Y, is defined as

$$F_{XY}(x, y) = P(X \leq x, Y \leq y)$$

Its properties are extensions of the properties of marginal cdf's. For example,

$$F_{XY}(-\infty, y) = F_{XY}(x, -\infty) = F_{XY}(-\infty, -\infty) = 0$$

and

$$F_{XY}(\infty, \infty) = 1$$

In addition, we have relations between the joint cdf and marginal cdf's:

$$F_{XY}(x, \infty) = F_X(x)$$

$$F_{XY}(\infty, y) = F_Y(y)$$

Finally, to find the joint probability of the random variables X and Y lying in a given rectangle, we use

$$P(x_1 \leq X \leq x_2, y_1 \leq Y \leq y_2) = P(X \leq x_2, Y \leq y_2) - P(X \leq x_1, Y \leq y_2)$$

$$- P(X \leq x_2, Y \leq y_1) + P(X \leq x_1, Y \leq y_1)$$
$$= F_{XY}(x_2, y_2) - F_{XY}(x_1, y_2) - F_{XY}(x_2, y_1) + F_{XY}(x_1, y_1)$$

2. The joint pdf of two random variables X and Y is defined as

$$f_{XY}(x, y) = \frac{\partial^2 F_{XY}(x, y)}{\partial x \partial y}$$

Its properties are extensions of the marginal pdf. For example, the volume between the joint pdf surface and the x–y plane is unity, and it is everywhere a nonnegative function. To get to a marginal pdf from a joint pdf, we "integrate out" the variable we don't want. For example,

$$f_X(x) = \int_{-\infty}^{\infty} f_{XY}(x, y)\, dy$$

$$f_Y(y) = \int_{-\infty}^{\infty} f_{XY}(x, y)\, dx$$

We get the joint cdf from the joint pdf by integration:

$$F_{XY}(x, y) = \int_{-\infty}^{x} \int_{-\infty}^{y} f_{XY}(u, v)\, du\, dv$$

and to find the probability of the random variables X and Y being in a certain rectangle, we use

$$P(x_1 \leq X \leq x_2, y_1 \leq Y \leq y_2) = \int_{x_1}^{x_2} \int_{y_1}^{y_2} f_{XY}(x, y)\, dx\, dy$$

3. Discrete random variable pairs can be described either by joint cdf's or by joint probability mass functions. The latter simply is the set of probabilities of the random variable pair taking on each possible set of pairs; it must sum to unity:

$$\sum_{i=1}^{m} \sum_{j=1}^{n} P_{ij} = 1$$

4. The conditional cdf of a random variable X, given a related random variable Y lying in a certain range R_y, is defined as

$$F_X(x \mid Y \in R_y) = P(X \leq x \mid Y \in R_y) = \frac{P(X \leq x, Y \in R_y)}{P(Y \in R_y)}$$

The conditional pdf of X given random variable Y is defined as

$$f_{X|Y}(x \mid y) = \frac{f_{XY}(x, y)}{f_Y(y)}$$

and similarly, the conditional pdf of Y given random variable X is defined as

$$f_{Y|X}(y \mid x) = \frac{f_{XY}(x, y)}{f_X(x)}$$

5. Two random variables are statistically independent if

$$F_{XY}(x, y) = F_X(x)F_Y(y)$$

or if

$$f_{XY}(x, y) = f_X(x)f_Y(y)$$

or if

$$f_{X|Y}(x \mid y) = f_X(x)$$

$$f_{Y|X}(y \mid x) = f_Y(y)$$

These statements are all equivalent.

6. The expectation or average of a function of two random variables is defined as

$$E[g(X, Y)] = \int_{-\infty}^{\infty}\int_{-\infty}^{\infty} g(x, y)f_{XY}(x, y)\,dx\,dy$$

where $f_{XY}(x, y)$ is their joint pdf. If X and Y are discrete random variables, then

$$E[g(X, Y)] = \sum_{i=1}^{m}\sum_{j=1}^{n} g(x_i, y_j)P(X = x_i, Y = y_j)$$

Two useful properties of joint expectations are:

(a) If $g(X, Y)$ is *a product of two functions,* one of which is a function of X alone and the other a function of Y alone, and if X and Y are *statistically independent,* then

$$E[g_1(X)g_2(Y)] = E[g_1(X)]E[g_2(Y)]$$

(b) If $g(X, Y) = a_1g_1(X, Y) + a_2g_2(X, Y)$ where $g(X, Y)$ and $g_2(X, Y)$ are separate functions of X and Y, and a_1 and a_2 are constants, then

$$E[a_1g_1(X, Y) + a_2g_2(X, Y)] = a_1E[g_1(X, Y)] + a_2E[g_2(X, Y)]$$

7. Special averages include the following:

(a) *Joint moments:*

$$m_{mn} = E(X^mY^n), \qquad m, n = 1, 2, \ldots$$

(b) *Joint central moments:*

$$\mu_{mn} = E[(X - \mu_X)^m(Y - \mu_Y)^n, \qquad m, n = 1, 2, \ldots$$

where μ_X and μ_Y are the means of X and Y, respectively.

(c) *Covariance:*

$$C_{XY} = E[(X - \mu_X)(Y - \mu_Y)] = E(XY) - \mu_X\mu_Y$$

(d) *Correlation coefficient:*

$$\rho_{XY} = \frac{C_{XY}}{\sigma_X\sigma_Y}$$

The correlation coefficient magnitude is bounded by unity. If it is zero, the random variables are said to be *uncorrelated.* Statistically independent random variables are uncorrelated, but the reverse is not necessarily true. If the random variables X and Y are linearly related, their correlation coefficient is either 1 or -1.

(e) *Characteristic function:*

$$M_{XY}(ju, jv) = E\left[e^{j(uX+vY)}\right] = E\left[e^{juX}e^{jvY}\right]$$

The characteristic function of statistically independent random variables factors:

$$M_{XY}(ju, jv) = E\left[e^{juX}\right]E\left[e^{jvY}\right] = M_X(ju)M_Y(jv)$$

The characteristic function of the sum of two statistically independent random variables is the product of their separate characteristic functions. Since the characteristic function is the Fourier transform of the pdf, this means that the pdf of a sum of two statistically independent random variables is the convolution of their separate pdf's.

8. The joint Gaussian pdf is given by

$$f_{XY}(x, y) = \frac{\exp\left\{-\dfrac{[(x-\mu_X)^2/\sigma_X^2] - [2r(x-\mu_X)(y-\mu_Y)/\sigma_X\sigma_Y] + [(y-\mu_Y)^2/\sigma_Y^2]}{2(1-r^2)}\right\}}{2\pi\sigma_X\sigma_Y\sqrt{1-r^2}}$$

where

$$\mu_X = E(X) = \text{mean of } X$$

$$\mu_Y = E(Y) = \text{mean of } Y$$

$$\sigma_X^2 = E[(X-\mu_X)^2] = \text{variance of } X$$

$$\sigma_Y^2 = E[(Y-\mu_Y)^2] = \text{variance of } Y$$

$$r = \frac{E[(X-\mu_X)(Y-\mu_Y)]}{\sigma_X\sigma_Y} = \text{correlation coefficient of } X \text{ and } Y$$

If two Gaussian random variables are uncorrelated, they are also statistically independent. This is easily seen by setting $r = 0$ in the pdf above and noting that it then factors.

9. Given a function of two random variables,

$$Z = g(X, Y)$$

the cdf of Z can be found from

$$F_Z(z) = P(Z \le z) = P(g(X, Y) \le z)$$

The pdf can be found by differentiation. The transformation

$$Z = \sqrt{X^2 + Y^2}$$

of two statistically independent Gaussian random variables with zero means and variances σ^2 results in the *Rayleigh* pdf:

$$f_z(z) = \frac{z}{\sigma^2}e^{-z/2\sigma^2}, \qquad z \ge 0$$

10. The pdf of the sum of two independent random variables is the convolution of their separate pdf's.
11. The pdf of the sum of two Gaussian random variables is Gaussian, whether or not the component random variables in the sum are independent or not.
12. Transformation of a pair of random variables into another pair of random variables by a transformation that is one-to-one results in the transformed pdf given by

$$f_{UV}(u,v) = f_{XY}(x,y)\left|J\begin{pmatrix} x & y \\ u & v \end{pmatrix}\right|_{\substack{x=g^{-1}(u,v)\\ y=h^{-1}(u,v)}}$$

where the transformation is denoted by

$$U = g(X,Y)$$
$$V = h(X,Y)$$

and the inverse transformation by

$$X = g^{-1}(U,V)$$
$$Y = h^{-1}(U,V)$$

The Jacobian is defined as

$$J\begin{pmatrix} x & y \\ u & v \end{pmatrix} = \begin{vmatrix} \dfrac{\partial x}{\partial u} & \dfrac{\partial x}{\partial v} \\ \dfrac{\partial y}{\partial u} & \dfrac{\partial y}{\partial v} \end{vmatrix}$$

where

$$\frac{\partial x}{\partial u} = \frac{\partial}{\partial u} g^{-1}(u,v),\quad \frac{\partial x}{\partial v} = \frac{\partial}{\partial v} g^{-1}(u,v),\quad \frac{\partial y}{u} = \frac{\partial}{\partial u} h^{-1}(u,v),\quad \frac{\partial y}{\partial v} = \frac{\partial}{\partial v} h^{-1}(u,v)$$

13. The central limit theorem states that a sum of statistically independent random variables of the form

$$\sum_{i=1}^{n} \frac{X_i - \mu_i}{\sqrt{n}\,\sigma_i}$$

where μ_i is the mean of X_i and σ_i^2 is its variance, tends to a zero-mean and unit-variance Gaussian pdf as the number of random variables in the sum increases, provided that the means and variances are finite and provided that no one random variable dominates the sum.

14. The weak law of large numbers states that the probability of the average of n independent values of a random variable X differs from $E(X)$ by more than an arbitrary $\epsilon > 0$ goes to zero as $n \to \infty$.

15. Extension of joint distribution functions from two to several random variables is straightforward. Several generalizations were given in Section 4-13. For the

case of N jointly Gaussian random variables, the joint pdf and characteristic function may be expressed in matrix form as

$$f_{\mathbf{x}}(\mathbf{x}) = |\mathbf{C}|^{-1/2}(2\pi)^{-N/2}\exp\left[-\frac{1}{2}(\mathbf{x}-\mathbf{m})^t\mathbf{C}^{-1}(\mathbf{x}-\mathbf{m})\right]$$

and

$$M_{\mathbf{X}}(j\mathbf{v}) = \exp(j\mathbf{m}^t\mathbf{v} - \tfrac{1}{2}\mathbf{v}^t\mathbf{C}\mathbf{v})$$

respectively, where

$$\mathbf{x} = \begin{bmatrix} x_1 \\ x_2 \\ \vdots \\ x_N \end{bmatrix}, \quad \mathbf{m} = \begin{bmatrix} m_1 \\ m_2 \\ \vdots \\ m_N \end{bmatrix}, \quad \mathbf{v} = \begin{bmatrix} v_1 \\ v_2 \\ \vdots \\ v_N \end{bmatrix}$$

and $\mathbf{C}$ is the matrix of covariances with elements $C_{ij} = E[(X_i - m_i)(X_j - m_j)]$.

16. The variance of a linear combination of random variables that are not necessarily statistically independent is

$$\operatorname{Var}\left(\sum_{i=1}^{n} a_i X_i\right) = \sum_{i=1}^{n} a_i^2\sigma_i^2 + \sum_{i=1}^{n}\sum_{\substack{l\neq 1 \\ i\neq j}}^{n} a_i a_j \sigma_i \sigma_j \rho_{ij}$$

where the σ_i^2 are their respective variances; ρ_{ij} is the correlation coefficient of X_i with X_j.

4–16 FURTHER READING

In addition to the references cited in Sections 1-5 and 2-10, Papoulis (1993) provides an alternative treatment of bivariate probability distributions at the graduate level. Kreyszig (1988) is a good reference on mathematics, in particular, the transformation of variables in multiple integrals.

4–17 PROBLEMS

Section 4-2

4-1. **(a)** Find the constant A to make the following function a proper cdf:

$$F(x,y) = A\frac{x}{2x+1}\frac{y}{y+1}u(x)u(y)$$

(b) Demonstrate that it does indeed approach the proper limits.
(c) Find the cdf's of X and Y alone.
(d) What is the probability of the joint event $-2 \le X \le 2$ and $1 \le Y \le 5$?

4-2. **(a)** Sketch the joint cdf

$$F(x,y) = \begin{cases} (1-e^{-3x})(1-e^{-2y}), & x, y \ge 0 \\ 0, & \text{otherwise} \end{cases}$$

(**b**) Demonstrate that it does indeed approach the proper limits.

Section 4-3

4-3. (**a**) Obtain the joint pdf corresponding to the cdf of Problem 4-1.
(**b**) Find and sketch the marginal pdf's corresponding to this joint pdf.

4-4. (**a**) Obtain the joint pdf corresponding to the cdf of Problem 4-2.
(**b**) Find and sketch the marginal pdf's corresponding to this joint pdf.

4-5. The joint pdf of the random variables X and Y is given by

$$f_{XY}(x, y) = Axe^{-x(y+1)}u(x)u(y)$$

(**a**) Find the constant A and plot $f_{XY}(x, y)$;
(**b**) Find the corresponding marginal pdf's.

4-6. You start for a certain truckstop and expect to arrive there between 30 and 45 min after starting out. Your friend starts from the opposite side of the truckstop at the same time and expects to arrive there between 20 and 40 min later. Assuming that both arrival times are equally likely to be anywhere in their respective intervals, what is the probability that you will arrive after your friend arrives? (*Hint:* First deduce the pdf's of your arrival time and that of your friend's. Since it is assumed that the two of you operate independently, the joint pdf of the two arrival times is the product of the marginal ones. The desired probability can then be found by integration over a properly chosen area under this joint pdf.)

Section 4-4

4-7. A fair coin and die are tossed simultaneously. The random variable X is 1 if a head comes up on the coin and is 0 if a tail comes up. The random variable Y is numerically equal to the number of spots up on the die. Provide a table defining the joint random variables and tabulate the probability mass function for this joint experiment.

Section 4-5

4-8. Rework Example 4-9 if the Gaussian pdf (4-24) is replaced by a Laplacian pdf:

$$f_X(x) = e^{-2|x-1000|}$$

4-9. Rework Example 4-8, but with the 2 in the exponent of (4-22a) replaced by a parameter, say S. Compute the probabilities given by (4-23a) and (4-23b) as a function of S. Compute numerical results for $S = 1, 4$, and 8. Let the threshold [upper limit on the integral in (4-23a) and lower limit on the integral in (4-23b)] be $S/2$.

Section 4-6

4-10. (**a**) Are the random variables with the joint cdf of Problem 4-1 statistically independent? Give a reason for your answer.
(**b**) Answer the same question for the joint pdf of Problem 4-2.

4-11. Are the random variables with the joint pdf of Problem 4-5 statistically independent? Why or why not?

Section 4-7

4-12. A pair of random variables has the joint pdf

$$f_{XY}(x, y) = e^{-2(|x|+|y|)}$$

Find averages of (**a**) the product, cos(X) sin(Y); (**b**) the sum, cos(X) + sin(Y).

Section 4-8

4-13. (**a**) Find the joint moments corresponding to the joint pdf of Problem 4-12. Show that they are the same as the joint central moments.
(**b**) Find the covariance of these random variables.
(**c**) Find their correlation coefficient.

4-14. Two random variables have joint pdf

$$f_{XY}(x, y) = \begin{cases} C, & x^2 + y^2 \leq 1, \ x > 0, \ y > 0 \\ 0, & \text{otherwise} \end{cases}$$

(**a**) Find the constant C.
(**b**) Find $E(XY)$, $E(X^2)$, and $E(Y^2)$.
(**c**) Are these random variables uncorrelated?

4-15. Find the joint moments for the experiment and random variable definition given in Problem 4-7. Tabulate them through $m = n = 2$. Give the correlation coefficient.

4-16. (**a**) Show that the characteristic function of a Gaussian random variable X with pdf given by (3-28) is given by

$$M_X(jv) = e^{j\mu v - (1/2)\sigma^2 v^2}$$

where μ is the mean and σ^2 is the variance.
(**b**) Show that the characteristic function of the sum of two statistically independent Gaussian random variables, X_1 and X_2, with means m_1 and m_2 and variances σ_1^2 and σ_2^2, respectively, is given by

$$M_{X+Y}(jv) = e^{j(m_1+m_2)v-(1/2)(\sigma_1^2+\sigma_2^2)v^2}$$

(**c**) From parts (b) and (a) conclude that the sum of two statistically independent Gaussian random variables is still Gaussian. Discuss how you would extend this to any number of statistically independent Gaussian random variables.

4-17. It is sometimes easier to compute the expectation of a function of two random variables using *conditional expectation*. That is, the expectation is found as

$$E[g(X, Y)] = E_Y\{E[g(X, Y)|Y]\}$$

where

$$E[g(X,Y)|Y] = \int_{-\infty}^{\infty} f_{X|Y}(x|y)g(x,y)\,dx$$

and $E_Y[\,\cdot\,]$ means the expectation with respect to the pdf of Y. Work the following problem using conditional expectations: projectiles are fired until a target is hit for the first time, after which firing ceases. Assume that the probability of a projectile's hitting the target is 0.7 and that the firings are independent of one another. Find the average number of projectiles fired at the target.

Section 4-9

4-18. A marksperson shoots at a target with the bullseye located at the origin. Because of wind (horizontal) and sighting error (vertical), the tendency is for the bullet to average 6 in. to the right (positive x direction) and 3 in. below (negative y direction) of the bullseye. Because of unsteadyness on the marksperson's part and other random influences, the standard deviations of the impact points in the x and y directions are 2 in. and 4 in., respectively. The errors in the two coordinate directions are uncorrelated. They are also known to be Gaussian. Write down an expression for the joint pdf of the impact points on the target.

4-19. Using the definition of a conditional pdf along with (4-55) and (3-28) with $m = 0$, show that the conditional Gaussian pdf is also Gaussian in form with conditional mean and variance given by

$$E(X \mid Y) = \mu_X + \frac{\rho\sigma_X}{\sigma_Y}(Y - \mu_Y) \quad \text{and} \quad \text{Var}(X \mid Y) = \sigma_X^2(1 - \rho_{XY})$$

respectively (no integration is necessary for this).

Section 4-10

4-20. Given the transformation of random variables

$$Z = X^2 + Y^2$$

where X and Y are statistically independent Gaussian random variables with zero means and variances σ^2. Find the cdf and pdf of Z.

4-21. Given the transformation $Z = XY$, show that the pdf of Z is given by

$$f_Z(z) = \int_{-\infty}^{\infty} f_{XY}\left(\frac{z}{y}, y\right)\frac{dy}{|y|}$$

Section 4-11

4-22. Repeat Example 4-20 for an arbitrary linear transformation, given by

$$U = aX + bY \quad \text{and} \quad V = cX + dY$$

where a, b, c, and d are constants. Let the joint pdf of X and Y be Gaussian with the means zero, but the correlation coefficient and variances arbitrary. Conclude the exponent of the joint pdf of U and V to be of the form

$$Au^2 + Buv + Cv^2$$

and that the result of a linear transformation on two jointly Gaussian random variables, uncorrelated or not, yields a pair of jointly Gaussian random variables. This an alternative derivation to that of Example 4-19.

Section 4-12

4-23. In surveying land, section markers are placed every square mile. Suppose that a mile can be measured to an accuracy of 5280 ft $\pm$ 1 %, uniformly distributed. Ten such markers are placed along a 10-mile strip, starting with a benchmark that is assumed to be accurate. Use the central limit theorem to compute the (approximate) probability that the tenth marker is within $\pm$ 1000 ft of where it should be.

Section 4-13

4-24. The purpose of this problem is to justify the mean and variance expressions for S_n used to obtain (4-101) from (4-99). Let $S_n = (1/n)\Sigma_{i=1}^{n} X_i$ where the X_i's are all distributed according to the same pdf and each has mean μ_X and variance σ^2_X.

(a) By extending (3-92) to n terms with each equal to an X_i and using (3-91), show that $E(S_n) = \mu_X$.

(b) Define $Y_i = X_i - \mu_X$ as a statistically independent set of random variables having zero means.

(i) If the variance of X_i is σ_X^2, show that this is also the variance of Y_i.

(ii) Write

$$S_n - \mu_X = \frac{1}{n}\sum_{i=1}^{n} X_i - \mu_X = \frac{1}{n}\sum_{i=1}^{n} X_i - \frac{1}{n}n\mu_X$$

$$= \frac{1}{n}\sum_{n=1}^{n} (X_1 - \mu_X) = \frac{1}{n}\sum_{i=1}^{n} Y_i$$

and note that

$$\text{Var}(S_n) = E[(S_n - \mu_X)^2]$$

$$= E\left\{\left[\frac{1}{n}\sum_{i=1}^{n} Y_i\right]^2\right\}$$

$$= \frac{1}{n^2}E\left(\sum_{i=1}^{n}\sum_{j=1}^{n} Y_iY_j\right)$$

$$= \frac{1}{n^2}\sum_{i=1}^{n}\sum_{j=1}^{n} E(Y_iY_j)$$

For $i \neq j$ note that $E(Y_iY_j) = E(Y_i)E(Y_j) = 0$ because of the independence of Y_i and Y_j and their zero means. For $i = j$, we get that $E(Y_iY_j) = E(Y_i^2) = \sigma_X^2$ as pointed out above. This should lead you through the proof that $\text{Var}(S_n) = \sigma^2_X/n$.

4-25. A fair die is tossed repeatedly. Using the weak law of large numbers, argue that the relative frequency of a 3 (or any number of spots up from 1 to 6) approaches 1/6.

Section 4-14

4-26. Three random variables X_1, X_2, and X_3 have variances 2, 5, and 3, respectively. The correlation coefficient of X_1 with X_2 is 0.2, of X_1 with X_3 is 0.1, and of X_2 with X_3 is 0.4.

(a) Find the variance of their sum, Z.

(b) Find the variance of the linear combination $W = 5X_1 + 2X_2 + X_3$.

4-18 COMPUTER EXERCISES

4-1. To generate two independent Gaussian pseudorandom variables, the following algorithm can be used [U and V are assumed uniform in (0,1)]:

$$R = \sqrt{-2\ln(U)}$$

$$X = R\cos(2\pi V)$$

$$Y = R\sin(2\pi V)$$

(**a**) Explain why this works by appealing to the results of Example 4-21.

(**b**) Write a program to generate a sequence of Gaussian pseudorandom numbers using this approach, and show histograms of X and Y to verify that they are Gaussian.

4-2. Using the results of Problem 4-19 and a Gaussian pseudorandom number generator (either the one available in MATLAB or the one discussed in Computer Exercise 4-1, design a Gaussian pseudorandom number generator that will provide a specified correlation coefficient between samples. Let the correlation coefficient between samples be

$$\rho(\tau) = e^{-\alpha|\tau|}$$

where α is a positive constant and τ is a parameter. Set $\tau = 1$ and choose various values of α to give high correlation and low correlation between adjacent pseudorandom variables. The sequence of operations will be as follows: (1) specify $\sigma = \sigma_X = \sigma_Y$; (2) generate a Gaussian pseudorandom number; (3) compute the conditional mean and variance of the next sample according to the formulas derived in Problem 4-19; (4) generate the next pseudo-Gaussian random number using the conditional mean and variance so computed; (5) continue until all the desired pseudorandom numbers are generated. Plot the sequences of pseudorandom numbers generated versus sample numbers to see the effect of the correlation.

5 Elementary Statistics, Empirical Probability Distributions, and More on Simulation

5-1 CONNECTING PROBABILITY WITH OBSERVATIONS OF DATA

In Chapters 2 to 4, we laid down the axioms of probability, defined several quantities, and developed various probabilistic relationships. For the most part we divorced ourselves from the real world and stayed in the world of mathematical models. Some of the problems that we considered had a touch of realism: for example, the development of testing procedures based on Bayes' theorem in Example 2-10. However, we still assumed that we knew the various probabilities that were required in computing the probability of type I and II errors.

In this chapter we want to get more involved with the real world. This is the realm of statistics, which we barely touch on in this chapter. We consider such things as the sample mean, variance, and standard deviation. In addition, we consider the empirical distribution of random data. In later chapters we consider estimation theory and decision making based on probabilistic concepts. Also considered in this chapter is more on computer simulation of random phenomena, which we began in Section 1-4.

5-2 SAMPLE MEAN AND SAMPLE VARIANCE

Suppose that we are taking a sample of manufactured items at the output of a production line for testing purposes, say resistors. The totality of all resistors manufactured is called the *population,* and the resistors selected for testing are called the *sample.* We perform a test, say measure the resistance of each resistor selected. We want to use the data collected to perform a judgment on the entire population. For example, suppose that we wish to infer whether or not the population of resistors is 1000 Ω ± 10%. We discuss more precise methods for making this inference in Chapter 7. For now, however, we consider two simple *statistics*, referred to as the *sample mean* and *sample variance*. The sample mean is simply the arithmetic average of all the sample values taken. If the sample values, or data, are denoted by x_i, $i = 1, 2, \ldots, n$, the sample mean is defined as

$$\bar{x} = \frac{1}{n}\sum_{i=1}^{n} x_i \tag{5-1}$$

Example 5-1.

Given the following resistance values in ohms: 900, 1013, 939, 1062, 1017, 996, 970, 1079, 1065, and 1049, find the sample mean.

Solution: Using (5-1), we compute

$$\bar{x} = 1009\ \Omega$$

Note that the sample mean is a random variable. It depends on the sample values selected from the population. We hope that it is close to the actual mean, an issue we explore later.

The next statistic we wish to consider is the sample variance. We recall from Chapter 2 that the variance was shown to be the expectation of the square of a random variable minus its mean squared; its square root, called the standard deviation, gives a measure of the spread of the realizations of the random variable around the mean. In analogy with the definition of the variance, we define the *sample variance* as

$$s_x^2 = \frac{1}{n-1}\sum_{i=1}^{n} (x_i - \bar{x})^2 \tag{5-2}$$

The reason for division by $n - 1$ rather than n will become apparent in a subsequent chapter when we consider estimators. If n is large, the difference is small. As was true for the sample mean, the sample variance is a random variable.

The square root of the sample variance is the *sample standard deviation*, which is

$$s_x = \sqrt{\frac{1}{n-1}\sum_{n=1}^{n} (x_1 - \bar{x})^2} \tag{5-3}$$

Example 5-2.

Obtain the sample standard deviation of the resistance samples given in Example 5-1.

Solution: Applying (5-3), we obtain $s_x = 58.7\ \Omega$.

For large n, the computations involved in (5-2) or (5-3) get tedious; we must first calculate the sample mean and then subtract it from every sample value before squaring and summing again. Furthermore, it is often unstable in that we are faced with taking the difference of two large numbers. A more convenient expression computationally can be obtained as follows. First, we expand the right side of (5-2) to obtain

$$s_x^2 = \frac{1}{n-1}\sum_{i=1}^{n} (x_i^2 - 2\bar{x}x_i + \bar{x}^2) \tag{5-4}$$

$$= \frac{1}{n-1}\left(\sum_{i=1}^{n} x_i^2 - 2\bar{x}\sum_{i=1}^{n} x_i + n\bar{x}^2\right)$$

Using (5-1), this becomes

$$s_x^2 = \frac{1}{n-1}\left[\sum_{i=1}^{n} x_i^2 - \frac{2}{n}\left(\sum_{i=1}^{n} x_i\right)^2 + \frac{1}{n}\left(\sum_{i=1}^{n} x_i\right)^2\right]$$

$$= \frac{1}{n-1}\left[\sum_{i=1}^{n} x_i^2 - \frac{1}{n}\left(\sum_{i=1}^{n} x_i\right)^2\right] \tag{5-5}$$

$$= \frac{n\sum_{i=1}^{n} x_i^2 - \left(\sum_{i=1}^{n} x_i\right)^2}{n(n-1)}$$

Note that by using this equation, we can compute the sum of the sample values and the sum of the squares of the sample values without having to go through the tedious step of subtracting the sample mean from each sample value.

Example 5-3.

The sum of the sample values for the resistance values of Example 5-1 is 10,090 and the sum of their squares is 10,211,806. Applying (5-5), we get 3444 for the sample variance, or 58.7 Ω for the sample standard deviation, which is the same as obtained previously.

Example 5-4.

The *residuals* are defined as

$$d_i = x_i - \bar{x} \tag{5-6}$$

Show that the sum of all the residuals is equal to zero.

Solution: From (5-1) we have

$$\sum_{i=1}^{n} x_i - n\bar{x} = 0$$

or

$$\sum_{i=1}^{n} (x_i - \bar{x}) = 0 \tag{5-7}$$

Using the definition of the residuals, we have

$$\sum_{i=1}^{n} d_i = 0 \tag{5-8}$$

5-3 REGRESSION TECHNIQUES

Suppose that we have data measurements taken in pairs. We know that there is some relationship between the data, but because of measurement error we are not able to determine it precisely. The simplest relationship would be a straight line. Thus, if $\{x_i, y_i,$

$i = 1, 2, \ldots, n\}$ are the data pairs, we assume that we can fit a straight line to them in some best sense[1]:

$$y = \alpha x + \beta \tag{5-9}$$

We choose the constants α and β so that the straight-line fit to the data is taken in the sense of minimum squared error:

$$\epsilon = \frac{1}{n-1}\sum_{i=1}^{n}(y_i - \alpha x_i - \beta)^2 \tag{5-10}$$

To find α and β making (5-10) as small as possible, we differentiate first with respect to α and then with respect to β and then set the resulting expressions equal to zero. Carrying out the differentiation, we obtain

$$\frac{\partial \epsilon}{\partial \alpha} = -\frac{2}{n-1}\sum_{i=1}^{n}(y_i - \alpha_o x_i - \beta_o)x_i = 0 \tag{5-11}$$

and

$$\frac{\partial \epsilon}{\partial \beta} = -\frac{2}{n-1}\sum_{i=1}^{n}(y_i - \alpha_o x_i - \beta_o) = 0 \tag{5-12}$$

where the subscripts "o" denote optimum. We can simplify (5-11) to

$$\left(\sum_{i=1}^{n} x_i^2\right)\alpha_o + \left(\sum_{i=1}^{n} x_i\right)\beta_o = \sum_{i=1}^{n} x_i y_i \tag{5-13}$$

and (5-12) to

$$\left(\sum_{i=1}^{n} x_i\right)\alpha_o + n\beta_o = \sum_{i=1}^{n} y_i \tag{5-14}$$

Thus we have two equations in two unknowns (α_o and β_o), which can be solved to yield

$$\alpha_o = \frac{n\sum_{n=1}^{n} x_i y_i - \sum_{n=1}^{n} x_i \sum_{n=1}^{n} y_i}{n(n-1)\, s_x^2} \tag{5-15a}$$

and

$$\beta_o = \bar{y} - \alpha_0 \bar{x} \tag{5-15b}$$

where s_x^2 is the sample variance of the x-values. The result for α_o can be simplified by defining the *sample covariance* as

$$c_{xy} = \frac{1}{n-1}\sum_{n=1}^{n}(x_i - \bar{x})(y_i - \bar{y}) \tag{5-16}$$

[1]Data dependencies other than linear dependencies can be handled if it is suspected that some other functional relationship is present. For example, a straight line could be fitted to the logarithm of the x values, or the y values, or both.

By expanding the product, it follows that this can also be written as

$$c_{xy} = \frac{n\sum_{n=1}^{n} x_i y_i - \sum_{n=1}^{n} x_i \sum_{n=1}^{n} y_i}{n(n-1)} \tag{5-17}$$

Thus

$$\alpha_o = \frac{c_{xy}}{s_x^2} \tag{5-18}$$

Using (5-15b) and (5-18), we may write the equation for the straight-line fit to the data as

$$y = \alpha_o x + \beta_o = \frac{c_{xy}}{s_x^2} x + \bar{y} - \frac{c_{xy}}{s_x^2} \bar{x} \tag{5-19}$$

This is called a *regression line* and the procedure used in getting it is called a *regression technique.* We can arrange (5-19) as

$$y - \bar{y} = \frac{c_{xy}}{s_x^2}(x - \bar{x}) \tag{5-20}$$

The *sample correlation coefficient* is defined as

$$r_{xy} = \frac{c_{xy}}{s_x s_y} \tag{5-21}$$

where s_x and s_y are the sample standard deviations of the data sets $\{x_i\}$ and $\{y_i\}$, respectively. In terms of the sample correlation coefficient, the regression line can be expressed as

$$\frac{y - \bar{y}}{s_y} = r_{xy} \frac{x - \bar{x}}{s_x} \tag{5-22}$$

which is a somewhat easier expression to remember.

If $r_{xy} = 0$, the regression line vanishes and we say that the data sets are *uncorrelated.* Another special case occurs when the data are linearly related as given by

$$y_i = mx_i + b \tag{5-23}$$

In this case it can be shown that

$$r_{xy} = \pm 1 \tag{5-24}$$

The proof of this is left to the problems. We may also show that

$$-1 \le r_{xy} \le 1 \tag{5-25}$$

which is also left to the problems.

The goodness of the fit is determined by the squared error given by (5-10). By the mathematics, we know that it has been minimized. However, the straight-line fit to the data may still not be very good, particularly if the correlation between data pairs is small.

Example 5-4.

The data pairs given in the following table are believed to be related linearly. Find their correlation coefficient and the regression line which is the least-squares fit to these points.

X_i	0.68	0.72	1.27	2.01	2.63	3.06	3.15	4.00	4.03	4.50
Y_i	12.45	9.93	6.64	10.14	8.93	13.34	11.56	16.72	19.62	15.03

Solution: The correlation coefficient is 0.71, the sample means are $\bar{x} = 2.6$ and $\bar{y} = 12.44$, and the sample standard deviations are $s_x = 1.39$ and $s_y = 3.88$. Thus the regression line, from (5-22), is

$$\frac{y - 12.44}{3.88} = 0.71\frac{x - 2.6}{1.39}$$

It is plotted in Figure 5-1 along with the data pairs.

5-4 EMPIRICAL DISTRIBUTION FUNCTIONS

As discussed in Chapter 3, the use of theoretical distribution functions assumes that a model for the data is available—we have some reason to use a Gaussian, or chi-square, or exponential distribution. In this section we discuss empirical distribution functions, which are based only on the data available.

Consider a random variable X with distribution function $F_X(x)$ which we do not know. However, we have a number of independent samples of the random variable, denoted $\{x_i, i = 1, 2, ..., n\}$. The *empirical cumulative distribution function,* or simply *empirical distribution,* is defined as

$$\tilde{F}_X(x|x_1, x_2, \cdots, x_n) = \frac{\text{number of samples } x_1, x_2, \cdots, x_n \text{ no greater than } x}{n} \tag{5-26}$$

Example 5-6.

Obtain the empirical distribution of the resistance samples given in Example 5-1.

Solution: The empirical distribution is obtained more easily by listing the samples in ascending order, which are 900, 939, 970, 996, 1013, 1017, 1049, 1062, 1065, and 1079. The empirical distribution function is plotted in Figure 5-2.

Note that the empirical distribution function has the properties of a cdf given in Section 2-2. That is, it has the limits zero and unity as its argument approaches $-$ and $+\infty$, respectively; it is a nondecreasing function of its argument and is continuous from the right. The empirical probability mass function can be obtained from empirical distribution function by plotting a line equal to the magnitude of each jump at the location on the abscissa where the jump occurs.

If the number of datums is large, a simplification in computing the empirical distribution function is handy. We take the range over which the data occur and divide it into a convenient number of intervals of equal length. The number of datums within each cell is then plotted as a bar graph called a *histogram.* The number of datums within each interval can be cumulatively summed to get the empirical cumulative distribution function. Example 5-6 illustrates this procedure.

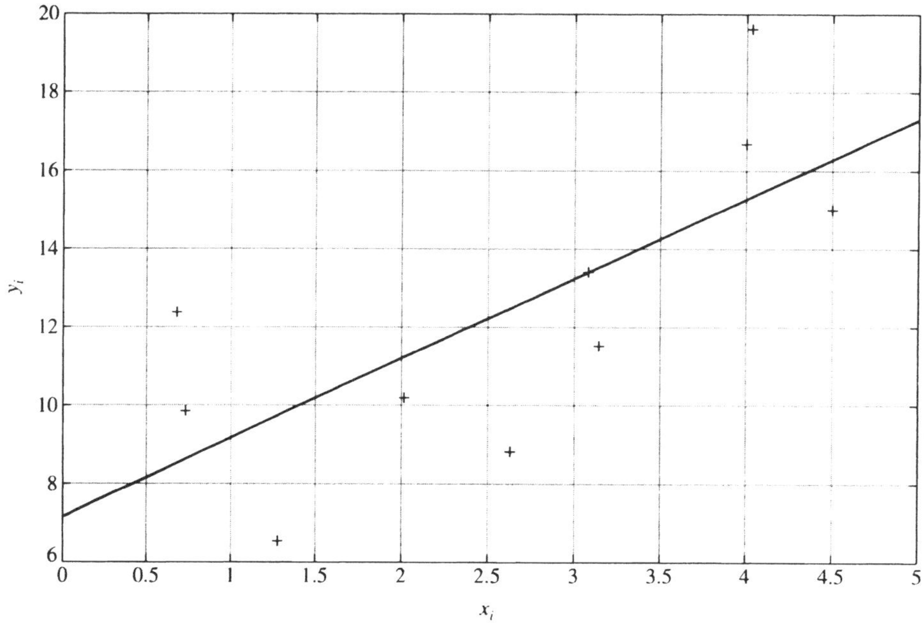

Figure 5-1 Data pairs and regression line for the random data set of Example 5-4.

Example 5-7.

The intervals between telephone calls arriving at a certain switching office are recorded in minutes. They are: 0.026, 0.977, 0.05, 0.183, 0.597, 0.426, 1.327, 0.017, 0.191, 0.938, 0.065, 0.098, 0.271, 0827, 0.863, 0.101, 0.372, 0.93, 0.343, 0.156, 0.451, 0.637, 0.282, 0.191, 0.14, 0.163, 0.372, 1.048, 0.5, 0.09, 1.675, 0.33, 0.206, 0.426, 1.128, 0.026, 0.041, 0.299, 0.531, 0.376, 0.49, 0.083, 0.575, 0.393, 0.651, 0.009, 0.606, 0.151, 0.283, and 0.815. Find the empirical cdf and the histogram of these time intervals.

Solution: The histogram and empirical cdf are shown in Figure 5-3. The histogram appears to be roughly exponentially decreasing.

The histogram obtained in Example 5-7 does not match the pdf for two reasons. To see why, we examine it in the light of (3-14), which is repeated below for convenience:

$$f_x(x)\,\Delta x = P(x < X \leq x + \Delta x) \simeq \frac{\text{number of data values in } (x, x + \Delta x)}{\text{total number of data values}}, \quad \Delta x \ll 1 \qquad (5\text{-}27)$$

The first reason is that we have not normalized the histogram blocks by the total number of datums. The second reason is apparent from (5-27); to get an approximation to the pdf, we must divide through by Δx. We replot Figure 5-3a in Figure 5-4 with these two corrections. Also plotted is the function

$$\tilde{f}_X(x) = 2e^{-2x}u(x) \qquad (5\text{-}28)$$

The agreement is surprisingly good. Indeed, the data values in Example 5-7 were generated by a random number generator using the technique discussed in Section 3-8 to generate exponentially distributed random variables.

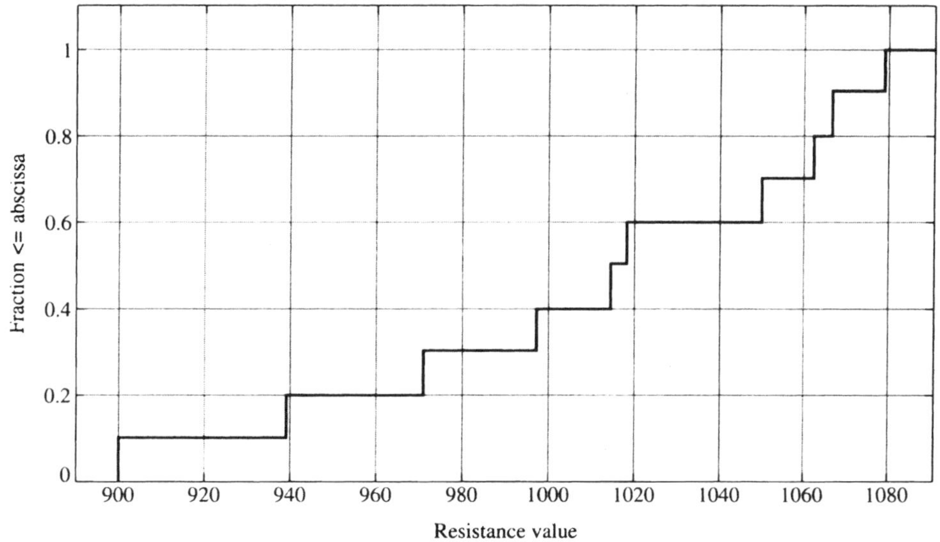

Figure 5-2 Empirical cumulative distribution function for the resistance samples given in Example 5-1.

Note that there is a trade-off between the number of bins used in constructing the histogram and the appearance of the histogram. Too many bins give too few samples per bin, and statistical irregularity causes a ragged-looking histogram. Too few bins give a low resolution on the abscissa of the histogram. Usually, selection of the number of bins will have to proceed by trial and error in the case of relatively few data samples, as is the case in these examples.

5-5 MORE ON MONTE CARLO SIMULATION

We discussed how to generate random variables by computer in Section 3-8. The purpose of this was to prepare for carrying out more complex computer simulations of systems undergoing random perturbations. This method is known as *Monte Carlo simulation.* We delve into this subject in this section, with more to come throughout the book. To illustrate the procedure, consider the following example.

Example 5-8.

Consider and RC circuit with random resistance and capacitance values that were a result of the manufacturing process. The 3-dB cutoff frequency of an RC filter, defined as

$$f_3 = \frac{1}{2\pi RC}$$

is of interest. For component values that are in a range of $\pm p\%$ centered on the nominal value, we can generate a histogram for f_3 to determine the influence of component variation on cutoff frequency. This gives much more information than an extreme-value analysis carried by computing the maximum and minimum values of f_3 through substitution of the minimum and maximum values of the resistance and capacitance. For example, we can estimate the most likely value by estimating the value of f_3 that corresponds to the maximum of the histogram of values for f_3.

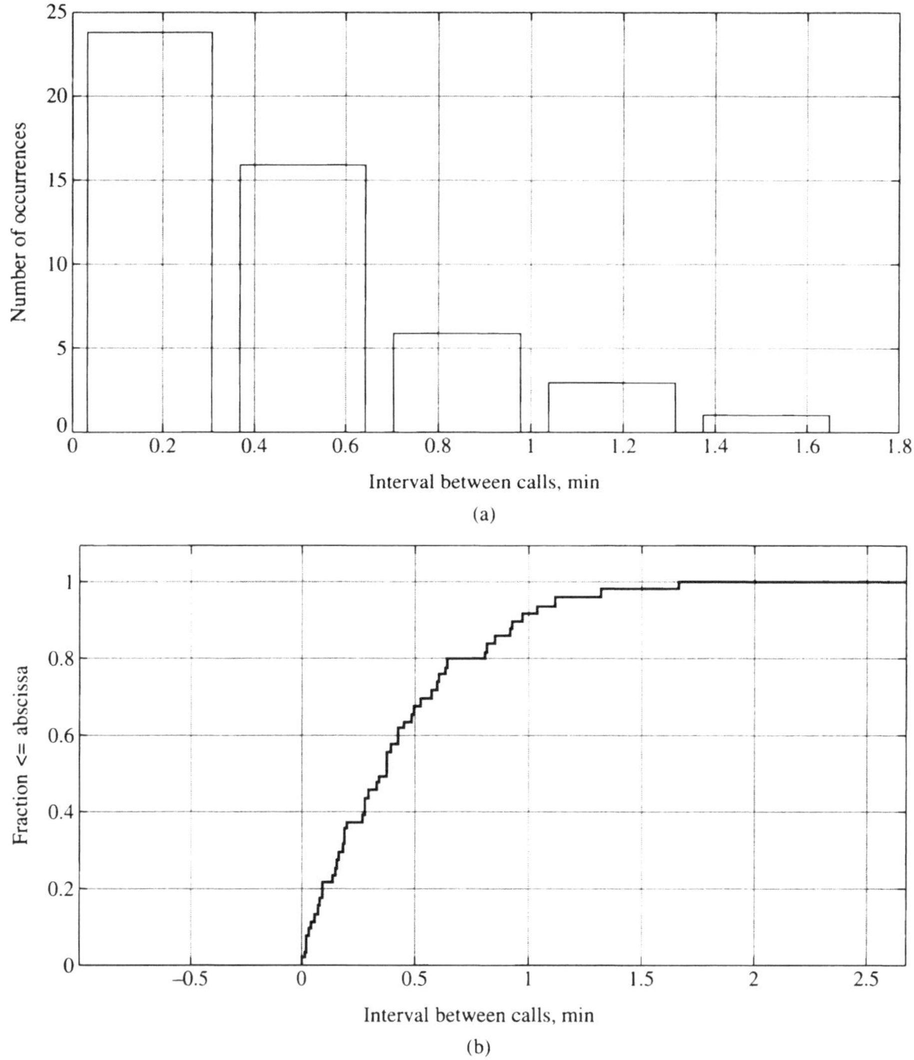

Figure 5-3. Empirical distributions for Example 5-7: (a) histogram; (b) empirical cdf.

To get the histogram of cutoff frequency values, we generate 10,000 resistance and capacitance values uniformly distributed within their allowed ranges about the nominal values, compute f_3 for each pair of values, and plot the histogram of cutoff frequencies. This procedure can be carried out by writing a computer program in a higher level language such as Pascal, Fortran, or C. A faster approach for this relatively simple problem is to use a mathematics package, in this case MATLAB. The histograms for the resistance, capacitance, and time constant values are shown in Figure 5-5. The MATLAB program statements are given in Table 5-1. In this case, the nominal value of $R = 1000\ \Omega$ and of $C = 1\ \mu F$. The tolerances were $\pm 10\%$ for the resistance value and $\pm 5\%$ for the capacitance values.

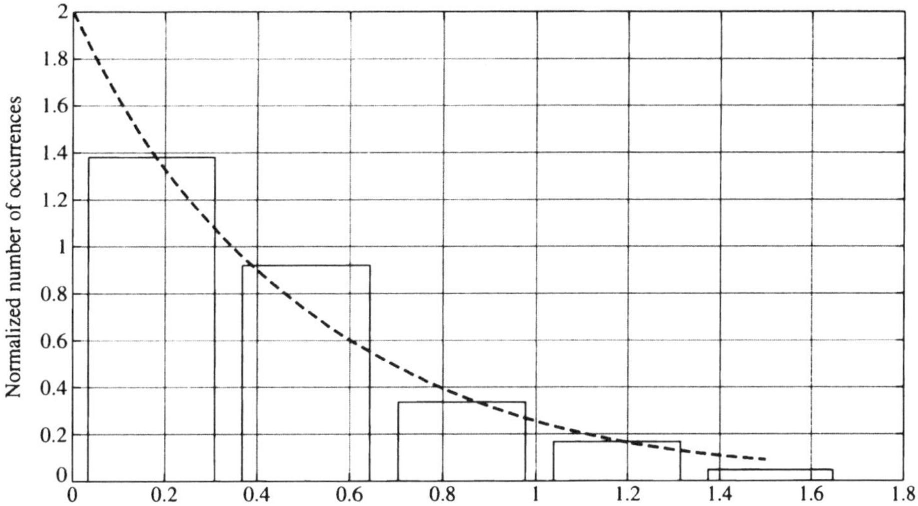

Figure 5-4. Histogram of Figure 5-3a normalized so as to approximate the theoretical pdf from which the random numbers were generated.

Table 5-1 MATLAB PROGRAM FOR GENERATING HISTOGRAMS

```
% MATLAB program for Example 5-8
%
N = input('Enter of repetitions for simulation')
tol_R = input('Enter per cent tolerance for resistors')
tol_C = input('Enter per cent tolerance for capacitors')
% Generate 1XN vectors of random numbers uniform in ( - 1,1)
delta1 = 2*rand(1,N) - 1;
delta2 = 2*rand(1,N) - 1;
% Form vector of random resistance values according to tolerance
R = 1000*(1 + (tol_R/100)*delta);
Rmax = max(R)
Rmin = min(R)
% Form vector of random capacitance values according to tolerance
C = 1*10^( - 6)*(1 + (tol_C/100)*delta2);
Cmax = max(C)
Cmin = min(C)
% Form vector of random 3-dB cutoff frequency values and plot
histograms
f3 = 1./(2*pi*R.*C);
subplot(3,1,1),hist(R,20),xlabel('Resistance, ohms'),ylabel('No. of
values'),grid
subplot(3,1,2),hist(C,20),xlabel('Capacitance, farads'),ylabel('No.
of values'),grid
subplot (3,1,3), hist(f3,20), xlabel('Cutoff frequency,Hz'),
ylabel('No. of values'),grid
```

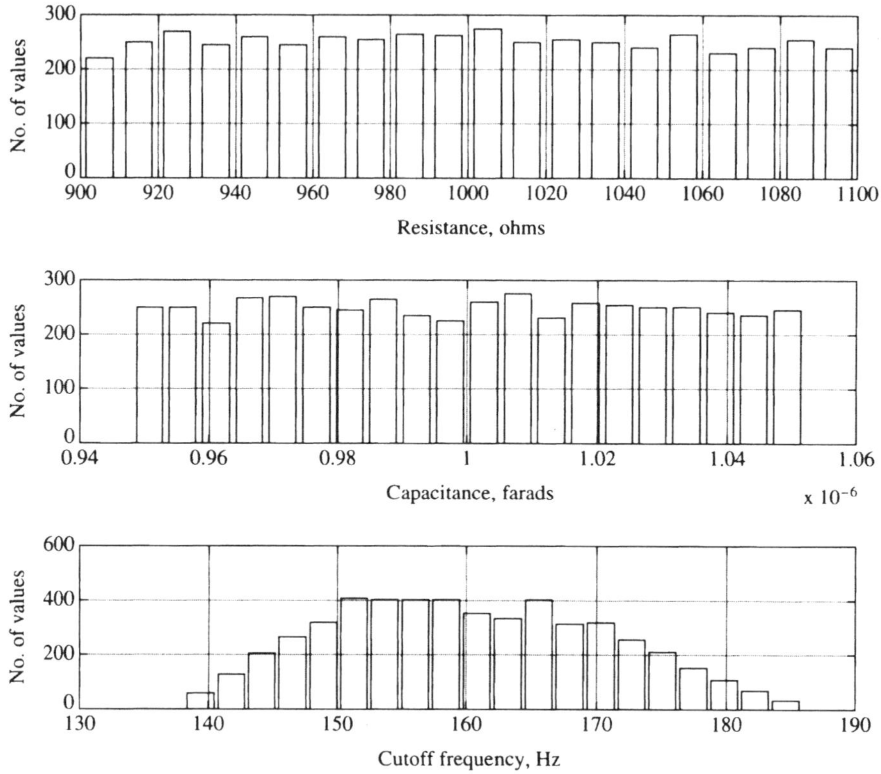

Figure 5-5. Histograms for the component values and 3-dB cutoff frequency of an *RC* filter generated by Monte Carlo simulation. Histograms of resistance values in ohms (a), capacitance values in farads (b), and cutoff frequency in hertz (c), 5000 repetitions.

Note that the data provided by Figure 5-5c are valuable in several ways. First, by simulating enough values of the time constant, we can find the maximum and minimum values of the time constant and carry out a worst-case design if the *RC* circuit is part of a larger system. Of course, in this case, we could have gotten these maximum and minimum values very easily since the relationship between the resistance and capacitance and the 3-dB cutoff frequency is very simple, and the resistance and capacitance values are bounded because of the assumption that within their tolerance regions their respective nominal values are uniformly distributed. With more complex functional relationships or probabilistic models for the component values, this might not be as simple.

A second use of the histogram of Figure 5-5c is to find the approximate probability that the 3-dB cutoff frequency values lie in a specified region. For example, a rough estimate from Figure 5-5c indicates that the probability of the cutoff frequency being between 140 and 150Hz is $900/5000 = 0.18$, whereas the probability of it being between 150 and 160 Hz, is about $1600/5000 = 0.32$. For this purpose, the empirical cdf would be more accurate.

A third use of data such as those provided by Figure 5-5 is to design a more suitable system. In this particular example, the only apparent solution to providing a more precise circuit is to tighten the tolerances on the resistor and capacitor. However, in more complex systems there may be a number of possible solutions that can be checked out rapidly by Monte Carlo simulation.

5–6 STATISTICAL PROCESS CONTROL

In a manufacturing process, there are usually several steps that take place in the process of producing the items being manufactured. In any one of these steps, a procedure can go bad and produce items that are not acceptable. The question arises as to how one might monitor the process and determine when things have gone so wrong that the manufactured items are no longer acceptable. One can then close down the process and look for the offending steps. One tool for monitoring such situations is the control chart. To illustrate, suppose that we are manufacturing transistors and that the gain of each is an item of concern. We measure the gain of each, divide the measurements into lots, and compute the sample mean of each lot. We also compute the sample mean of the sample means, which is the sample mean of the entire sample. Upper and lower control limits are then established as $\pm$ 3 sample standard deviations of the lot sample means from the sample mean of the entire sample. If the sample mean of a lot falls outside these control limits, it is assumed that something is wrong with the process and it is closed down while the cause is sought. An example will illustrate the procedure.

Example 5-9.

Transistors are manufactured according to a certain process. For 25 lots of size 5, the sample means of the current gains of the lots (i.e. the sample means of transistor gains taken five at a time) are as shown in Figure 5-6 by the $\times$'s. The sample mean of the entire population of the $5 \times 25 = 125$ transistor gains is 98.8, and the sample standard deviation of the lot's sample means is 6.95. We establish upper and lower control limits of

$$\text{UCL} = 98.81 + 3 \times 6.95 = 119.65$$

$$\text{LCL} = 98.81 - 3 \times 6.95 = 77.96$$

The sample means of the 25 lots of five samples are plotted versus lot number, as shown in Figure 5-6, along with the upper and lower control limits. This is called a *control chart.* If the sample mean of any lot exceeds or goes below the UCL or LCL, respectively, the process is said to be *out of control.* If out of control, the cause of the excursion outside the control limit band should be sought and corrected. We note that for this particular process we are well within the control limits. The MATLAB program is given in Table 5-2.

5–7 CONVERGENCE OF THE SAMPLE MEAN TO THE MEAN

In Chapter 3 we derived Chebyshev's inequality, given by (3-104). The sample mean, expressed by (5-1), is hopefully a good estimate of the true mean of our population. We can get a bound on how good it is by applying Chebyshev's inequality. First, if μ_X is the true mean of the samples drawn from the population, it was shown in Problem 4-24a that

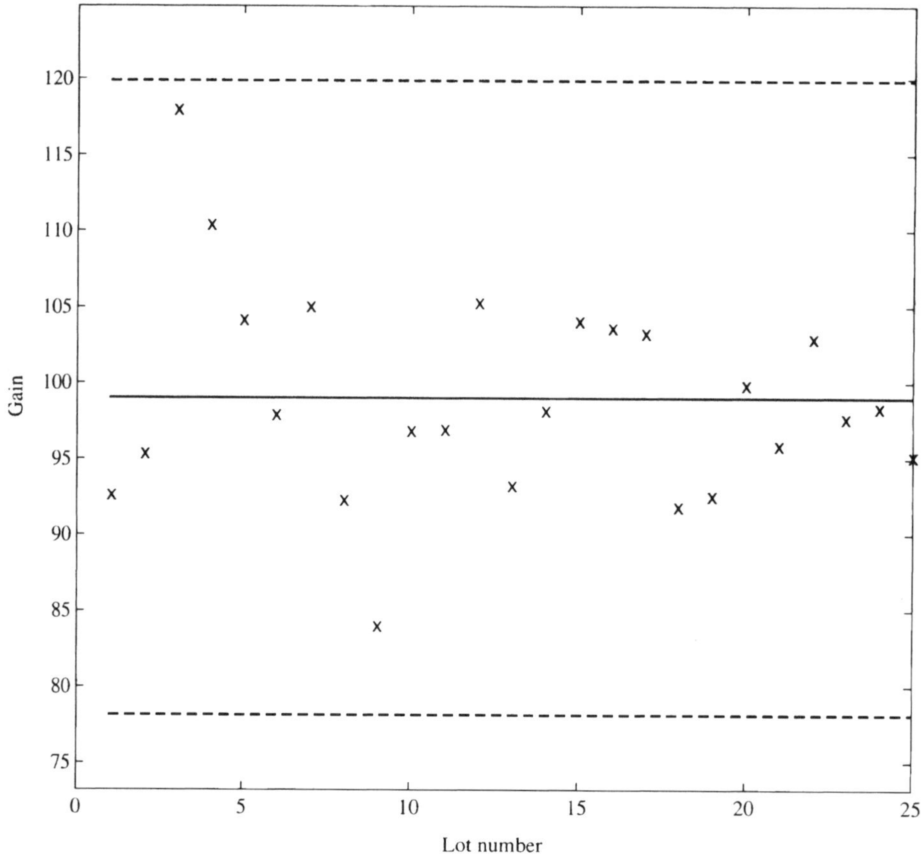

Figure 5-6. Control chart for transistor gains: 25 lots of five transistors each.

$$E(\bar{x}) = \mu_X \tag{5-30}$$

This is one ingredient in Chebyshev's in equality. The other is the variance of the sample mean. From the definitions of sample mean and variance, it is

$$\begin{aligned}
\mathrm{Var}(\bar{x}) &= E\{[\bar{x} - E(\bar{x})]^2\} = E\left[\left(\frac{1}{n}\sum_{i=1}^{n} X_i - \frac{1}{n}\sum_{i=1}^{n} \mu_X\right)^2\right] \\
&= E\left\{\frac{1}{n^2}\left[\sum_{n=1}^{n} (X_i - \mu_X)\right]^2\right\} = \frac{1}{n^2} E\left[\sum_{i=1}^{n}\sum_{j=1}^{n} (X_i - \mu_X)(X_j - \mu_Y)\right] \\
&= \frac{1}{n^2}\left\{\sum_{i=1}^{n}\sum_{j=1}^{n} E[(X_i - \mu_X)(X_j - \mu_Y)]\right\} = \frac{1}{n^2}\sum_{i=1}^{n} E[(X_i - \mu_X)^2] \\
&= \frac{1}{n^2}\sum_{i=1}^{n} \mathrm{Var}(X_i) = \frac{\mathrm{Var}(X_i)}{n} = \frac{\sigma_X^2}{n}
\end{aligned} \tag{5-31}$$

Table 5-2 MATLAB PROGRAM FOR GENERATING A CONTROL CHART

```
% MATLAB program for Example 5-9
%
m = input('Enter lot size');
n = input('Enter number of lots');
gain = input('Enter nominal transistor gain');
std_dev_gain = input('Enter standard deviation of transistor gain');
X = randn(m,n);
unit = ones(m,n);
gain_meas = std_dev_gain*X + gain*unit;
lot_mean = mean(gain_meas);
mean_mean = mean(lot_mean);
std_dev_lot_mean = std(lot_mean);
lot_no = 1:n;
UCL = mean_mean + 3*std_dev_lot_mean*ones(1,n);
LCL = mean_mean - 3*std_dev_lot_mean*ones(1,n);
fprintf('UCL = %f; LCL = %f; sample mean = %f; std dev of lot
mean = %f\n',...
UCL(1),LCL(1),mean_mean,std_dev_lot_mean)
clg
plot(lot_no,lot_mean,'x'),xlabel('Lot number'),ylabel('Gain')
hold
plot(lot_no,mean_mean*ones(1,n),'r')
plot(lot_no,UCL'-')
plot(lot_no,LCL,'-')
```

where σ_X^2 is the variance of each sample from the population and uppercase letters are used for the X_i's since they are not specific sample values in this context (i.e., they are random variables). (This is a special case of the result of Problem 4-24b.) Thus Chebyshev's inequality as given by (3-104) becomes

$$P\left(\left|\frac{1}{n}\sum_{i=1}^{n} x_i - \mu_X\right| \geq \frac{k\sigma_X}{\sqrt{n}}\right) \leq \frac{1}{k^2} \tag{5-32}$$

Unfortunately, it is usually true that we also do not know the variance of the samples, so (5-32) is not applicable to true estimation problems. We take up this subject again in Chapter 6.

Example 5-10.

Samples drawn from a population are known to have standard deviations of 2. We want the probability that the absolute value of the difference between their sample mean and the true mean is greater than 0.5 to be less than 1%. How many samples should be drawn?

Solution: From (5-32) we want

$$\frac{1}{k^2} = 0.01 \quad \text{or} \quad k^2 = 100 \quad \text{or} \quad k = 10 \tag{5-33}$$

Also, from (5-32), we want

$$\frac{k\sigma_X}{\sqrt{n}} = \frac{10 \times 2}{\sqrt{n}} = 0.5 \quad \text{or} \quad \sqrt{n} = \frac{10 \times 2}{0.5} = 40 \quad \text{or} \quad n = 1600 \tag{5-34}$$

This is a fairly large number of samples.

If we know something about the distribution of the sample mean, this can be reduced. We saw in Chapter 4, that under fairly broad restrictions, the sample mean is approximately Gaussian for a large number of samples (central limit theorem). Using this assumption, the left-hand side of (5-32) becomes

$$2\int_{k\sigma_X/\sqrt{n}}^{\infty} \frac{e^{-v^2/(2\sigma_X^2/n)}}{\sqrt{2\pi\sigma_X^2/n}}\,dx = 2Q(k) = 0.1 \tag{5-35}$$

where the substitution $u = n^{1/2}v/\sigma_X$ has been used in the integrand to get the Q-function expression. Using a table of Q-function values or a rational approximation, we find that $k = 1.645$ satisfies (5-35). Since k must be integer, we round it up to 2. From the argument of (5-32)

$$\frac{k\sigma_X}{\sqrt{n}} = 0.5 \quad \text{or} \quad \frac{2 \times 2}{\sqrt{n}} = 0.5 \quad \text{or} \quad \sqrt{n} = \frac{2 \times 2}{0.5} = 8 \quad \text{or} \quad n = 64 \tag{5-36}$$

which is considerably less than the result found using Chebyshev's inequality.

5-8 SUMMARY

In this chapter we considered certain topics that relate probability to the real world. For the most part, these topics come from the realm of *statistics*. The topics considered included the sample mean, variance, and standard deviation, the sample covariance and correlation coefficient, regression techniques, empirical distribution functions including histograms and the empirical cdf, an introduction to Monte Carlo simulation, and control charts.

1. A *sample* consists of the selection of a certain number of random variable realizations from the universe, called the *population*. In the context of statistics, we are usually attempting to make measurements on the sample to infer something about the population. The measures we compute from the sample are called *statistics*.

2. Given a sample or data set, $\{x_i, i = 1, 2, ..., n\}$, the sample mean is defined as

$$\bar{x} = \frac{1}{n}\sum_{i=1}^{n} x_i$$

3. The *sample variance* of a sample is defined as

$$s_x^2 = \frac{1}{n-1}\sum_{i=1}^{n} (x_i - \bar{x})^2$$

4. The *sample standard deviation* is defined as the square root of the sample variance.

5. A handier expression for computing the sample variance is

$$s_x^2 = \frac{n\sum_{i=1}^{n} x_i^2 - \left(\sum_{i=1}^{n} x_i\right)^2}{n(n-1)}$$

6. The *sample covariance* of a sample where the data are grouped in pairs $\{x_i, y_i, i = 1, 2, ..., n\}$ is defined as

$$c_{xy} = \frac{1}{n-1}\sum_{n=1}^{n}(x_i - \bar{x})(y_i - \bar{y})$$

where, as before, the overbars denote sample means.

7. The sample covariance is more conveniently computed from the equivalent expression

$$c_{xy} = \frac{n\sum_{n=1}^{n} x_i y_i - \sum_{n=1}^{n} x_i \sum_{n=1}^{n} y_i}{n(n-1)}$$

8. The *sample correlation coefficient* is defined as

$$r_{xy} = \frac{c_{xy}}{s_x s_y}$$

Data set pairs for which $r_{xy} = 0$ are said to be *uncorrelated.*

9. If the $\{y_i\}$ are plotted versus the $\{x_i\}$ and a straight line is fit to this scatter plot such that the total squared error between the data points and the line is minimized we obtain a *regression line.* It is given by

$$\frac{y - \bar{y}}{s_y} = r_{xy}\frac{x - \bar{x}}{s_x}$$

10. The *empirical cumulative distribution function* for a sample is defined as

$$\tilde{F}_X(x|x_1, x_2, \cdots, x_n) = \frac{\text{number of samples } x_1, x_2, \cdots, x_n \text{ no greater than } x}{n}$$

11. While the foregoing definition for the empirical cdf works well for small samples, it is difficult to compute for large sample sizes. For easier computation, we divide the range of values for the sample into n equal-length intervals or cells, denoted by Δx, and compute the *histogram,* defined as

$$\frac{\text{number of data values in } (x_i, x_i + \Delta x)}{\text{total number of data values}} \approx P(x_i < X \le x_i + \Delta x),$$

$$i = 1, 2, \cdots, \text{number of cells}$$

where x_i is the left-hand edge of each interval. The left-hand side of the equation above is known as the histogram. According to the relative frequency definition of probability, it is approximately equal to the right-hand side, which is the probability of the random variable lying in the range Δx as long as the number of samples in each bin is large. We explore this approximation more in a subsequent chapter.

12. To get the approximate empirical cdf from the histogram, simply sum the elements in each cell from $i = 1$ to x.

13. In *Monte Carlo simulation,* we have a system described in terms of input–output equations. The input or system parameters have random aspects (e.g., random noise at the input, unknown component values, etc.). We can get a probabilistic description of the system behavior by modeling the random quantities with random number generators on a computer, repeatedly solving the system equations as we draw different realizations for the random quantities and computing some statistic of the desired response(s) or output(s).

24. *Statistical process control* can be effected by taking the output of the process (say, transistor gains) as a sample, dividing this sample up into lots, computing the sample mean of the parameter of interest of each lot, and comparing these sample means of the lots to the mean of the entire sample. If the sample mean of any lot deviates from the overall sample mean by more than, say 3 standard deviations, it is assumed that the process has gone out of control and the cause is looked into. The plot of sample means of the lots is called a *control chart.* The limits with which the sample means of the lots are compared are called *upper* and *lower control limits.*

25. Use of Chebyshev's inequality or the weak law of large numbers shows that the sample mean converges to the mean (i.e. the probability of the sample mean differing from the true mean by more than an arbitrary $\epsilon > 0$ goes to zero as the number of samples becomes large).

5–9 FURTHER READING

The references cited in Sections 1-5 and 2-10 are appropriate reading for this chapter as well. In particular, Breipohl (1970) has a good treatment of empirical distribution functions, and Williams (1991) has a discussion of process control using control charts with different examples than used here. Control charts are also discussed in Walpole and Meyers (1993).

5–10 PROBLEMS

Section 5-2

5-1. Given that each datum is drawn from a population with mean m, show that the sample mean defined by (5-1) has expectation

$$E(\bar{x}) = E\left(\frac{1}{n_i}\sum_{i=1}^{n} x_i\right) = m$$

5-2. Given that each datum is drawn from a population with mean m and second moment $E(X_i^2) = K$, independent of sample number i, show that the sample variance, defined by (5-2), has expectation

$$E(s_x^2) = \frac{n}{n-1}[E(X_i^2) - m^2] = \frac{n}{n-1}\text{Var}(X_i) = \frac{n}{n-1}(K - m^2)$$

5-3. Given the following set of measurements of transistor gains at the output of a production line: 112, 77, 113, 83, 95, 105, 102, 120, 73, and 95. Compute the sample mean and sample standard deviation.

Section 5-3

5-4. Go through the details in deriving equations (5-15a) and (5-15b).

5-5. Show that (5-16) and (5-17) are equivalent.

5-6. Show that (5-19), (5-20) and (5-22) are equivalent.

5-7. Give the steps in the derivation of (5-25).

5-8. Given the following pairs of measurements, which are suspected to be linearly related, do a regression analysis and comment on the degree of certainty that you have that they are related.

X_i	0.77	4.39	4.11	2.91	0.56	0.89	4.09	2.38	0.78	2.52
Y_i	14.62	22.21	20.12	19.42	14.69	15.23	24.48	16.88	8.56	16.24

5-9. Given the following pairs of measurements, which are suspected to be linearly related, do a regression analysis, and comment on the degree of certainty that you have that they are related.

X_i	0.77	4.39	4.11	2.91	0.56	0.89	4.09	2.38	0.78	2.52
Y_i	9.35	1.40	6.21	7.66	5.27	2.06	10.53	10.24	6.50	4.49

5-10. An exponential relationship between the following pairs of measurements is suspected to exist, with the x_i's as the independent variables. Do a regression analysis to find this exponential relationship.

X_i	1.14	4.55	7.52	6.86	5.43	0.74	4.37	2.02	6.96	2.90
Y_i	10.61	190.6	3702	1914	464.5	5.79	162.7	24.70	2118	45.7

Section 5-4

5-11. Obtain the empirical cdf for the data of Problem 5-3.

5-12. Obtain the empirical cdf's for both sets of data given in Problem 5-9.

Section 5-6

5-13. Given the data of Problem 5-3, establish upper and lower control limits. Test the data given for being within these control limits by dividing it into lots of 3. (*Note:* This problem is meant merely to provide you with practice in setting up a control chart; there are insufficient data to perform a reasonable test.)

Section 5-7

5-14. One hundred samples are drawn from a population where it is known that the true variance of the samples is unity. The sample mean is computed. Make up a table giving a bound, α, on the probability that the sample mean deviates by more than β from the true mean, where several pairs of α and β are shown.

5-11 COMPUTER EXERCISES

5-1. **(a)** Design a random number generator to generate random numbers according to the Laplacian pdf

$$f_X(x) = \frac{\alpha}{2} e^{-\alpha|x|}$$

(b) Test your number generator by computing a histogram for 1000 random numbers generated by it. Normalize this histogram properly so that you can compare it with the pdf given above. Choose $\alpha = 2$.

(c) Also compute the empirical cdf corresponding to the histogram of part (b).

5-2. Repeat Computer Exercise 5-1 for the Rayleigh random variable considered in Problem 3-38.

5-3. An oscillator has a frequency given by

$$f_0 = \frac{1}{2\pi\sqrt{LC}} \quad \text{Hz}$$

where L and C are the inductance and capacitance, respectively, in a tank circuit. A frequency of 10,000 Hz is desired. Assume that the inductors and capacitors available have $\pm P\%$ tolerance around their nominal values, uniformly distributed in this range (this should be a program input). Design a Monte Carlo simulation to determine the histogram for the oscillator frequency for several combinations (i.e. at least three) of L and C with nominal values chosen to give the desired frequency (the number of repetitions should be a program input). Answer the following questions.

(a) From the tolerance ranges for L and C, determine the minimum and maximum resonant frequency bounds for a random selection of L and C within their tolerance ranges.

(b) Obtain the sample mean of f_0 for each choice of L and C. Does the sample mean appear to be a good estimate for f_0? (f_0 means the value of resonant frequency obtained when the nominal values of L and C are used.)

(c) Obtain the sample standard deviation for the generated resonant frequencies. Does it appear to be dependent on your selections for the nominal values of L and C?

(d) From the histogram of the resonant frequencies, what can you say about the probabilities of resonant frequencies below the nominal f_0 (using the nominal values of L and C) and above the nominal f_0 (i.e. which is more probable—frequencies above or below the nominal f_0)?

5-4. Rivets that are supposed to have a nominal diameter of 5 mm are manufactured. Design a computer simulation to give diameters of 5 mm $\pm$ $P\%$, uniformly distributed. Use the data generated to construct a control chart for a batch of 1000 rivets, tested against the control limits in lots of size 5, 10, and 20 and for $P = 5$ and 10%. Does the process ever go out of control? Does it appear to depend on P? What lot size appears most reasonable? You might determine this by plotting the rivet diameters versus the control limits and seeing if this curve goes outside the control limits. If so, does the control chart for lot size 20 detect this event (or events)? Repeat for lot sizes of 5 and 10.

6

Estimation Theory and Applications

6–1 INTRODUCTION

In Chapter 5 we considered the sample mean and sample variance, which were written down in a more-or-less intuitive fashion, of a set of observations from the overall population. The hope was that these estimates would be close to the actual mean and variance of the population. We had no reason to believe that this is indeed the case, except for the application of Chebyshev's inequality to the sample mean, which indicates that the probability that the sample mean deviates from the true mean by a certain amount can be made as small as desired through taking a large enough sample size. This fact is summarized by the weak law of large numbers. Since estimates depend on the samples taken from the population, we refer to them as *statistics.*

In this chapter, we address the subject of estimates, their properties, and measures of their goodness. Any procedure that provides an estimate of a parameter is called an *estimator.* We consider two types of estimators: point estimators and interval estimators. An example of a *point estimator* is the sample mean, for it provides a single number which, we hope, is a good estimate of the true mean.[1] Measures of the quality of a point estimator include its bias and variance, ideas that we explore in the next section.

An interval estimator involves a statement like

$$P(\theta_1 < \theta < \theta_2) = 1 - \alpha, \qquad \alpha > 0 \tag{6-1}$$

[1]There is no guarantee that an estimate given by a point estimator, such as the sample mean, will not deviate, perhaps by a great deal, from the true value. Later, when we take up interval estimators, we formulate the accuracy of estimators in probabilistic terms.

where θ is the unknown parameter and $1 - \alpha$ is a given constant, called the *confidence coefficient.* The parameter α is called the *confidence level.* The numbers θ_1 and θ_2 are called *confidence limits.* They depend on the data or samples taken from the population and are therefore random variables. If we choose α close to zero, we expect with near certainty that the parameter θ is in the *confidence interval* (θ_1, θ_2). The objective of interval estimation is to find θ_1 and θ_2 such that the length of the confidence interval, $\theta_2 - \theta_1$, is minimized subject to the constraint (6-1). In many applications, this is not an easy problem. We consider a few examples in this chapter for which answers can be obtained with relative ease.

6–2 DESIRABLE QUALITIES OF POINT ESTIMATORS

If we take a series of samples from a population and compute the sample mean, we hope that this estimate is close to the true mean of the population. One measure of the goodness of an estimate is its expectation. We note that the estimator is a random variable, because it is a function of random variables. For example, consider the sample mean, given by[2]

$$\bar{x} = \frac{1}{n}\sum_{i=1}^{n} X_i \tag{6-2}$$

Using the fact that the expectation of a sum of terms is the sum of the expectations of each term, we obtain for the expectation of the sample mean

$$E(\bar{x}) = E\left(\frac{1}{n}\sum_{i=1}^{n} X_i\right) = \frac{1}{n}\sum_{i=1}^{n} E(X_i) = \frac{1}{n}\sum_{i=1}^{n} \mu_x = \mu_x \tag{6-3}$$

That is, the expectation of the estimate of the parameter we are trying to estimate is equal to the parameter itself. Such an estimator is said to be *unbiased.* In general, if θ is the parameter being estimated, we say an estimator is unbiased if

$$E(\hat{\theta}) = \theta \tag{6-4}$$

where $\hat{\theta}$ is the estimate. If

$$E(\hat{\theta}) = \theta + B \tag{6-5}$$

where B is a constant, the estimator is said to be *biased with bias B.* It is clearly desirable to have an unbiased estimator, although if we know the bias, it can be subtracted from the estimate.

Another measure of goodness of an estimator is its *variance,* defined as

$$\text{Var}(\hat{\theta}) = E\{[\hat{\theta} - E(\hat{\theta})]^2\} \tag{6-6}$$

It gives a measure of the imprecision of the estimator (i.e., the spread of values provided by the estimator about the average estimate). If an estimator is biased, this spread of values is around the true value of the parameter being estimated plus the bias as given by (6-5). Thus, the total average error of the estimator has two components: one due to the bias, and

[2]In Chapter 3, lowercase letters were used for sample values because we thought of them in terms of numbers selected from the population. Depending on the sample chosen, they will vary from sample to sample. In short, they are really random variables. We therefore use our convention for random variables in this chapter—namely, that the samples are denoted by capital letters.

the other due to the random spread about the mean of the estimator, characterized by the square root of the variance, or *standard deviation* of the estimator.

A measure of goodness that takes into account both the bias and imprecision of the estimator is the *mean-squared error*, given by

$$\text{MSE} = E[(\hat{\theta} - \theta)^2] \tag{6-7}$$

Substituting from (6-5), we have

$$\begin{aligned}\text{MSE} &= E\{[\hat{\theta} - E(\hat{\theta}) + B]^2\}\\ &= E\{[\hat{\theta} - E(\hat{\theta})]^2 + 2B[\hat{\theta} - E(\hat{\theta})] + B^2\}\\ &= E\{[\hat{\theta} - E(\hat{\theta})]^2\} + 2BE\{[\hat{\theta} - E(\hat{\theta})]\} + B^2\\ &= \text{Var}(\hat{\theta}) + B^2\end{aligned} \tag{6-8}$$

Clearly, the smaller the mean-squared error, the more accurate the estimator, both in terms of precision (variance or standard deviation) and bias. An estimator whose variance and bias go to zero as the number of observations go to infinity is called *consistent*. From (6-8) it follows that the mean-squared error also goes to zero for a consistent estimator as the number of observations goes to infinity. Such an estimator can be called an *accurate* estimator because not only are the values of successive estimates clustered around the true value of the parameter (zero bias), but the spread of these estimates about the true value goes to zero as the number of observations increases without bound.

Example 6-1

Show that the sample mean is a consistent estimator (note that it is assumed that the samples drawn from the population are statistically independent).

Solution: We have already shown that the bias is zero by (6-3). The variance is

$$\begin{aligned}\text{Var}(\bar{x}) &= E\left[\left(\frac{1}{n}\sum_{i=1}^{n} X_i - \mu_X\right)^2\right]\\ &= E\left[\left[\frac{1}{n}\sum_{i=1}^{n}(X_i - \mu_X)\right]^2\right]\\ &= \frac{1}{n^2}E\left[\sum_{i=1}^{n}\sum_{j=1}^{n}(X_i - \mu_X)(X_j - \mu_X)\right]\\ &= \frac{1}{n^2}\sum_{i=1}^{n}\sum_{j=1}^{n}E[(X_i - \mu_X)(X_j - \mu_X)]\\ &= \frac{1}{n^2}\sum_{i=1}^{n}\text{Var}(X_i) = \frac{\text{Var}(X_i)}{n}\end{aligned} \tag{6-9}$$

where, for statistically independent samples, we have used the fact that

$$E[(X_i - \mu_X)(X_j - \mu_X)] = \begin{cases}\text{Var}(X_i), & i = j\\ E(X_i - \mu_X)E(X_j - \mu_X) = 0, & i \neq j\end{cases} \tag{6-10}$$

Thus the variance goes to zero as the number of samples goes to infinity, and the estimator is consistent.

6–3 METHOD FOR OBTAINING ESTIMATORS: THE MAXIMUM LIKELIHOOD TECHNIQUE

The sample mean used as an example in Section 6-2 is what we might call an *intuitive* estimator: It appears reasonable and we simply use it. In this section we look at a more logical way of obtaining estimators, called the *maximum likelihood technique*. As before, we denote the parameter to be estimated by θ. We take n statistically independent samples from the population, $\{X_1, X_2, \ldots, X_n\}$, which are random variables. Let the conditional pdf on the ith sample given the parameter be denoted by $f_i(x_i \mid \theta)$. The joint conditional pdf of all samples given the parameter, because of the statistical independence of the samples, is given by

$$L(\theta) = f_n(x_1, x_2, \ldots, x_n \mid \theta) = \prod_{i=1}^{n} f_i(x_i \mid \theta) \tag{6-11}$$

The "L" denotes *likelihood function,* for we interpret (6-11) as follows: once the samples are drawn, they are known. The parameter θ is, of course, unknown because we are trying to estimate it. Given the samples, we ask what value of θ is most likely to have produced them. It is that value of $\theta = \hat{\theta}$ that maximizes (6-11) (i.e., the most probable). Thus we call it the *maximum likelihood estimate* of the parameter θ.

A condition for finding the maximum likelihood estimate is obtained by differentiating the likelihood function, given by (6-11), setting the resulting equation equal to zero, and solving for θ. It is often easier to differentiate the natural logarithm of the likelihood function, and this is permissible since the logarithm is a monotonically increasing function of its argument: The value of the independent variable maximizing the logarithm of a function will also maximize the function itself. Thus a condition that the maximum likelihood estimate of a parameter θ must satisfy is

$$\frac{d}{d\theta} \ln L(\theta) \Big|_{\theta=\hat{\theta}} = \frac{1}{L(\theta)} \frac{dL(\theta)}{d\theta}\Bigg|_{\theta=\hat{\theta}} = 0 \tag{6-12}$$

Example 6-2

Find the maximum likelihood estimator for the mean of n statistically independent samples which are known to be Gaussian. The variance of each sample is assumed known.

Solution: We denote the unknown mean of each sample by μ_X and the known variance of each sample by $\sigma_X{}^2$. Thus from (3-28) and statistical independence of the samples, we have for the likelihood function

$$L(\mu_X) = \prod_{i=1}^{n} \frac{e^{-(X_i - \mu_X)^2/2\sigma_X^2}}{\sqrt{2\pi\sigma_X^2}} = \frac{e^{-\sum_{i=1}^{n}(X_i - \mu_X)^2/2\sigma_X^2}}{(2\pi\sigma_X^2)^{n/2}} \tag{6-13}$$

Denoting the log-likelihood function by a ℓ, we have

$$\ell(\mu_X) = -\sum_{i=1}^{n} \frac{(X_i - \mu_X)^2}{2\sigma_X^2} - \frac{n}{2} \ln(2\pi\sigma_X^2) \tag{6-14}$$

Differentiating with respect to μ_X and setting the result equal to zero, we obtain

$$\frac{d\ell(\hat{\mu}_X)}{d\hat{\mu}_X} = \sum_{i=1}^{n} \frac{2(X_i - \hat{\mu}_X)}{2\sigma_X^2} = 0$$

or

$$\sum_{i=1}^{n} (X_i - \hat{\mu}_X) = 0$$

or

$$\sum_{i=1}^{n} X_i - n\hat{\mu}_X = 0 \tag{6-15}$$

Solving for the unknown mean, we obtain

$$\hat{\mu}_X = \frac{1}{n} \sum_{i=1}^{n} X_i \tag{6-16}$$

which is the sample mean. The circumflex in (6-15) and (6-16) reminds us that this is an estimator for the mean. Thus the intuitive estimator for the mean is also the maximum likelihood estimate. However, we had to know the distribution of the samples to get the maximum likelihood estimator.

It has not yet been shown that (6-16) maximizes (6-14). Another way to demonstrate that (6-16) provides an extremum of (6-14), and eventually a maximum, is to rewrite (6-14) as

$$\begin{aligned} I(\mu_X) = 2\sigma_X^2\ell(\mu_X) &= -\sum_{i=1}^{n} (X_i - \mu_X)^2 - n\sigma_X^2 \ln(2\pi\sigma_X^2) \\ &= -\sum_{i=1}^{n} (X_i^2 - 2X_i\mu_X + \mu_X^2) - n\sigma_X^2 \ln(2\pi\sigma_X^2) \\ &= -(A - 2B\mu_X + n\mu_X^2) \end{aligned} \tag{6-17}$$

The maximum of $I(\mu_X)$ will, of course, correspond to the maximum of $\ell(\mu_X)$. For purposes of the maximization with respect to μ_X, A and B are constants given by

$$A = \sum_{i=1}^{n} X_i^2 - n\sigma_X^2 \ln(2\pi\sigma_X^2) \quad \text{and} \quad B = \sum_{i=1}^{n} X_i \tag{6-18}$$

By completing the square in μ_X in (6-17), we obtain

$$I(\mu_X) = -\left(\mu_X - \frac{B}{n}\right)^2 + \left(\frac{B^2}{n} - A\right) \tag{6-19}$$

Since the first term is nonpositive, the maximum value of (6-19) with respect to μ_X is $B^2/n - A$, which occurs for $\mu_X = B/n$ [the same as obtained in (6-16)]; any other value for μ_X makes the first term more negative. Clearly, $\mu_X = B/n$ provides an absolute maximum if the value of μ_X is unconstrained. If $a \leq \mu_X \leq b$ and the sample mean lies outside the interval $[a, b]$, the maximum is either a or b, whichever gives the largest value of (6-19).

Example 6-3.

Using (6-14) for the log-likelihood function, find the maximum likelihood estimator for the variance of Gaussian samples with the mean known.

Solution: We differentiate (6-14) with respect to σ_X^2 considering μ_X to be a constant, and obtain

$$\frac{d\ell(\sigma_X^2)}{d\sigma_X^2} = \sum_{i=1}^{n} \frac{(X_i - \mu_X)^2}{2(\sigma_X^2)^2} - \frac{n}{2}\frac{1}{2\pi\sigma_X^2}2\pi = 0 \tag{6-20a}$$

Note that once we set the derivative equal to zero, we have the condition for the maximum likelihood estimator, which we indicate by the circumflex. Solving, we obtain

$$\widehat{\sigma_X^2} = \frac{1}{n}\sum_{i=1}^{n} (X_i - \mu_X)^2 \tag{6-20b}$$

A second differentiation shows that this indeed provides a maximum. Here the circumflex again reminds us that this is an estimator of the variance, not the true variance. Note that to carry out this estimation procedure for the variance, we must know the true mean of the samples.

If we wish to find maximum likelihood estimators for the mean and the variance, with neither known, we would differentiate the log-likelihood function partially with respect to each parameter, set the resulting two equations equal to zero, and then solve for the parameters simultaneously. Examining (6-15), we easily convince ourselves that the estimate for the mean is the same as that obtained in (6-16). However, the condition (6-20a) with the circumflex over the mean parameter because it is unknown becomes

$$\sum_{i=1}^{n} (X_i - \hat{\mu}_X)^2 - n\widehat{\sigma_X^2} = 0 \tag{6-21}$$

Solution of (6-21) for the maximum likelihood estimator of σ_X^2 results in

$$\widehat{\sigma_X^2} = \frac{1}{n}\sum_{i=1}^{n} (X_i - \hat{\mu}_X)^2 \tag{6-22a}$$

which is a *biased* estimator of σ_X^2, because the expectation of (6-22a) can be shown to be $(n - 1)\sigma_X^2/n$. To get an unbiased estimator for the variance with the mean unknown, we use

$$\widehat{\sigma_X^2} = \frac{1}{n-1}\sum_{i=1}^{n} (X_i - \hat{\mu}_X)^2 \tag{6-22b}$$

Example 6-4

The following samples are drawn from a population that is known to be Gaussian: 7.31, 10.8, 11.27, 11.91, 5.51, 8.0, 9.03, 14.42, 10.24, and 10.91. Find estimates for the mean and variance.

Solution: Applying (6-16) and (6-22b), we obtain

$$\hat{\mu}_X = 9.94 \quad \text{and} \quad \widehat{\sigma_X^2} = 6.51$$

The samples were drawn from a population with true mean of 10 and true variance of 4. Note that the estimates for the mean and the variance are random variables. If we draw another 10 samples, these estimates will almost surely be different from the values above. In fact, by drawing other batches of 10 samples and computing the estimates for the mean and variance, we get the results in Table 6-1.

What is the distribution of the mean and variance estimates? For the mean estimate, the answer is easy (after a little thought). From (6-16) we note that the estimate for the mean depends linearly on the samples drawn; or, in other words, the mean estimate is a linear combination of the samples, which are assumed Gaussian. From Chapter 4, we remember that the sum of Gaussian random variables is Gaussian. We have already found the true

Table 6-1 SAMPLE MEANS AND VARIANCES FOR BATCHES DRAWN FROM A GAUSSIAN POPULATION

Batch No.	Sample Mean	Sample Variance
1	10.24	9.42
2	9.32	4.90
3	11.44	4.16
4	9.65	5.23
5	10.98	4.85
6	10.03	8.30
7	9.72	4.36
8	10.01	3.89
9	10.39	5.28
10	9.84	2.70

mean and variance of the sample mean, which are given by (6-3) and (6-9), respectively. Thus we can write down the pdf of the sample mean. In the case of Example 6-4, where the samples were drawn from a population with true mean of 10 and variance of 4, we get the following for the pdf of a 10-sample batch:

$$f_{\bar{X}}(\alpha) = \frac{e^{-(\alpha-10)^2/0.8}}{\sqrt{0.8\pi}} \quad \text{(10-sample batches)} \tag{6-23a}$$

Since the variance decreases inversely with the number of samples according to (6-9), this pdf becomes narrower as the number of samples drawn becomes larger. For example, for 100-sample batches, we obtain

$$f_{\bar{X}}(\alpha) = \frac{e^{-(\alpha-10)^2/0.08}}{\sqrt{0.08\pi}} \quad \text{(100-sample batches)} \tag{6-23b}$$

These pdf's are plotted in Fig. 6-1, which shows that the estimate for the mean probably gets more and more accurate as the number of samples increases, a result we would have expected intuitively. In the next section, we put a more quantitative measure on how much more accurate our estimates become as we take a larger number of samples. We postpone considering the distribution of the sample variance until then.

6-4 FINDING INTERVAL ESTIMATES FOR THE MEAN AND VARIANCE

Recall (6-1) for the definition of an interval estimate. In this section, we apply that definition to the mean and variance estimates found in Section 6-3. We consider the sample mean first.

Confidence Intervals for the Sample Mean

In Figure 6-1, sketches are given for the pdf of the sample mean assuming Gaussian samples. Also shown are two vertical lines that represent the confidence limits of (6-1). They have been placed symmetrically about the true mean. We know that the expectation of the

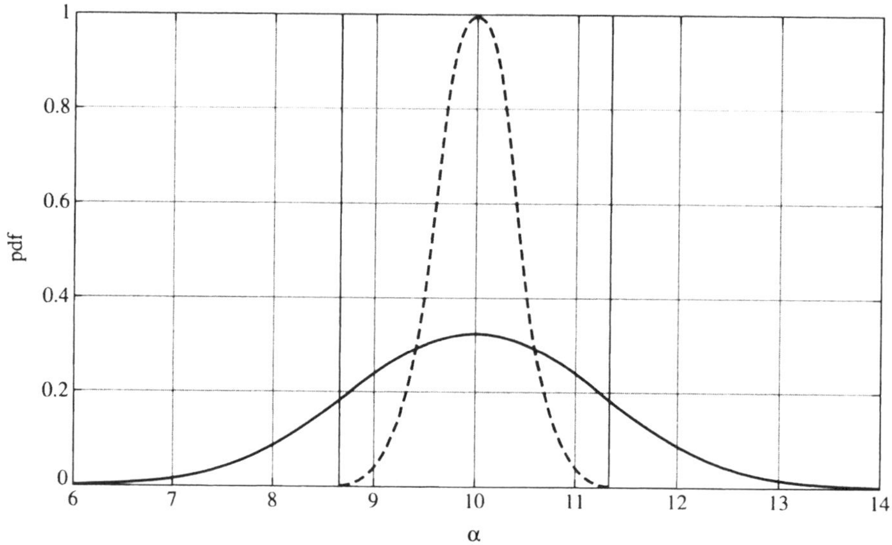

Figure 6-1 Probability density functions of the sample mean for 10-sample batches (solid curve) and 100-sample batches (dashed curve). Vertical lines refer to the confidence intervals for the sample mean discussed in Section 6-4.

sample mean is the true mean itself, and from Example 6-1 we know that the variance of the sample mean is the variance of each sample divided by *n*, the number of samples. Therefore,

$$Z = \frac{\sqrt{n}(\bar{x} - \mu_X)}{\sigma_X} \tag{6-24}$$

is a zero-mean, unit-variance random variable which will be used for convenience. Since it is a linear combination of Gaussian random variables (the sample mean is a sum), it is also Gaussian. Using a rational approximation for the Q-function or a set of tables, we can find z_c such that

$$P(-z_c < Z < z_c) = 1 - \alpha = \int_{-z_c}^{z_c} \frac{e^{-u^2/2}}{\sqrt{2\pi}} du = 1 - 2Q(z_c) \tag{6-25}$$

Some values for z_c for various values of $1 - \alpha$ are given in Table 6-2.

We may now compute the confidence interval for the sample mean using the following series of steps:

$$P\left(-z_c < Z = \frac{\sqrt{n}(\hat{\mu}_X - \mu_X)}{\sigma_X} < z_c\right) = 1 - \alpha$$

or

$$P\left(-\frac{\sigma_X z_c}{\sqrt{n}} < \hat{\mu}_X - \mu_X < \frac{\sigma_X z_c}{\sqrt{n}}\right) = 1 - \alpha$$

or

TABLE 6-2 ENDPOINT VALUES FOR CONFIDENCE INTERVALS FOR THE MEAN VERSUS CONFIDENCE LEVEL

$(1 - \alpha)$ %	z_c
90	1.645
93	1.812
95	1.96
98	2.326
99	3.09

$$P\left(\hat{\mu}_X - \frac{\sigma_X z_c}{\sqrt{n}} < \mu_X < \hat{\mu}_X + \frac{\sigma_X z_c}{\sqrt{n}}\right) = 1 - \alpha \tag{6-26}$$

The last equation of (6-26) expresses the $1 - \alpha$ confidence interval for the mean where μ_X and σ_X are the true mean and variance, respectively, and n is the number of samples.

Suppose that we want a confidence level of 95% or 0.95. Thus we set the final equation in the series given by (6-26) equal to 0.95. From Table 6-2, we have that this corresponds to $z_c = 1.96$. Given the samples so that we can compute the sample mean, their number, and their common variance, we can then compute the interval within which the true mean lies with 95% confidence.

Example 6-5

Find the confidence limits for a 99% confidence level for the first batch of data in Example 6-4. If the number of samples is increased to 100, how do the 99% confidence limits change?

Solution: The sample mean is 9.94 for 10 samples. The samples were drawn from a population with variance of each sample of 4. From Table 6-2, for a confidence level of 99%, we have z_c of 3.09. From the last equation of (6-26), the confidence interval is

$$P\left(\hat{\mu}_X - \frac{\sigma_X z_c}{\sqrt{n}} < \mu_X < \hat{\mu}_X + \frac{\sigma_X z_c}{\sqrt{n}}\right) = 1 - \alpha$$

or

$$P\left(9.94 - \frac{\sqrt{4}(3.09)}{\sqrt{10}} < \mu_X < 9.94 + \frac{\sqrt{4}(3.09)}{\sqrt{10}}\right) = 0.99$$

or

$$P(7.986 < \mu_X < 11.894) = 0.99 \tag{6-27}$$

Because the square root of the number of samples appears in the denominator of the endpoints, 100 samples will mean that this number is smaller by a factor of about 3.3, which means a tighter confidence interval for 100 samples than for 10.

Confidence Limits for the Variance: Mean Known

What about confidence limits for the variance? Unfortunately, we don't yet know the pdf of the sample variance with the mean known. However, we can extend the results of Chapters 3 and 4 on transformations of random variables to obtain this pdf. First, recall Example 3-16 where the pdf of the square of a Guassian random variable of mean zero and variance σ^2 was found to be

$$f_Y(y) = \frac{e^{-y/2\sigma^2}}{\sqrt{2\pi\sigma^2 y}}, \quad y > 0 \tag{6-28}$$

Second, recall from Example 4-18 that the pdf of the sum of two independent random variables is the convolution of their separate pdf's. Let us find the pdf of the sum of two independent random variables with the pdf of (6-28). The convolution integral is

$$\begin{aligned} f_{Z_2}(z) &= \int_{-\infty}^{\infty} f_Y(y) f_Y(z-y)\, dy \\ &= \int_0^z \frac{e^{-y/2\sigma^2}}{\sqrt{2\pi\sigma^2 y}} \frac{e^{-(z-y)/2\sigma^2}}{\sqrt{2\pi\sigma^2(z-y)}}\, dy \\ &= \frac{e^{-z/2\sigma^2}}{2\pi\sigma^2} \int_0^z \frac{dy}{\sqrt{y(z-y)}}, \quad z > 0 \end{aligned} \tag{6-29}$$

Now make the change of variables $y = zu$ in the last integral, which gives

$$f_{Z_2}(z) = \frac{e^{-z/2\sigma^2}}{2\pi\sigma^2} \int_0^1 \frac{du}{\sqrt{u(1-u)}}, \quad z > 0 \tag{6-30}$$

A table of integrals provides the definite integral[3]

$$B(\alpha, \beta) = \int_0^1 u^{\alpha-1}(1-u)^{\beta-1} du = \frac{\Gamma(\alpha)\Gamma(\beta)}{\Gamma(\alpha+\beta)}, \quad \alpha > 0,\ \beta > 0 \tag{6-31}$$

where $\Gamma(\bullet)$ is the gamma function defined by (3-35). Using this integral in (6-30) and recalling that $\Gamma(1/2) = \pi^{1/2}$ from (3-36), we obtain

$$f_{Z_2}(z) = \Gamma^2(\tfrac{1}{2}) \frac{e^{-z/2\sigma^2}}{2\pi\sigma^2} = \frac{e^{-z/2\sigma^2}}{2\sigma^2}, \quad z > 0 \tag{6-32}$$

This is an exponential pdf. Because it resulted from the sum of the squares of two independent Gaussian random variables of zero mean, it is not hard to show that the mean and variance are given by

$$\mu_{Z_2} = 2\sigma^2 \quad \text{and} \quad \sigma_{Z_2}^2 = 4\sigma^4 \tag{6-33}$$

respectively.

Let us continue by adding another random variable to the sum, so that we have the sum of the squares of three independent Gaussian random variables. However, to simplify matters, let $\sigma = 1$. Thus we convolve (6-28) and (6-32) with $\sigma = 1$. The convolution integral is

$$\begin{aligned} f_{Z_3}(z) &= \int_0^z \frac{e^{-u/2}}{\sqrt{2\pi u}} \frac{e^{-(z-u)/2}}{2}\, du \\ &= \frac{e^{-z/2}}{2\sqrt{2\pi}} \int_0^z \frac{du}{\sqrt{u}} \end{aligned}$$

[3]This is called the *beta integral*.

$$= \frac{\sqrt{z}e^{-z/2}}{\sqrt{2\pi}}, \quad z > 0 \tag{6-34}$$

This is a chi-square pdf of three degrees of freedom (order three). We could continue and show that for the sum of m squares of independent Guassian random variables of mean zero and unit variance, the pdf is

$$f_{Z_m}(z) = \frac{z^{m/2-1}e^{-z/2}}{\Gamma(m/2)2^{m/2}}, \quad z > 0, \quad m = 1, 2, 3, \ldots \tag{6-35}$$

which is a chi-square pdf of order m. The mean and variance of Z_m can be shown to be

$$\mu_{Z_m} = m \quad \text{and} \quad \sigma_{Z_m}^2 = 2m \tag{6-36}$$

respectively.

How does this apply to estimation of the variance of a sample from a population of Gaussian random variables? Recall (6-20b), which we rewrite as

$$V = \frac{n\hat{\sigma}_X^2}{\sigma_X^2} = \sum_{i=1}^{n} \left(\frac{X_i - \mu_X}{\sigma_X}\right)^2 \tag{6-37}$$

This may seem like cheating since we don't know σ_X^2. However, everything will work out in the end. Since we have subtracted each random variable's mean and divided by its standard deviation, it follows that the sum is composed of the squares of zero-mean, unit-variance Gaussian random variables. Thus we apply the results of the demonstration above that the pdf of a sum of squares of n zero-mean, unit-variance Gaussian random variables is chi-square with n degrees of freedom.

The confidence interval for the variance is defined by

$$P\left(\gamma < V = \frac{n\hat{\sigma}_X^2}{\sigma_X^2} < \delta\right) = 1 - \alpha \tag{6-38}$$

where the sketch in Figure 6-2 shows the relationship among γ, δ, and α. We want the two tail areas to be equal. Using the definition for V and rearranging inequalities, we obtain

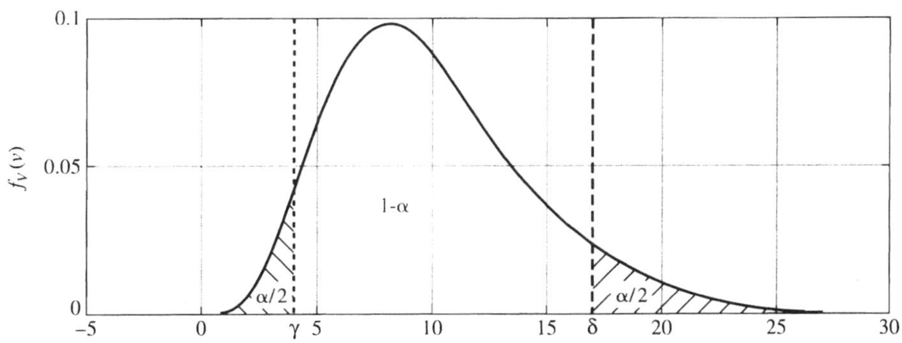

Figure 6-2 Chi-square pdf showing the tails to be integrated in finding the confidence intervals for the variance with the mean known.

$$P\left(\frac{n\widehat{\sigma_X^2}}{\delta} < \sigma_X^2 < \frac{n\widehat{\sigma_X^2}}{\gamma}\right) = 1 - \alpha \tag{6-39}$$

Example 6-6

Find 95% confidence intervals for estimation of the variance of the samples given in Example 6-4 with the mean known to be 10.

Solution: A calculation using (6-20b) and the data of Example 6-4 gives

$$\widehat{\sigma_X^2} = 5.87 \tag{6-40}$$

For a confidence level of 95%, α in (6-39) is 0.05, which we split between the left and right tails of the chi-square pdf for the sample variance. From Table C-6 of Appendix C, using the $n = 10$ row we have

$$\begin{aligned} P(V > \delta) &= 0.025 \Rightarrow \delta = 20.483 \\ P(V > \gamma) &= 1 - 0.025 = 0.975 \Rightarrow \gamma = 3.247 \end{aligned} \tag{6-41}$$

Putting these into (6-39) along with (6-40), we have

$$P\left(\frac{10(5.92)}{20.483} < \sigma_X^2 < \frac{10(5.92)}{3.247}\right) = 0.95 \tag{6-42}$$

or

$$P(2.89 < \sigma_X^2 < 18.23) = 0.95 \tag{6-43}$$

Finally, we address confidence intervals for the mean and variance assuming that both are unknown. Those for the variance are surprisingly easy once we have gone through the mathematics for finding confidence intervals for the variance with the mean known.

Confidence Limits for the Variance: Mean Unknown

The unbiased estimator for the variance with mean unknown is given by (6-22b), but we normalize it and consider the random variable

$$W = \frac{(n-1)\widehat{\sigma_X^2}}{\sigma_X^2} = \sum_{i=1}^{n}\left(\frac{X_i - \mu_X}{\sigma_X}\right)^2 \tag{6-44}$$

With the X_i's Guassian, the random variable W can be shown to be chi-square with $n - 1$ degrees of freedom. It is well beyond the scope of this book to prove this, but a demonstration of this fact may be found in Walpole and Meyers (1993, pp. 224-225).

Proceeding as in the case of confidence intervals for the variance with mean known, we rewrite (6-39) as

$$P\left[\frac{(n-1)\widehat{\sigma_X^2}}{\delta} < \sigma_X^2 < \frac{(n-1)\widehat{\sigma_X^2}}{\gamma}\right] = 1 - \alpha \tag{6-45}$$

Example 6-7

Find 95% confidence intervals for the variance of the samples of Example 6-4 if the mean is unknown.

Solution: Again using Table C-6 of Appendix C, but with $n - 1 = 9$ degrees of freedom, we have

$$P(W > \delta) = 0.025 \Rightarrow \delta = 19.023$$

$$P(W > \gamma) = 1 - 0.025 = 0.975 \Rightarrow \gamma = 2.7 \tag{6-46}$$

Putting these results into (6-45), we obtain

$$P\left(\frac{9(5.96)}{19.023} < \sigma_X^2 < \frac{9(5.96)}{2.7}\right) = 0.95 \tag{6-47}$$

or

$$P(2.82 < \sigma_X^2 < 19.87) = 0.95 \tag{6-48}$$

Confidence Limits for the Mean: Variance Unknown

Finally, we turn to confidence intervals for the mean with the variance unknown. We define the random variable

$$T = \frac{\sqrt{n}(\hat{\mu}_X - \mu_X)}{\hat{\sigma}_X} \tag{6-49a}$$

where $\hat{\sigma}_X = \sqrt{\widehat{\sigma_X^2}}$. Equation (6-49a) is the ratio of two random variables: the sample mean and the sample standard deviation. In terms of our previously defined random variables, Z and W, we have

$$T = \frac{Z}{\sqrt{W/(n-1)}} \tag{6-49b}$$

where it will be recalled that Z is a Gaussian random variable of unit variance and zero mean and W is a chi-square random variable. It can be shown that Z and W are statistically independent, and the method of transformation of multiple random variables outlined in Chapter 4 can be used to find the pdf of T. This procedure is fairly complex and would detract from the development of confidence intervals here. Therefore, we simply state that T is distributed according to a Student's t pdf[4], given by

$$f_T(t) = \frac{\Gamma[(r+1)/2]}{\sqrt{\pi r}\Gamma(r/2)(1 + t^2/r)^{(r+1)/2}}, \quad r = n - 1 \text{ integer} \tag{6-50}$$

Note that $f_T(t)$ is an even function of t. The confidence interval is defined by

$$P\left[-t_c < T = \frac{\sqrt{n}(\hat{\mu}_X - \mu_X)}{\hat{\sigma}_X} < t_c\right] = 1 - \alpha \tag{6-51a}$$

Rearranging the inequalities, we have

$$P\left(\hat{\mu}_X - \frac{t_c\hat{\sigma}_X}{\sqrt{n}} < \mu_X < \hat{\mu}_X + \frac{t_c\hat{\sigma}_X}{\sqrt{n}}\right) = 1 - \alpha \tag{6-51b}$$

Example 6-8

Find 95% confidence intervals for the mean of the samples of Example 6-4 if the variance is unknown.

[4]First published in 1908 in a paper by W. S. Gosset under the pseudonym "Student" because the Irish brewery by which he was employed did not allow publication of research by its employees.

Solution: Using Table C-8 with $n - 1 = 9$, we have

$$P(T > t_c) = P(T < -t_c) = \frac{\alpha}{2} = 0.025 \Rightarrow t_c = 2.263 \tag{6-52}$$

In Example 6-4, it was found that

$$\hat{\mu}_X = 9.94 \quad \text{and} \quad \widehat{\sigma_X^2} = 6.51$$

so (6-51b) becomes

$$P\left(9.94 - \frac{2.263\sqrt{6.51}}{\sqrt{10}} < \mu_X < 9.94 + \frac{2.263\sqrt{6.51}}{\sqrt{10}}\right) = 0.95 \tag{6-53}$$

or

$$P(8.11 < \mu_X < 11.77) = 0.95 \tag{6-54}$$

Confidence Limits for the Relative Frequency of Bernoulli Trials

We wish to measure the probability, p, of the occurrence of an event A that either may or may not occur upon performance of the experiment. To do so, we repeat the experiment n times and note the number K of occurrences of A. The random variable K is binomial and has mean np and variance npq, as found in Chapter 3 [see (3-96) and (3-97)]. Consider $X = K/n$ as an estimate of p. Since K is a binomial random variable, the expectation of $X = K/n$ is $np/n = p$ and its variance is $\text{Var}(K/n) = \text{Var}(K)/n^2 = npq/n^2 = pq/n$. For large n we can approximate the binomially distributed sample mean of X as Gaussian, and the confidence limits are the same as found previously for the sample mean. We have

$$P\left(p - \beta < \bar{x} = \frac{K}{n} < p + \beta\right) = 1 - \alpha \tag{6-55a}$$

In terms of the approximate Gaussian pdf of $\bar{x}$ we have

$$\begin{aligned} P(p - \beta < \bar{x} < p + \beta) &= \int_{p-\beta}^{p+\beta} \frac{e^{-(v-p)^2/(2pq/n)}}{\sqrt{2\pi pq/n}}\,dv \\ &= 2\int_0^{\beta\sqrt{n/pq}} \frac{e^{-u^2/2}}{\sqrt{2\pi}}\,du \\ &= 2\left[0.5 - Q\left(\beta\sqrt{\frac{n}{p(1-p)}}\right)\right] = 1 - \alpha \end{aligned} \tag{6-55b}$$

or

$$Q\left(\beta\sqrt{\frac{n}{p(1-p)}}\right) = \frac{\alpha}{2} \tag{6-55c}$$

Since we don't know p, we replace it by its estimate $\hat{p} = K/n$. Thus the confidence interval is found by solving (6-55c) for β given a particular α, or vice versa.

Example 6-9

Voters are polled on how they are going to vote on a certain issue, with 280 of a sample of 500 saying they are going to vote yes. Find the 95% confidence interval of the probability p that the issue will pass.

Solution: In this example,

$$n = 500 \quad \text{and} \quad \hat{p} = \frac{K}{n} = \frac{280}{500} = 0.56$$

Thus we solve

$$Q\left(\beta\sqrt{\frac{500}{0.56(0.44)}}\right) = \frac{0.05}{2} = 0.025 \tag{6-56a}$$

or

$$Q(45.05\beta) = 0.025 \quad \text{or} \quad 45.05\beta = 1.96 \quad \text{or} \quad \beta = 0.0435 \tag{6-56b}$$

Thus, solving the inequalities on the left side of (6-55b), we have that the probability p that the issue will pass is between $0.56 - 0.0435 = 0.5165$ and $0.56 + 0.0435 = 0.6035$ with a confidence of 95%.

6–5 MORE ON MAXIMUM LIKELIHOOD ESTIMATORS

Maximum likelihood estimators have several nice properties that will be summarized in this section. In the estimates of the sample variance with mean known and unknown and the estimation of the mean with the variance unknown, we saw in Section 6-4 that the pdf of the estimator was fairly complex to obtain. It can be shown that the pdf of a maximum likelihood estimate tends to a Gaussian pdf as the number of samples tends to infinity. Its mean is equal to the parameter being estimated and the variance is given by

$$\text{Var}(\hat{\theta}) = \frac{1}{nI} \tag{6-57a}$$

where

$$I = E\left[\left|\frac{\partial}{\partial\theta}\ell\,(X_1, X_2, \ldots, X_n; \theta)\right|^2\right] \tag{6-57b}$$

in which $\ell(X_1, \ldots, X_n : \theta)$ is the log-likelihood function. Using integration by parts, it can be shown that (6-57b) can also be written as

$$I = -E\left[\frac{\partial^2}{\partial\theta^2}\ell\,(X_1, X_2, \ldots, X_n; \theta)\right] \tag{6-57c}$$

The proof that the maximum likelihood estimate is Gaussian distributed is based on the central limit theorem.

Maximum likelihood estimates can also be shown to be asymptotically unbiased. Furthermore, there is a bound on the minimum variance of a maximum likelihood estimate called the Cramér–Rao bound. Maximum likelihood estimates can be shown to satisfy this lower bound asymptotically as the number of samples becomes large.

The Cramér–Rao inequality can be expressed as follows. Given an unbiased estimator, the variance of the estimate is no smaller than

$$\text{Var}(\hat{\theta}) = E[\hat{\theta}(X_1, X_2, \ldots, X_n)^2] \geq \frac{1}{nI} \tag{6-58}$$

where I is defined by (6-57b) or (6-57c). Equality holds if, and only if,

$$\frac{\partial \ell(X_1, X_2, \ldots, X_n; \theta)}{\partial \theta} = nI[g(X_1, X_2, \ldots X_n) - \theta] \tag{6-59}$$

If an estimator satisfies (6-58) with equality, it is called *efficient*.

It can be shown that distributions of exponential type for the samples lead to efficient estimators. An exponential class estimator is of the form

$$f(x_1, x_2, \ldots, x_n; \theta) = h(x_1, x_2, \ldots, x_n)e^{a(\theta)g(x_1, \ldots, x_n) - b(\theta)} \tag{6-60}$$

where the functions $a(\theta)$ and $b(\theta)$ are functions of the unknown parameter only, and the functions $h(\bullet)$ and $g(\bullet)$ are functions of the samples only.

Example 6-10

Show that estimation of the parameter of an exponential pdf is of exponential type and therefore efficient.

Solution: The joint pdf of n independent samples, but of course dependent on the parameter, is

$$f_X(x_1, x_2, \ldots, x_n; \theta) = \prod_{i=1}^{n} f(x_i; \theta) = \theta^n e^{-\theta \sum_{i=1}^{n} x_i} = e^{-\theta \sum_{i=1}^{n} x_n + n \ln \theta} \tag{6-61}$$

The log-likelihood function is (uppercase X_i's are now used because they represent data)

$$\ell(\theta) = -\theta \sum_{i=1}^{n} X_i + n \ln \theta \tag{6-62}$$

Clearly, the joint pdf is of exponential type. The maximum likelihood estimator for θ is found as

$$-\sum_{i=1}^{n} X_i + \frac{n}{\hat{\theta}} = 0 \quad \text{or} \quad \hat{\theta} = \left(\frac{1}{n}\sum_{i=1}^{n} X_i\right)^{-1} \tag{6-63}$$

The second derivative can be used in (6-57c) to give the variance via (6-58):

$$\frac{\partial^2 \ell}{\partial \theta^2} = -\frac{n}{\theta^2} \quad \text{or} \quad I = -E\left(\frac{\partial^2 \ell}{\partial \theta^2}\right) = \frac{1}{\theta^2} \tag{6-64}$$

Because the pdf is of the exponential type, the estimator is efficient with variance

$$\text{Var}(\hat{\theta}) = \frac{1}{nI} = \frac{\theta^2}{n} \tag{6-65}$$

6-6 CONDITIONAL MEAN AS AN ESTIMATOR

Suppose that we wish to estimate some characteristic of a population—say, the amount of rainfall in a year in a certain country. We take samples by placing rain gauges in certain places and base our estimate on the sample mean of the readings of these rain gages over the course of a year. This gives us an estimate that is representative of the average over the whole country. However, certain regions of the country are dry and others are humid. If we have additional information, such as locations of the rain gauges as northeast, southeast, northwest, and southwest, we can form estimates of the average rainfall that are more representative with respect to region.

Let us couch this in more mathematical terms. Given two random variables X and Y, which we assume to be dependent on each other, with joint pdf $f_{XY}(x, y)$, we wish to form an estimate of Y, with knowledge of an observation of X, say x. As previously, we designate the estimator by θ, but since it depends on our knowledge of X, it is now a function of the observation. The criterion we shall use to choose the estimate is minimum mean-squared error, given by

$$e = E\{[Y - \theta(X)]^2\} = \int_{-\infty}^{\infty}\int_{-\infty}^{\infty} [y - \theta(x)]^2 f_{XY}(x, y)\, dx\, dy \tag{6-66}$$

That is, we want to find the function $\theta(x)$ to minimize this expression. To do so, we rewrite the joint pdf in terms of the conditional and marginal pdf's as

$$f_{XY}(x, y) = f_{Y|X}(y\,|\,x) f_X(x) \tag{6-67}$$

so that (6-66) becomes

$$e = E\{[Y - \theta(X)]^2\} = \int_{-\infty}^{\infty}\int_{-\infty}^{\infty} [y - \theta(x)]^2 f_{Y|X}(y|x)\, f_X(x)\, dx\, dy$$

$$= \int_{-\infty}^{\infty} f_X(x) \int_{-\infty}^{\infty} [y - \theta(x)]^2 f_{Y|X}(y|x)\, dy\, dx \tag{6-68}$$

Now both integrands in (6-68) are nonnegative. The pdf of X is given. Hence the best we can do is minimize the inner integral. It follows that this integral will be a minimum if

$$\theta(x) = \int_{-\infty}^{\infty} y\, f_{Y|X}(y\,|\,x)\, dy = E(Y\,|\,x) \tag{6-69}$$

To show this another way, consider the expression

$$E[(Y - c)^2] = E\{[\mu_Y - c) + (Y - \mu_Y)]^2\}$$

$$= E[\mu_Y - c)^2] + 2E[(\mu_Y - c)(Y - \mu_Y)] + E[(Y - \mu_Y)^2]$$

$$= (\mu_Y - c)^2 + \sigma_Y^2 \tag{6-70}$$

where $c = \theta(x)$ and the expectation is with respect to the conditional pdf $f_{Y|X}(y \mid x)$. It is clear that the right-hand side of the last equation of (6-70) is a minimum of $\mu_Y = c$ or if (6-69) holds.

Example 6-11

Consider the estimation of a signal, Y, in Gaussian noise. The signal component is assumed Gaussian with mean m and variance σ^2. A single sample of the observed data is of the form

$$X_k = Y + N_k, \quad k = 1, 2, \ldots, n \tag{6-71}$$

where the N_k's are independent, Gaussian random variables with zero means and variances σ_n^2, and n samples are taken. Find the conditional mean estimator for Y and the resulting mean-squared error.

Solution: The conditional pdf of the data given the parameter Y is

$$f_{X|Y}(x_1, x_2, \ldots, x_n \mid Y) = \prod_{k=1}^{n} \frac{\exp[-(x_k - Y)^2/2\sigma_n^2]}{\sqrt{2\pi\sigma_n^2}}$$

$$= \frac{\exp\left[\sum_{k=1}^{n} (x_k - Y)^2/2\sigma_n^2\right]}{(2\pi\sigma_n^2)^{n/2}} \tag{6-72}$$

After some algebra it can be shown that

$$f_{Y|X}(y \mid \mathbf{x}) = (2\pi\sigma_P^2)^{-1/2} \exp\left\{\frac{-[y - \sigma_P^2(n\bar{x}/\sigma_n^2 + m/\sigma^2)]}{2\sigma_p^2}\right\} \tag{6-73}$$

where $\mathbf{x} = x_1, x_2, \ldots, x_n)$, $\bar{x}$ is the sample mean of the data, and

$$\frac{1}{\sigma_P^2} = \frac{n}{\sigma_n^2} + \frac{1}{\sigma^2} \tag{6-74}$$

From the form of (6-73) it should be clear that the conditional mean is just

$$E(Y \mid \mathbf{x}) = \sigma_P^2\left(\frac{n\bar{x}}{\sigma_n^2} + \frac{m}{\sigma^2}\right)$$

$$= \frac{n\sigma^2/\sigma_n^2}{1 + n\sigma^2/\sigma_n^2}\bar{x} + \frac{1}{1 + n\sigma^2/\sigma_n^2} m \tag{6-75}$$

The conditional variance is σ_P^2. Since it is independent of the data, this is also the mean-squared error. We note that as $n\sigma^2/\sigma_n^2$ gets large, the conditional mean estimate approaches $\bar{x}$ or the sample mean of the samples, $X_1, X_2, \ldots, X_n$. That is, the ratio of signal variance to noise variance becomes large, the best strategy is to rely entirely on the observed data. On the other hand, as $n\sigma^2/\sigma_n^2$ gets small, the conditional mean estimate approaches m, or the mean of the parameter Y that we are trying to estimate. In other words, if the noise dominates, we should rely on the prior knowledge about the signal, and the best estimate we have of it is its mean.

6–7 ORTHOGONALITY PRINCIPLE

A useful observation for minimum mean-square estimates is that the estimation error is orthogonal, in a statistical sense, to any function $g(X)$ of the data. That is, if $\theta(X) = E(Y \mid X)$, then

$$E\{[Y - \theta(X)]g(X)\} = 0 \tag{6-76}$$

That this is the case can be proved by making use of conditional expectations as discussed in Problem 4-17. The expectation on the left-hand side can be written as

$$E_X\{g(X)E_Y[Y - \theta(X)|X]\} \tag{6-77}$$

where the subscript denotes the random variable, respectively, with respect to which the expectation is taken. The inside expectation is zero, for

$$E_Y[Y - \theta(X) \mid X] = E(Y \mid X) - \theta(X) = 0 \tag{6-78}$$

by definition of the minimum mean-square estimate (6-69). Thus the overall expectation (6-78) is zero and the assertion (6-76) is true.

Example 6-12

Consider the estimation of a random variable Y in terms of a single observation of another, related random variable X in the form of a linear estimator given by

$$\theta(X) = a_0 + a_1 X$$

where the constants a_0 and a_1 are to be found. Assume that the mean and variance of X are μ_X and σ_X^2, respectively, that the mean of Y is μ_Y, and that the covariance of X and Y is C_{XY}.

Solution: From (6-76) with $g(X)$ replaced by 1 and X in turn, we obtain the equations

$$E(Y - a_0 - a_1 X) = 0 \quad \text{or} \quad a_0 + \mu_X a_1 = \mu_Y$$

$$E[(Y - a_0 - a_1 X) X] = 0 \quad \text{or} \quad \mu_X a_0 + (\sigma_X^2 + \mu_X^2) a_1 = C_{XY}$$

Solving for a_0 and a_1, we obtain

$$a_0 = \mu_Y - \frac{(C_{XY} - \mu_X \mu_Y)\mu_X}{\sigma_X^2}$$

$$a_1 = \frac{C_{XY} - \mu_X \mu_Y}{\sigma_X^2}$$

Thus the conditional mean estimate is given by

$$\hat{Y} = \theta(X) = (\mu_Y - a_1 \mu_X) + a_1 X$$

If X and Y are uncorrelated, $C_{XY} = \mu_X \mu_Y$ and $a_1 = 0$, so that the optimum estimate is μ_Y (i.e., we don't use the observation in making the estimate).

6–8 SUMMARY

In this chapter, the estimation of parameters from samples from a population of data has been considered. Two types of estimators were introduced: point and interval. Measures of their goodness were discussed. A procedure, called maximum likelihood, for finding point estimators if the distribution of the samples is known, was discussed. Some examples of finding interval estimators were then given, and finally, asympototic properties in the limit of large sample sizes of maximum likelihood estimators were discussed.

1. A point estimator provides a single value for the estimator of a parameter for each batch of samples taken from the population. An example is the sample mean, defined as

$$\hat{\theta}(X_1, X_2, \ldots, X_n) = \bar{x} = \frac{1}{n} \sum_{i=1}^{n} X_i$$

Since an estimator is a function of random variables (the samples), it is itself a random variable.

2. Measures of goodness for point estimators include their *expectation, variance,* and *mean-squared error.* If the expectation of a point estimator is not equal to the parameter being estimated, it is said to be *biased.* The variance of an estimator is a measure of its precision, or spread of values of the estimates about the mean of the estimates. The mean square error is a measure of its total error: precision and bias.
3. The *maximum likelihood technique* for finding an estimate answers the question: What value of the parameter is most likely in the light of the samples taken from the population? It is that value of the parameter that maximizes the joint pdf of the samples, given the parameter, with a particular set of samples substituted in this joint pdf. It is found by maximizing the *likelihood function,* which is the conditional pdf of the samples given the parameter:

$$L(\theta) = f_n(x_1, x_2, \ldots, x_n \mid \theta) = \prod_{i=1}^{n} f_i(x_i \mid \theta)$$

In obtaining the product form on the right, it is assumed that the samples are statistically independent. Quite often, it is easier to maximize the logarithm of the likelihood function, which is permissible since the logarithm is a monotonically increasing function. Thus a necessary condition for the maximum likelihood estimate of a parameter is

$$\left.\frac{d}{d\theta} \ln L(\theta)\right|_{\theta=\hat{\theta}} = \left.\frac{1}{L(\theta)} \frac{dL(\theta)}{d\theta}\right|_{\theta=\hat{\theta}} = 0$$

4. An interval estimate satisfies a probability statement such as

$$P(\theta_1 < \theta < \theta_2) = 1 - \alpha, \qquad \alpha > 0$$

In this equation, $1 - \alpha$ is called the *confidence coefficient,* α is called the *confidence level,* and the range (θ_1, θ_2) is called the *confidence interval,* with the endpoints of this interval known as the *confidence limits.*
5. To find the confidence limits for a given confidence level, we must be able to find the pdf of the estimator. This is often difficult to do. We were able to do it for the sample mean, the sample variance knowing the mean, the sample variance with the mean unknown, and the sample mean with the variance unknown.
6. Maximum likelihood estimators have the following properties in the limit of large sample sizes:
   (a) They are Gaussian.
   (b) They are efficient, or satisfy the Cramér–Rao lower bound for variance with equality:

$$\text{Var}\,(\hat{\theta}) = E[\hat{\theta}(X_1, X_2, \ldots, X_n)^2] \geq \frac{1}{nI}$$

where

$$I = E\left[\left|\frac{\partial}{\partial\theta} \ell(X_1, X_2, \ldots, X_n; \theta)\right|^2\right]$$

or

$$I = -E\left[\frac{\partial^2}{\partial\theta^2}\ell(X_1, X_2, \ldots, X_n; \theta\right]$$

where $\ell(\bullet)$ denotes the log-likelihood function.

7. Any estimator satisfying the Cramér–Rao inequality with equality is called an *efficient estimator*. If the joint pdf of the samples given the parameter is of the exponential type, the maximum likelihood estimate will be efficient. An exponential type pdf is of the form

$$f(x_1, x_2, \ldots, x_n; \theta) = h(x_1, x_2, \ldots, x_n)e^{a(\theta)g(x_1, x_2, \ldots, x_n) - b(\theta)}$$

where the functions $a(\theta)$ and $b(\theta)$ are functions of the unknown parameter only, and the functions $h(\bullet)$ and $g(\bullet)$ are functions of the samples only.

8. Given two random variables X and Y that are dependent, the *minimum mean-squared-error estimate* of Y based on an observation of X, denoted x, is

$$\theta(x) = E(Y|x) = \int_{-\infty}^{\infty} y f_{Y|x}(y|x)\,dy$$

9. Minimum mean-squared-error estimates satisfy the *orthononality condition*

$$E\{[Y - \theta(X)]\,g(X)\} = 0$$

where $g(X)$ is any function of the random variable X.

6–9 FURTHER READING

The books by Williams (1991) and Papoulis (1990) are particularly appropriate as additional reading for the material presented in this chapter. A careful treatment of interval estimators at about this level but with broader scope may be found in Walpole and Meyers (1993). Proof of the Cramér–Rao lower bound can be found in Cramér (1946) or Van Trees (1968). These are more advanced books than those referenced so far. In particular, the proof makes use of Schwartz's inequality, which although not hard to understand, has not been introduced in this book.

6–10 PROBLEMS

Section 6-2

6-1. The median is often a better estimator for data sets that have a very few samples that are far away from the majority of samples. Such samples are called outliers.

(a) Compare the sample mean and median for the following data set. Recall that to find the median, the samples are placed in ascending order and the middle sample is chosen as the median. Comment on your results.

i	1	2	3	4	5	6	7	8	9	10	11	12	13
X_i	2.0	13.4	10.4	7.1	10.9	13.6	8.4	1.0	12.8	10.7	10.3	6.8	9.4

(b)Add the sample 1000 to the data set of part (a) and recompute the sample mean and the median. Which appears to be more representative of the average data sample?

6-2. (a) The variance of a certain estimate is 4 and its bias is 2. What is its mean-squared error?
(b) The mean-squared error of an estimate is 10 and its bias is 3. What is its variance? Its standard deviation?

6-3. The standard deviation of the sample mean for 10 samples is 10. How many samples must be taken to make the standard deviation 0.1?

6-4. The median m is defined by the equation

$$F_X(m) = 0.5$$

where $F_X(x)$ is the cdf. Using the definition of the empirical cdf defined in Chapter 5, find an estimator for the median.

6-5. Show that the median m minimizes

$$E(|X - m|)$$

by showing that

$$E(|X - a|) = E(|X - m|) + 2\int_a^m (x - a) f_X(x)\,dx, \qquad a < m$$

and

$$E(|X - a|) = E(|X - m|) + 2\int_a^m (a - x) f_X(x)\,dx, \qquad a > m$$

6-6. Show that the estimator for the variance given by (6-22b) is unbiased, assuming statistically independent samples.

6-7. Show that the following estimator for the covariance, assuming statistically independent samples, is unbiased:

$$\hat{C}(X, Y) = \frac{n \sum_{i=1}^{n} X_i Y_i - \sum_{i=1}^{n} X_i \sum_{i=1}^{n} Y_i}{n(n-1)}$$

6-8. Annual salaries for a certain small company are as follows:
20 assemblers at \$15,000 each
5 junior engineers at \$35,000 each
1 senior engineer at \$60,000
1 owner at \$300,000
In the annual report, a "typical" salary is given. Should the sample mean or median salary be used for this purpose?

6-9. The number of customers in a checkout line at a supermarket is suspected to obey a Poisson distribution given by

$$P_T(k) = \frac{(\lambda T)^k}{k!} e^{-\lambda T}, \qquad k = 0, 1, 2, \ldots$$

What would an intuitive estimator for the parameter λ of the distribution be? Indicate specifically how you would estimate it.

Section 6-3

6-10. Referring to Problem 6-9, given that 10 customers arrive in line in 15 minutes, what is a maximum likelihood estimate for λ?

6-11. A coin is flipped 100 times in 10 different experiments. It is not known whether or not the coin is fair. The following numbers of heads are obtained in these 10 experiments of 100 flips each.

(a) Find the maximum likelihood estimate for the probability p of a head based on the data for each experiment.

(b) Compute the sample mean of the estimates.

(c) Compute the sample variance according to (6-22b).

Experiment	Number of Heads
1	57
2	63
3	59
4	70
5	65
6	52
7	61
8	49
9	55
10	67

6-12. Find the bias of the estimator for Problem 6-11. Find its variance.

6-13. From the result of Problem 6-6, conclude that the bias for the estimator of (6-22b) is zero. Therefore, the estimator of (6-22a) cannot be unbiased. What is the bias for the estimator of (6-22a)?

6-14. Obtain and plot the empirical cdf's for the sample mean and sample variance for the data of Example 6-4 based on the statistics given in Table 6-1. Does either appear Gaussian? Which, if either?

6-15. The following data were obtained from a population known to have a mean of 15 and variance of 9.

(a) Compute the sample mean.

(b) What is the expectation of the sample mean? What is its variance?

i	1	2	3	4	5	6	7	8	9	10
X_i	8.81	15.91	20.39	8.63	13.80	21.73	27.23	7.32	14.83	20.26

Section 6-4

6-16. Refer to the samples of Problem 6-15, which are known to have come from a Gaussian population. Find the 95% confidence interval for the mean.

6-17. Repeat Problem 6-16 for the variance assuming that the mean is known.

6-18. Repeat Problem 6-16 for the variance assuming that the mean is unknown.

6-19. Repeat Problem 6-16 for the mean assuming that the variance is unknown.

6-20. Of 500 voters asked whether they are going to vote Republican, 260 said yes.
(a) Find the 95% confidence interval of the probability p that the vote for a certain office will be for the Republican candidate.
(b) Suppose that 475 of 500 voters say they will vote Democrat. Repeat part (a).

Section 6-5

6-21. Is the maximum likelihood estimator found in Problem 6-10 unbiased? Is it efficient?

6-22. Is the maximum likelihood estimator for the variance of a Gaussian pdf with the mean known (Example 6-3) efficient?

Section 6-6

6-23. In a certain classroom of 35 students, the weights and heights of all students are recorded. Their heights are modeled by the relation

$$H = \frac{W}{16} + \delta \text{ feet}, \qquad W \text{ in pounds}$$

where δ is a Gaussian random variable with the pdf

$$f_\Delta(\theta) = \frac{e^{-\delta^2\ 0.4}}{\sqrt{0.4\pi}}$$

(a) Given the weight of a student, find the conditional pdf of his or her height.
(b) A student weighs 90 pounds. Find the minimum mean-squared error estimate for the student's height.
(c) What is the mean-square error of the estimate found in part (b)? What is the root mean-squared error?

6-24. Refer to the statement of Example 6-11 to work this problem. The signal Y has mean 10 and variance 5. The noise has variance 25. How many independent samples should be taken to make the mean-squared error of the estimate 1? Give an expression for the conditional mean estimate for the parameters given here and the number of samples that you found. Is it more dependent on the sample mean or on the mean of the signal?

Section 6-7

6-25. Given the following moments for two random variables: $E(X) = 2$, $E(X^2) = 5$, $E(Y) = 1$, and $E(XY) = -3$. A minimum mean-square estimate of the form $\theta(X) = AX + B$ is to be found. With the values for the moments given, what should the values of A and B be to make it so?
Hint: Use the orthogonality principle, first with $g(X) = 1$ and then with $g(X) = X$, to get two equations in the two unknowns A and B.

6–11 COMPUTER EXERCISES

6-1. Write a computer program to generate N samples according to a Gaussian distribution with mean m and variance σ^2 (these should all be inputs to the program). Perform the following estimates:
(a) The mean
(b) The variance assuming the mean is known

(c) The variance assuming the mean is unknown

Do this for several choices of the parameters and draw any conclusions possible.

6-2. Write for a computer program to generate results such as those shown in Table 6-1. Replicate the computation of the sample mean several times and plot its histogram. Do for 10 and 100 sample batches and compare the histogram with Figure 6-1 after proper normalization by the total number of samples and bin width.

6-3. Design a Monte Carlo simulation to check the 95% confidence intervals for the sample mean as computed in Example 6-4. That is, your program should repeatedly generate the sample mean for 10 sample batches and determine the number of times that the sample mean falls within the limits as computed following (6-28). Use a mean of 10 and variance of 4 for the samples as in Table 6-1.

6-4. Repeat Computer Exercise 6-3 for the variance with mean known and check the confidence limits found in Example 6-5.

6-5. Repeat Computer Exercise 6-3 for the variance with mean unknown and check the confidence limits found in Example 6-7.

6-6. Repeat Computer Exercise 6-3 for the mean with variance unknown and check the confidence limits found in Example 6-8.

6-7. Design a Monte Carlo simulation of Example 6-11. All parameters given in the example should be inputs to your program. The equation implemented should be (6-75). By doing an estimate for the variance (mean-squared error) of the estimate, determine if it is close to σ_p^2 as given by (6-74).

7

Engineering Decisions

7-1 INTRODUCTION

Given a set of possible alternatives, it is important to make an informed decision as to the course to follow. Supposedly, each alternative will have a cost associated with it. Obviously, the decision maker desires to make the decision that costs the least on average. This is the type of problem that we address in this chapter. We begin the chapter by formalizing a decision-making process, in probabilistic terms, that minimizes the average cost of making a decision. This formal development, called the *Bayes decision strategy,* is followed with an example based on detection of a signal in noise to illustrate the decision-making methodology.

Another type of decision problem, deciding between acceptance or rejection of a parameter value, is referred to as *classical decision theory.* Its purpose is similar to the theory developed in Chapter 6, except that now we only want to decide between a parameter value or an alternative hypothesis that rejects the hypothesized parameter value.

In addition to the goodness criterion of minimum average cost, the Bayes decision strategy requires prior probabilities for the various hypotheses, an enumeration of the possible choices, and costs associated with each possible decision given the prior state of affairs. In many cases one or more of these ingredients may be missing. An alternative decision-making strategy that does not require the prior probabilities of the hypotheses, the *Neyman–Pearson criterion,* is discussed next. Another decision strategy not requiring the prior probabilities is the maximum likelihood technique. This approach is especially easy to generalize to more than two hypotheses. The chapter ends with a discussion of the maximum likelihood approach to decision making.

7–2 BAYES' DECISION STRATEGY

Several ingredients are necessary for decision making:

1. The possible actions or hypotheses from which to choose
2. The possible ways that outside influences may enter the problem
3. A criterion to be used in choosing the best course of action to follow

For example, in deciding whether or not a signal is present in a noisy background, the possible hypotheses from which to choose may be "signal not present" or "signal present." The outside influence present in such a problem is the noise component, which obscures the signal. A possible criterion for choosing the course of action to follow is minimum average cost. This assumes that costs can be assigned to each possible hypothesis–decision pair. The procedure of minimizing average cost leads to what is called a *decision rule.* In the signal detection example just mentioned, the decision rule is "choose the hypothesis 'no signal' if the receiver output is less than the threshold, and choose the hypothesis 'signal present' if the receiver output is greater than the threshold."

In this section, we formalize mathematically the procedure for making a choice between two alternatives using the criterion of minimum average cost. We refer to these two alternatives as hypothesis 0 (H_0) and hypothesis 1 (H_1). The resulting decision strategy is called *Bayes' hypothesis testing.*

We define four costs of making a decision:

c_{00}: the cost of deciding in favor of H_0 when H_0 is actually true
c_{01}: the cost of deciding in favor of H_0 when H_1 is actually true
c_{10}: the cost of deciding in favor of H_1 when H_0 is actually true
c_{11}: the cost of deciding in favor of H_1 when H_1 is actually true

Given that H_0 was actually true, the conditional average cost of making a decision, $C(D|H_0)$, is

$$C(D|H_0) = c_{00}P(\text{decide } H_0|H_0 \text{ true}) + c_{10}P(\text{decide } H_1|H_0 \text{ true}) \tag{7-1}$$

Similarly, given that H_1 was actually true, the conditional average cost of making a decision, $C(D|H_1)$, is

$$C(D|H_1) = c_{01}P(\text{decide } H_0|H_1 \text{ true}) + c_{11}P(\text{decide } H_1|H_1 \text{ true}) \tag{7-2}$$

We represent the data that we observe by the random variable Z with pdf $f_Z(z|H_i)$, $i = 0, 1$, depending on H_0 or H_1 being true, respectively. Also, we suppose that we make the decision in favor of H_0 if Z falls into the region R_0 on the real line; conversely, we make the decision in favor of H_1 if Z falls into the region R_1 on the real line. Given that H_0 is really true, we may write the probability of making a correct decision as

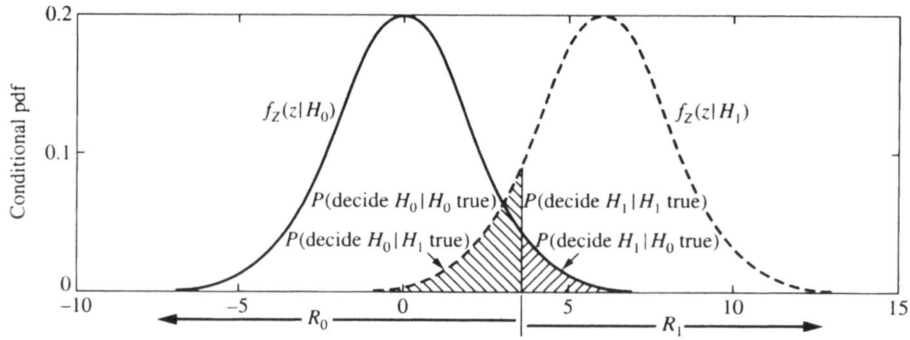

Figure 7-1 Conditional pdf's $f_Z(z|H_1)$ and $f_Z(z|H_1)$, along with the various probabilities of making right and wrong decisions.

$$P(\text{decide } H_0|H_0 \text{ true}) = \int_{R_0} f_Z(z|H_0)\,dz \tag{7-3}$$

where $f_Z(z|H_0)$ is the pdf of Z given H_0 was true. This conditional pdf and the corresponding probability given by (7-3) are shown in Figure 7-1.

On the other hand, the probability of making a wrong decision, given that H_0 is really true, is

$$P(\text{decide } H_1|H_0 \text{ true}) = \int_{R_1} f_Z(z|H_0)\,dz \tag{7-4}$$

Now either Z must lie in the region R_0 or the region R_1 since we are forced to make a decision. Thus (7-4) can be written as

$$P(\text{decide } H_1|H_0 \text{ true}) = \int_{R_1} f_Z(z|H_0)\,dz = 1 - \int_{R_0} f_Z(z|H_0)\,dz \tag{7-5}$$

Similarly, given that hypothesis H_1 is true, we can write that

$$P(\text{decide } H_0|H_1 \text{ true}) = \int_{R_0} f_Z(z|H_1)\,dz \tag{7-6}$$

for the probability of making a wrong decision, and

$$P(\text{decide } H_1|H_1 \text{ true}) = \int_{R_1} f_Z(z|H_1)\,dz = 1 - \int_{R_0} f_Z(z|H_1)\,dz \tag{7-7}$$

for the probability of making a correct decision. The conditional pdf $f_Z(z|H_1)$ and the probabilities (7-5) to (7-7) are illustrated in Figure 7-1.

Assume that hypothesis H_0 happens with probability p and that hypothesis H_1 happens with probability $q = 1 - p$. The overall average cost of making a decision is the average of the conditional costs given by (7-1) and (7-2), which is

$$C(D) = p[c_{00}P(\text{decide } H_0|H_0 \text{ true}) + c_{10}P(\text{decide } H_1|H_0 \text{ true})]$$

$$+ q[c_{01}P(\text{decide } H_0|H_1 \text{ true}) + c_{11}P(\text{decide } H_1|H_1 \text{ true})] \tag{7-8}$$

When (7-4) to (7-7) are substituted, (7-8) becomes

$$\begin{aligned} C(D) &= p\left\{c_{00}\int_{R_0} f_Z(z \mid H_0) + c_{10}\left[1 - \int_{R_0} f_Z(z \mid H_0)\, dz\right]\right\} \\ &+ q\left\{c_{01}\int_{R_0} f_Z(z|H_1)\, dz + c_{11}\left[1 - \int_{R_0} f_Z(z|H_1)\, dz\right]\right\} \end{aligned} \tag{7-9}$$

Collection of all terms involving integration over R_0 results in the expression

$$C(D) = (pc_{10} + qc_{11}) + \int_{R_0} \{[q(c_{01} - c_{11})f_Z(z|H_1)] - [p(c_{10} - c_{00})f_Z(z|H_0)]\}\, dz \tag{7-10}$$

The first term is fixed. We assume that wrong decisions are more costly than right decisions, which is only reasonable. Thus $c_{01} > c_{11}$ and $c_{10} > c_{00}$, and the two bracketed terms inside the integral of (7-10) are positive because probabilities and pdf's are positive. To minimize the overall expression, we adopt the following strategy: if a realization of the random variable Z is such that the first bracketed term in (7-10) is smaller than the second bracketed term, it is left in region R_0, whereas if it is such that the first bracketed term in (7-10) is larger than the second bracketed term, it is left out of region R_0 (i.e., assigned to R_1). Thus the values left in the integral make it negative and therefore minimize the overall expression. This procedure can be summarized by the *decision rule*

$$q(c_{01} - c_{11})f_Z(Z|H_1) \underset{H_0}{\overset{H_1}{\gtrless}} p(c_{10} - c_{00})f_Z(Z|H_0) \tag{7-11}$$

When rearranged, this becomes

$$\frac{f_Z(Z|H_1)}{f_Z(Z|H_0)} \underset{H_0}{\overset{H_1}{\gtrless}} \frac{p(c_{10} - c_{00})}{q(c_{01} - c_{11})} \tag{7-12}$$

This is interpreted as follows: if an observed value for Z (in this instance, a lower case z would be used) results in the left-hand ratio of pdf's being greater than the right-hand ratio of constants, choose H_1; if not, choose H_0. The left-hand ratio of pdf's is called the *likelihood ratio,* and the right-hand ratio of constants is called the *threshold* of the test. This testing procedure is called *Bayes' decision making.*

7-3 DECISION-MAKING EXAMPLES

In this section we consider two related examples to illustrate the decision-making process according to Bayes' hypothesis testing.

Example 7-1

A signal of 10 V is sent to a distant receiver. In the process, a noise component, which is represented by a Gaussian random variable with mean zero and variance 4 V^2, is added to the signal. The signal is detected by comparing the received signal plus noise with a threshold of 5 V. If greater than the threshold, it is decided that the signal was indeed sent. If less than the threshold, it is decided that the signal was not sent. The cost of making a decision is 0 if the correct decision is made (i.e., decide signal present when it is indeed present, or decide signal absent when it is indeed absent). The cost of making a decision is 1 if the incorrect decision is made (i.e., decide signal is present when it is indeed absent, or decide signal absent when it is indeed present). The probabilities of sending or not sending a signal are equally likely, or $\frac{1}{2}$. Find the average cost of making a decision. Can it be made smaller by using a different threshold?

Solution A general expression for the average cost of making a decision is given by

$$C = \frac{c_{00}}{2} P\,(\text{say no signal} \mid \text{no signal sent}) + \frac{c_{01}}{2} P\,(\text{say no signal} \mid \text{signal sent}) + \frac{c_{10}}{2} P\,(\text{say signal} \mid \text{no signal sent}) + \frac{c_{11}}{2} P\,(\text{say signal} \mid \text{signal sent}) \tag{7-13a}$$

Because of the costs given, this simplifies to

$$C = \frac{1}{2} P\,(\text{say no signal} \mid \text{signal sent}) + \frac{1}{2} P\,(\text{say signal} \mid \text{no signal sent}) \tag{7-13b}$$

The received signal plus noise is

$$Z = 10 + N \tag{7-14}$$

where N is a Gaussian random variable representing the added noise, having a mean of 0 and a variance of 4. Thus its pdf is

$$f_N(x) = \frac{e^{-x^2/8}}{\sqrt{8\pi}} \tag{7-15}$$

The probability of deciding no signal when a signal was indeed sent is the probability that $10 + N < 5$, where 5 is the threshold, or $N < -5$. In terms of the pdf given by (7-15), this is

$$P\,(\text{say no signal} \mid \text{signal sent}) = \int_{-\infty}^{-5} \frac{e^{-x^2/8}}{\sqrt{8\pi}}\,dx = \int_{5}^{\infty} \frac{e^{-x^2/8}}{\sqrt{8\pi}}\,dx \quad \text{(by symmetry)}$$
$$= \int_{2.5}^{\infty} \frac{e^{-u^2/2}}{\sqrt{2\pi}}\,du = Q(2.5) \tag{7-16}$$

Similarly, the probability of deciding signal when no signal was sent is the probability that $N > 5$, which gives the same expression as (7-16). Thus the average cost is

$$C = Q(2.5) = 0.00621 \tag{7-17}$$

Letting the threshold be a variable, T, we obtain

$$P\,(\text{say no signal} \mid \text{signal sent}) = \int_{-\infty}^{T-10} \frac{e^{-x^2/8}}{\sqrt{8\pi}}\,dx = \int_{(10-T)/2}^{\infty} e^{-u^2/2}\,du = Q\left(\frac{10 - T}{2}\right) \tag{7-18}$$

and

$$P\,(\text{say signal} \mid \text{no signal sent}) = \int_{T}^{\infty} \frac{e^{-x^2/8}}{\sqrt{8\pi}}\,dx = \int_{T/2}^{\infty} \frac{e^{-u^2/2}}{\sqrt{2\pi}}\,du = Q\left(\frac{T}{2}\right) \tag{7-19}$$

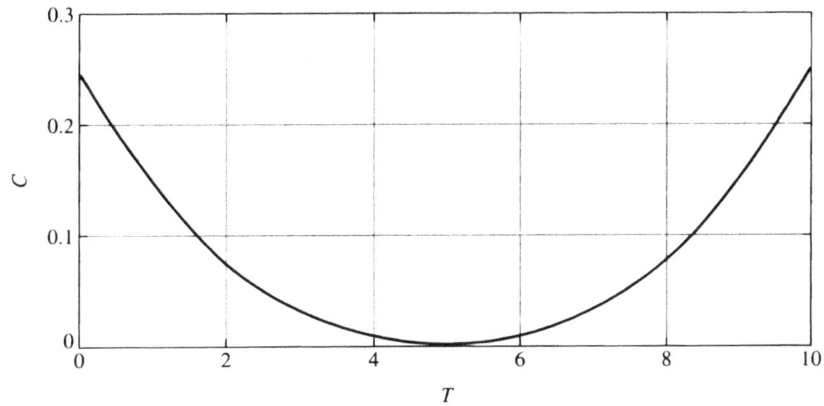

Figure 7-2 Average cost per decision versus the threshold for a signal detection problem (costs for correct decisions are 0 and costs for incorrect decisions are equal).

The average cost of making a decision is

$$C(T) = \frac{1}{2}Q\left(\frac{10 - T}{2}\right) + \frac{1}{2}Q\left(\frac{T}{2}\right) \tag{7-20}$$

This average cost is plotted as a function of the threshold, T, in Figure 7-2. Note that $T = 5$ gives the minimum average cost.

Example 7-2

We now generalize Example 7-1 to the case where we are trying to detect a signal of arbitrary amplitude, A, in additive Gaussian noise, N, with mean zero and variance σ^2. The costs are the same as in Example 7-1 (i.e., $c_{00} = c_{11} = 0$ and $c_{01} = c_{10} = 1$). If signal is present, the input to the receiver is

$$Z = A + N \tag{7-21}$$

which is Gaussian with mean A and variance σ^2. Thus the conditional pdf's appearing in (7-12) are

$$f_Z(z \mid H_1) = \frac{e^{-(z-A)^2/2\sigma^2}}{\sqrt{2\pi\sigma^2}} \tag{7-22}$$

if signal is present, and

$$f_Z(z \mid H_0) = \frac{e^{-z^2/2\sigma^2}}{\sqrt{2\pi\sigma^2}} \tag{7-23}$$

if signal is absent [note that it is immaterial which hypothesis we identify as H_0 and which we identify as H_1, but recall that $p = P(H_0)$]. The likelihood ratio is

$$\frac{f_Z(Z \mid H_1)}{f_Z(Z \mid H_0)} = \frac{e^{-(Z-A)^2/2\sigma^2}/\sqrt{2\pi\sigma^2}}{e^{-Z^2/2\sigma^2}/\sqrt{2\pi\sigma^2}} = e^{(2AZ-A^2)/2\sigma^2} \tag{7-24}$$

We leave the prior probabilities of the two possible hypotheses arbitrary as p (for $Z = N$) and $q = 1 - p$ (for $Z = A + N$) for now. Bayes' test then becomes

$$e^{(2AZ-A^2)/2\sigma^2} \underset{H_0}{\overset{H_1}{\gtrless}} \frac{p}{q} \tag{7-25a}$$

We can take the natural logarithm of both sides and not change the test since the logarithm is a monotonically increasing function of its argument. Doing this, (7-25a) becomes

$$\frac{2AZ - A^2}{2\sigma^2} \underset{H_0}{\overset{H_1}{\gtrless}} \ln\frac{p}{q}$$

or

$$Z \underset{H_0}{\overset{H_1}{\gtrless}} \frac{\sigma^2}{A} \ln\frac{p}{q} + \frac{A}{2} \tag{7-25b}$$

If $p = q = \frac{1}{2}$, the logarithm is zero and we get the result shown in Figure 7-2, where it was demonstrated that the best threshold is one-half the value of the signal amplitude. As p becomes larger (and q smaller), the decision boundary moves to the right, reflecting that the signal-absent hypothesis should be chosen with greater probability, since the signal-absent case is more probable before an observation of the data is made.

The average cost of making a decision is obtained by generalizing (7-2) to

$$C(D) = q\,P\,(\text{say no signal} \mid \text{signal sent}) + p\,P\,(\text{say signal} \mid \text{no signal sent}) \tag{7-26}$$

which is a special case of the average cost (7-8) because $c_{00} = c_{11} = 0$ and $c_{01} = c_{10} = 1$. The probability of saying that no signal is present when one is, in fact, present is the probability that Z is less than the right-hand side of (7-25b) given the pdf (7-22). This is called the probability of a miss and can be expressed as

$$P_{\text{miss}} = \int_{-\infty}^{T} \frac{e^{-(z-A)^2/2\sigma^2}}{\sqrt{2\pi\sigma^2}}\,dz = \int_{(A-T)/\sigma}^{\infty} \frac{e^{-u^2/2}}{\sqrt{2\pi}}\,du = \begin{cases} Q\left(\dfrac{A-T}{\sigma}\right), & T < A \\ 1 - Q\left(\dfrac{T-A}{\sigma}\right), & T > A \end{cases} \tag{7-27}$$

where T is given by the right-hand side of (7-25b) and the change of variables

$$u = \frac{z - A}{\sigma} \tag{7-28}$$

has been made to get from the first integral to the second one. The symmetry of the Gaussian pdf about zero was used in simplifying (7-27). The probability of saying that the signal was present when no signal was, in fact, present is the probability that Z exceeds the right-hand side of (7-25b) given the pdf (7-23). This is termed the *probability of a false alarm* and is given by

$$P_{FA} = \int_{T}^{\infty} \frac{e^{-z^2/2\sigma^2}}{\sqrt{2\pi\sigma^2}}\,dz = \int_{T/2\sigma}^{\infty} \frac{e^{-u^2/2}}{\sqrt{2\pi}}\,du = \begin{cases} Q\left(\dfrac{T}{\sigma}\right), & T > 0 \\ 1 - Q\left(\dfrac{|T|}{\sigma}\right), & T < 0 \end{cases} \tag{7-29}$$

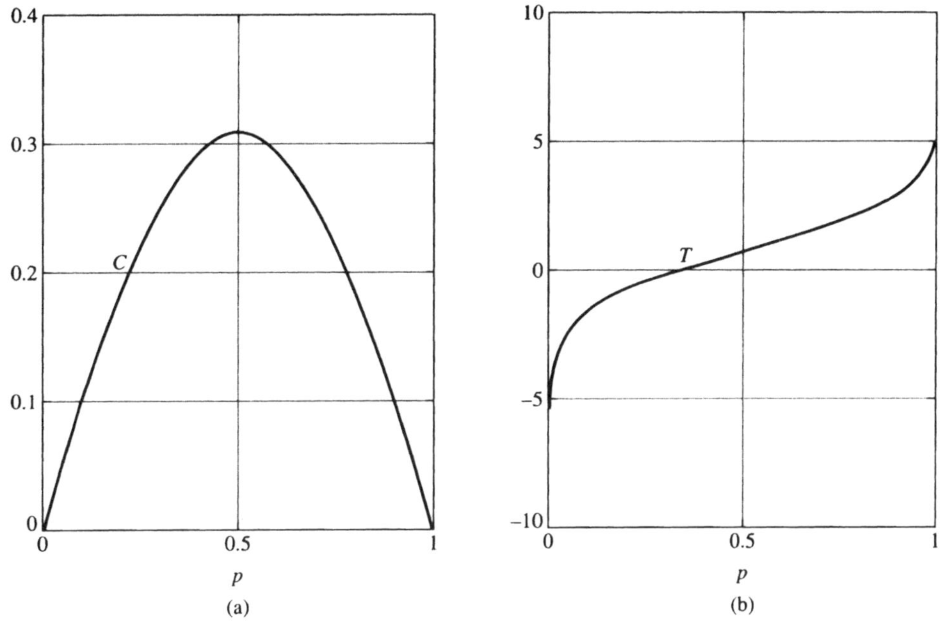

Figure 7-3 Average cost of a decision and optimum threshold for the signal detection problem of Example 7-2.

The average cost of a decision, which is the average probability of making an error in this example, is plotted in Figure 7-3a as a function of p along with a plot of the threshold shown in Figure 7-3b. Note that the average cost is a maximum for $p = \frac{1}{2}$ and that the optimum threshold for this value of p is one-half the signal amplitude, which was 1 in this case [see the right-hand side of (7-25b)]. This graphically verifies the result of Example 7-1 shown in Figure 7-2, which can also be shown by differentiation of (7-26) with respect to T with (7-27) and (7-29) substituted.

7-4 CLASSICAL DECISION THEORY

We now look at hypothesis-testing procedures where we wish to make a decision regarding the reasonableness of a statement about the value of a parameter. Two types of errors are possible in such a decision process:

1. We can say that the hypothesis is false when it is really correct (this is the probability of a miss in Examples 7-1 and 7-2).
2. We can say that the hypothesis is correct when it is really false (this is the probability of a false alarm in Examples 7-1 and 7-2).

The first is called a *type I error* and its probability is denoted by α; the second is called a *type II error* and its probability is given the symbol β.

The test is based on the underlying pdf of the statistic used to estimate the parameter and may be of the form

$$\begin{aligned} H_0&: \Theta = \theta_0 \\ H_1&: \Theta \neq \theta_0 \end{aligned} \tag{7-30}$$

where H_0 is known as the *null hypothesis* and H_1 is called the *alternative hypothesis.* In this case the alternative hypothesis is called a *two-sided alternative hypothesis* because it specifies values that Θ could be either greater than or less than θ_0. We may also formulate *one-sided alternative hypotheses* of the form

$$\begin{aligned} H_0&: \Theta = \theta_0 \\ H_1&: \Theta < \theta_0 \end{aligned} \tag{7-31a}$$

or

$$\begin{aligned} H_0&: \Theta = \theta_0 \\ H_1&: \Theta > \theta_0 \end{aligned} \tag{7-31b}$$

It is important to emphasize that hypotheses are statements about the population under consideration, not the sample.

In terms of the null and alternative hypotheses, the probabilities of the type I and type II errors discussed above can be expressed as

$$\begin{aligned} \alpha &= P\,(\text{type I error}) = P\,(\text{reject } H_0|H_0 \text{ is true}) = P\,(\text{accept } H_1|H_0 \text{ is true}) \\ \beta &= P\,(\text{type II error}) = P\,(\text{accept } H_0|H_0 \text{ is false}) = P\,(\text{accept } H_0|H_1 \text{ is true}) \end{aligned} \tag{7-32}$$

Sometimes α is called the *significance level* or *size* of the test, and $1 - \beta = P(\text{accept } H_1|H_1 \text{ true})$ is called the *power* of the test. The value of β depends on the true parameter value and is therefore a conditional probability. The power of the test plotted against the true parameter value is called the *operating characteristic* (OC) *curve.*

Example 7-3

Given the random variable X with mean μ. Design a test for the hypotheses

$$\begin{aligned} H_0&:\mu = \mu_0 \\ H_1&:\mu \neq \mu_0, \quad \mu > \mu_0 \text{ or } \mu < \mu_0 \end{aligned} \tag{7-33}$$

Assume that we take n samples from the population, which are assumed Gaussian with known variance σ^2.

Solution Form the test statistic

$$Y = \frac{\bar{x} - \mu_0}{\sigma/\sqrt{n}} \tag{7-34}$$

where $\bar{x}$ is the sample mean, defined as

$$\bar{x} = \frac{1}{n}\sum_{i=1}^{n} x_i \tag{7-35}$$

Under hypothesis H_0, the random variable Y is Gaussian with mean zero and unit variance. We set the probability of the type I error at some low acceptable value, denoted by α, and find the boundaries such that

$$P(c_1 < Y < c_2) = 1 - \alpha = 2P(0 \le Y < c)$$
$$= 2[1 - P(Y \ge c)] = 1 - 2Q(c) \tag{7-36}$$

where it is assumed that $c_2 = |c_1| = c$. The region bounded by c_1 and c_2 is called the *critical region.* Using the fact that Y is a zero-mean, unit-variance Gaussian, random variable (7-36) becomes

$$\int_0^c \frac{e^{-u^2/2}}{\sqrt{2\pi}}\,du = 0.5 - Q(c) = \frac{1-\alpha}{2}$$

or

$$c = Q^{-1}\left(\frac{\alpha}{2}\right) \tag{7-37}$$

To find the OC curve, assume that hypothesis H_0 is false (H_1 is true). In this case, $\bar{x}$ is Gaussian with mean $\mu \neq \mu_0$ and variance σ^2/n, or, from (7-34), Y is Gaussian with mean

$$\frac{\mu - \mu_0}{\sigma/\sqrt{n}}$$

and unit variance. Thus

$$\beta(\mu) = P(c_1 < Y < c_2 | H_1) = P(-c < Y < c | H_1) = \int_{-c}^{c} \frac{e^{-(1/2)[y-(\mu-\mu_0)/(\sigma/\sqrt{n})]^2}}{\sqrt{2\pi}}\,dy \tag{7-38}$$

Making the change of variables

$$\zeta = y - \frac{\mu - \mu_0}{\sigma/\sqrt{n}} \tag{7-39}$$

and using the definition of the Q-function, (3-29), we obtain

$$\beta(\mu) = \int_{-c-(\mu-\mu_0)/(\sigma/\sqrt{n})}^{c-(\mu-\mu_0)/(\sigma/\sqrt{n})} \frac{e^{-\zeta^2/2}}{\sqrt{2\pi}}\,d\zeta = \int_{-\infty}^{c-(\mu-\mu_0)/(\sigma/\sqrt{n})} \frac{e^{-\zeta^2/2}}{\sqrt{2\pi}}\,d\zeta - \int_{-\infty}^{-c-(\mu-\mu_0)/(\sigma/\sqrt{n})} \frac{e^{-\zeta^2/2}}{\sqrt{2\pi}}\,d\zeta$$
$$= \int_{-\infty}^{c-(\mu-\mu_0)/(\sigma/\sqrt{n})} \frac{e^{-\zeta^2/2}}{\sqrt{2\pi}}\,d\zeta - \int_{c+(\mu-\mu_0)/(\sigma/\sqrt{n})}^{\infty} \frac{e^{-\zeta^2/2}}{\sqrt{2\pi}}\,d\zeta \tag{7-40}$$
$$= 1 - Q\left(c - \frac{\mu - \mu_0}{\sigma/\sqrt{n}}\right) - Q\left(c + \frac{\mu - \mu_0}{\sigma/\sqrt{n}}\right)$$

where symmetry of the Gaussian pdf has been used.

Example 7-4

Suppose a manufacturing line produces rods that are supposed to be 10 cm in length. Ten rods are taken from the production line and measured, with the following results: 9.86, 9.90, 9.93, 9.95, 9.96, 9.97, 9.98, 10.01, 10.02, and 10.04 cm. Assume that the measurements are the actual length, μ, of the bar plus a Gaussian measurement error with mean zero and variance 0.1. Test the hypothesis that the bar is 10 cm long versus the hypothesis that it is not 10 cm long at a 5% significance level.

Solution The hypotheses are

$$H_0\text{: } \mu = \mu_0 = 10 \text{ cm} \tag{7-41}$$

$$H_1: \mu \neq 10 \text{ cm}$$

The sample mean is found to be

$$\bar{x} = 9.962 \tag{7-42}$$

and the normalized random variable Y is

$$Y = \frac{\bar{x} - \mu_0}{\sigma/\sqrt{n}} = -1.202 \tag{7-43}$$

For $\alpha = 0.05$, the boundary for the confidence interval is

$$c = 1.86 \tag{7-44}$$

Since $-c < Y < c$ we accept the hypothesis H_0 that the rod is 10 cm long. The OC curve, from (7-40), is

$$1 - \beta(\mu) = Q\left(1.96 - \frac{\mu - 10}{\sqrt{0.1}/\sqrt{10}}\right) + Q\left(1.96 + \frac{\mu - 10}{\sqrt{0.1}/\sqrt{10}}\right) \tag{7-45}$$

A plot of the OC curve is given in Figure 7-4. Note that the probability that we accept H_1 given H_1 is true $(1 - \beta)$ is low (≈ 0.5) for the true value of the parameter μ close to 10. However, we aren't as concerned with making a type II error as long as the mean is close to the hypothesized value as we are in detecting large differences between the true mean and the value specified in the null hypothesis. The probability of accepting H_1 given that H_1 is true becomes larger rapidly (i.e., approaches 1) as μ departs from the hypothesized value.

From (7-40) we note that the type II error depends on the sample size n as well. This is illustrated in Figure 7-5, where the OC curve for Example 7-4 is shown for a sample size of 50. With the larger sample size, the type II error decreases more rapidly than for the smaller size as the true mean departs from the hypothesized value, μ_0.

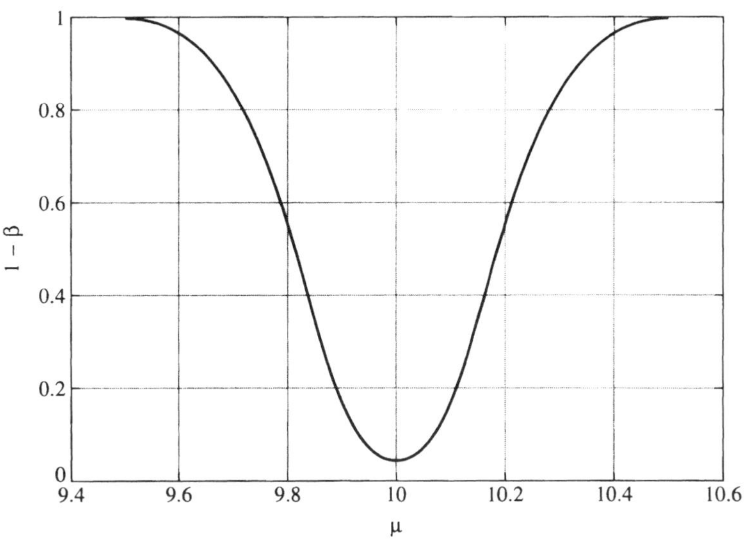

Figure 7-4 OC curve for Example 7-4 with a sample size of 10.

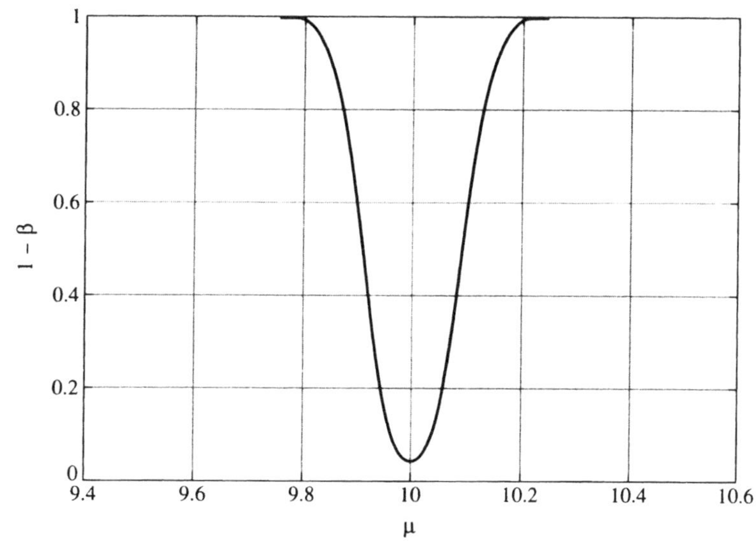

Figure 7-5 OC curve for Example 7-4 with a sample size of 50.

The probability of the type I error can be controlled by the analyst by choosing the parameter c. Hence, wrongly rejecting H_0 is called a *strong conclusion.* On the other hand, the probability of a type II error depends on the true value of the parameter and the sample size. For this reason, acceptance of H_0 is often referred to as a *weak conclusion* unless it is known that β is acceptably small. Often, instead of saying that H_0 is accepted, the phrase "fail to reject H_0" is used. The latter phraseology does not necessarily mean that there is a high probability that H_0 is true but that more data are needed to reach a strong conclusion.

In constructing a hypothesis, the null hypothesis will always be stated as an equality. This allows the type I error to be set to a specific value. The alternative hypothesis might be two-sided, as in Examples 7-3 and 7-4, or one-sided. The one-sided alternative hypothesis statement is used if it is desired to make statements such as "exceeds" or "at least" or "less than" if H_0 is rejected. For example, a one-sided alternative hypothesis statement in Example 7-4 would have led to the conclusion that "the bar is less than 10 cm" or "the bar is greater than 10 cm" had H_0 been rejected, depending on how the alternative hypothesis had been set up.

Example 7-5

Restate the test of Example 7-4 as a one-sided test to provide a decision on whether the bar is 10 cm or shorter than 10 cm. Find an expression for the OC curve versus the true mean for a significance level of α.

Solution The hypotheses are

$$\begin{aligned} H_0&: \mu = \mu_0 = 10 \text{ cm} \\ H_1&: \mu < 10 \text{ cm} \end{aligned} \tag{7-46}$$

We again use the test statistic

$$Z = \frac{\mu - \mu_0}{\sigma/\sqrt{n}} \tag{7-47}$$

In establishing the critical region for the test, we observe that a positive value for the test statistic Z would never lead to the conclusion that $\mu = \mu_0$ is false. Thus we would reject H_0 if $Z < -c$, and we would accept H_0 ($\mu = \mu_0$) if $Z \geq -c$. Since the test is one sided, we calculate c from $Q(c) = \alpha$ or $c = Q^{-1}(\alpha)$. For $\alpha = 0.05$ (5% significance), $c = 1.643$. The OC of the test is computed by recalling that $\beta = P(\text{accept } H_0|H_1 \text{ true})$ or $1 - \beta = P(\text{accept } H_1|H_1 \text{ true})$. Using the normalized statistic (7-47), this may be computed from

$$\begin{aligned} 1 - \beta(\mu) = P(\text{accept } H_1 \,|\, H_1 \text{ true}) &= \int_{-\infty}^{-c} \frac{e^{-(1/2)[z-(\mu-\mu_0)/(\sigma/\sqrt{n})]^2}}{\sqrt{2\pi}}\,dz \\ &= \int_{-\infty}^{-c-(\mu-\mu_0)/(\sigma/\sqrt{n})} \frac{e^{-u^2/2}}{\sqrt{2\pi}}\,du = Q\left(c + \frac{\mu - \mu_0}{\sigma/\sqrt{n}}\right) \end{aligned} \tag{7-48}$$

where the evenness of the integrand has been used to go from the last integral to the Q-function. The OC curve for a sample size of 10 and variance of 0.1 is shown in Figure 7-6.

7-5 OTHER DECISION STRATEGIES

In finding a Bayes' decision rule, several items were required, including the costs and prior probabilities. What if these are not known?

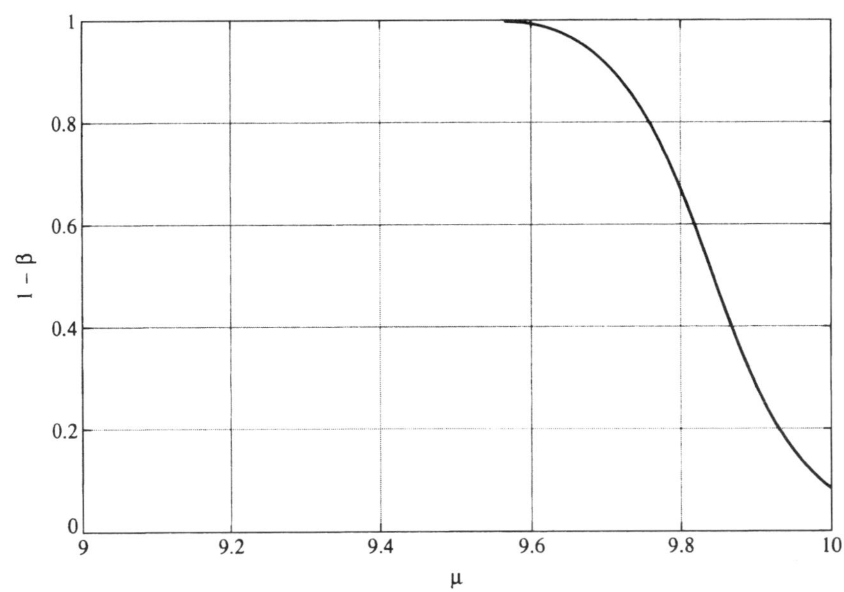

Figure 7-6 OC curve for the one-sided test of Example 7-5.

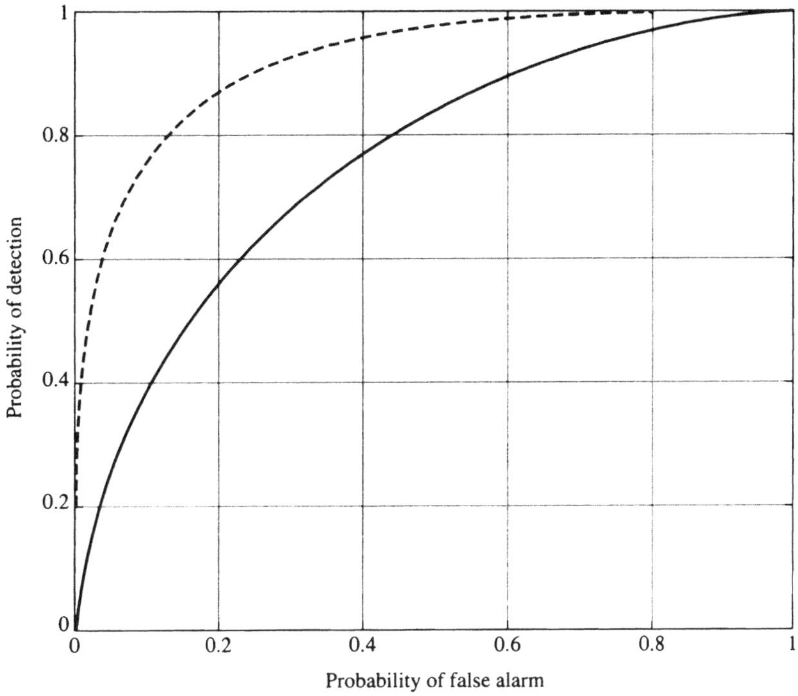

Figure 7-7 Receiver operating characteristic for the test of Example 7-2; upper curve is for $A/\sigma = 2$ and lower curve is for $A/\sigma = 1$.

Neyman–Pearson Strategy

A strategy that can be used if the prior probabilities and the costs are not known is the *Neyman–Pearson rule.* It is derived by keeping the probability of false alarm below some acceptable value and minimizing the probability of a miss. It can be shown that the Neyman–Pearson test amounts to a likelihood ratio test whose threshold has been fixed by the allowable value of the probability of a false alarm.

The performance of a decision-testing strategy, such as Bayes' or Neyman–Pearson, is sometimes portrayed by a *receiver operating characteristic* (ROC) for the test, which is a plot of the probability of detection (one minus the probability of a miss) versus the probability of false alarm. The ROC for the test derived in Example 7-2 is shown in Figure 7-7. A property of the ROC that can be proved is the fact that the slope of the curve at any given point (P_{FA}, $P_D = 1 - P_{\text{miss}}$) is the threshold for the Bayes' test [see (7-12)].

Similarly, the OC curves of Example 7-6 can be plotted in this fashion. In this case, $1 - \beta$ is plotted versus α, or the probability of accepting H_1 given H_1 is true versus the probability of accepting H_1 given H_0 is true. Examples of the OC curves for Example 7-4 plotted in this way are shown in Figure 7-8.

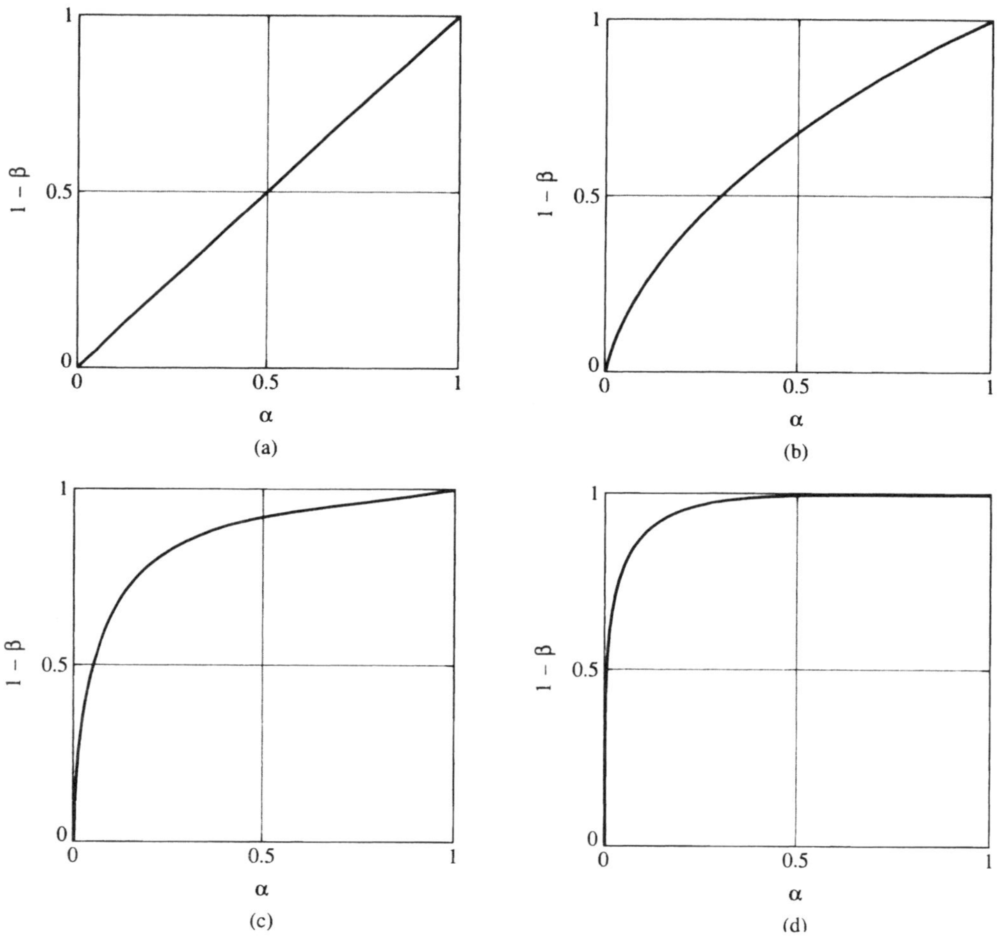

Figure 7-8 OC curves for Example 7-4 replotted as ROCs: (a) $\mu = 10$; (b) $\mu = 10.1$; (c) $\mu = 10.2$; (d) $\mu = 10.3$.

Maximum a Posteriori and Maximum Likelihood Decision Strategies

Returning to the Bayes' decision criterion expressed by the set of inequalities (7-12), we note that if right decisions cost zero and either type of wrong decision costs the same, the ratio of costs on the right-hand side is unity and (7-12) can be rewritten as

$$f_Z(Z|H_1)P(H_1) \underset{H_0}{\overset{H_1}{\gtrless}} f_Z(Z|H_0)P(H_0) \tag{7-49}$$

where the definitions of p and q have been used. We divide both sides of (7-30) by $f_Z(z)$ and use Bayes' theorem to rewrite (7-49) as

$$\frac{f_Z(Z/H_1)P(H_1)}{f_Z(Z)} = P(H_1|Z) \underset{H_0}{\overset{H_1}{\gtrless}} \frac{f_Z(Z|H_0)P(H_0)}{f_Z(Z)} = P(H_0|Z) \tag{7-50}$$

This revised decision rule amounts to choosing the hypothesis that is most probable given the data observed. It is called the *maximum a posteriori decision rule.*

An easier decision rule to apply occurs if $P(H_0) = P(H_1)$. In this case we may generalize the decision rule given by (7-50) to several hypotheses, say $H_0, H_1, \ldots, H_{M-1}$. After canceling the probabilities $P(H_i)$, which are assumed equal, our decision rule amounts to observing a datum, computing the conditional probability density functions

$$p(Z \mid H_0), p(Z \mid H_1), p(Z \mid H_2), \ldots, p(Z \mid H_{M-1}) \tag{7-51}$$

and then choosing the hypothesis corresponding to the largest conditional pdf as the true hypothesis. This is called the *maximum likelihood decision rule.*

Example 7-6

In a game of chance, a die is rolled; a zero-mean, unit-variance Gaussian random number is added to the number on the up face before giving the result to the player. In a given playing of the game, suppose that the datum given to the player is 4.3. What should the player guess in accordance with a maximum likelihood decision rule?

Solution The required conditional probabilities are

$$p(Z|H_i) = \frac{e^{-(Z-i)^2/2}}{\sqrt{2\pi}}, \quad i = 1, 2, 3, 4, 5, 6$$

Substituting the given datum, we have

$$p(4.3|H_i) = \frac{e^{-(4.3-i)^2/2}}{\sqrt{2\pi}}, \quad i = 1, 2, 3, 4, 5, 6$$

which is maximum for $i = 4$. Thus the maximum likelihood guess is four spots up on the die.

7–6 SUMMARY

In this chapter we have considered statistical approaches to making engineering decisions. The minimization of average cost led to the idea of a Bayes' decision strategy, which required prior probabilities of the hypotheses and costs. If both the costs and prior probabilities are unknown, the Neyman–Pearson strategy can be used, wherein the probability of false alarm is fixed to be less than some tolerable value and the probability of detection is maximized. In classical decision theory, we test a hypothesis against an alternative.

1. In making a *minimum cost decision* between two hypotheses, a general expression for the average cost of making a decision is

$$C(D) = p[c_{00}P(\text{decide } H_0 | H_0 \text{ true}) + c_{10}P(\text{decide } H_1 | H_0 \text{ true})]$$
$$+ q[c_{01}P(\text{decide } H_0 \mid H_1 \text{ true}) + c_{11}P(\text{decide } H_1 \mid H_1 \text{ true})]$$

where H_0 and H_1 are the two hypotheses, c_{ij} are the costs of deciding in favor of hypothesis i when in fact hypothesis j was true, and $p = P(H_0) = 1 - q = 1 - P(H_1)$, where p and q are known as the *a priori*, or *prior, probabilities* of the hypotheses.

2. The strategy that minimizes the above-average cost of making a decision is expressed by the inequalities

$$\frac{f_Z(Z \mid H_1)}{f_Z(Z \mid H_0)} \underset{H_0}{\overset{H_1}{\gtrless}} \frac{p(c_{10} - c_{00})}{q(c_{01} - c_{11})}$$

In this expression, $f_Z(Z \mid H_0)$ and $f_Z(Z \mid H_1)$ are the conditional pdf's of the observed data given the respective hypothesis. This decision strategy is known as a *Bayes' test.* The left-hand side is called the *likelihood ratio,* and the right-hand side involving constants required to make the test is called the *threshold* of the test.

3. In classical decision making, we are testing the hypothesis whether a parameter is within a certain range versus the alternative that it is not. Two types of errors are possible in such a decision process:
   (a) We can say that the hypothesis is false when it is really correct.
   (b) We can say that the hypothesis is correct when it is really false.

   The first is called a type I error and is denoted by α, and the second is called a type II error and is given the symbol β.

4. The value of β depends on the true parameter value and is therefore a conditional probability. When $1 - \beta$ is plotted as a function of the parameter value, we have an *operating characteristic* (OC) *curve.*

5. Several specializations of the Bayes' test are possible, depending on our state of prior knowledge. In making a Bayes' test, we need several ingredients, including the conditional pdf's, the costs, and the prior probabilities, $p = P(H_0)$ and $q = P(H_1)$. If neither the prior probabilities nor the costs are known, we can set the probability of deciding in favor of H_0, when, in fact, H_1 was true (called the probability of a false alarm) below some tolerable level. This fixes the threshold of the test. If the probability of a miss is minimized (or detection maximized), where the probability of a miss is the probability of deciding in favor of H_1 when, in fact, H_0 was true, we obtain what is referred to as a *Neyman–Pearson decision rule.* A plot of the probability of detection versus the probability of false alarm for a Neyman–Pearson test provides what is referred to as the operating characteristic for the test.

6. Another specialization of the Bayes' test is a *maximum a posteriori decision rule* that is implemented by testing the inequalities.

$$\frac{f_Z(Z|H_1)P(H_1)}{f_Z(Z)} = P(H_1|Z) \underset{H_0}{\overset{H_1}{\gtrless}} \frac{f_Z(Z|H_0)P(H_0)}{f_Z(Z)} = P(H_0|Z)$$

This amounts to a Bayes' test using zero costs for right decisions and costs of one for wrong decisions.

7. A *maximum likelihood decision rule* is implemented by finding the maximum of the pdf's

$$p(Z|H_0), p(Z|H_1), p(Z|H_2), \ldots, p(Z|H_{M-1})$$

and choosing that hypothesis as the true hypothesis. Such a decision rule is a maximum a posteriori decision rule with equal a priori probabilities.

7-7 FURTHER READING

The book by Papoulis (1990, Chapter 10) is appropriate supplementary reading for the material appearing in this chapter. The material on Bayes' decision theory is treated in several advanced-level texts. A simplified treatment is provided in Ziemer and Tranter (1995, Chapter 9). An excellent chapter on classical decision theory at the undergraduate level is found in Montgomery and Runger (1994, Chapter 8).

7-8 PROBLEMS

Sections 7-2 and 7-3

7-1. Given the following conditional pdf's for hypotheses H_0 and H_1, respectively:

$$f_Z(z|H_0) = \frac{\exp(-z^2/32)}{\sqrt{32\pi}} \quad \text{and} \quad f_Z(z|H_1) = \frac{\exp[-(z-10)^2/32]}{\sqrt{32\pi}}$$

The prior probabilities are $p = \frac{1}{3}$ and $q = \frac{2}{3}$, and the costs are $c_{00} = 1$, $c_{11} = 2$, $c_{01} = 5$, and $c_{10} = 15$.

(a) Find the likelihood ratio.

(b) Compute the threshold for the Bayes' test.

(c) Put the Bayes' test in the form of comparing a datum value with a modified threshold such as

$$Z \underset{H_0}{\overset{H_1}{\gtrless}} T$$

7-2. Design a Bayes' test to determine which of the following pdf's random variable Z is distributed according to:

$$H_0 : f_Z(z|H_0) = 10e^{-10z}u(z)$$

$$H_1 : f_Z(z|H_1) = \tfrac{1}{10}[u(z) - u(z-10)]$$

Assume that $c_{00} = c_{11} = 0$ and $c_{01} = c_{10} = 1$ and $p = q = \frac{1}{2}$. Find the average cost of a decision.

7-3. Consider a Bayes' test where the conditional pdf's of the data given either hypothesis are

$$f_Z(z|H_0) = \frac{\exp(-z^2/2)}{\sqrt{2\pi}} \quad \text{and} \quad f_Z(z|H_1) = \tfrac{1}{2}\exp(-|z|)$$

(a) Obtain the likelihood ratio.

(b) For an arbitrary threshold, find the decision regions R_0 and R_1. That is, give ranges for Z that define R_0 and R_1. You will obtain different cases depending on the value chosen for the threshold.

(c) Find the average cost of making a decision.

Section 7-3

7-4. Work Example 7-4 for a 5% significance level. Plot the OC curve assuming 10 samples.

7-5. The weight of a gold bar is measured 10 times with the following results: 60.6, 60.8, 59.9, 60.1, 59.7, 60.4, 60.3, 60.05, 59.7, and 59.8 oz. Assume that the measurements are the actual weight, μ, plus a Gaussian measurement error with mean zero and variance 0.64 oz^2.

(a) Test the hypothesis that the bar is 60 oz versus the hypothesis that is not 60 oz at a 5% significance level.

(b) Find and plot the OC curve.

7-6. Design a one-sided test for the situation of Problem 7-5 to test whether the bar is 60 oz or greater than 60 oz at a 2% significance level. Based on the data given, is the hypothesis that the bar is 60 oz accepted or rejected?

Section 7-5

7-7. Referring to Example 7-2 suppose that the costs and prior probabilities are unknown. From the formula for the false alarm probability,

$$P_{FA} = \int_T^\infty \frac{e^{-z^2/2\sigma^2}}{\sqrt{2\sigma^2}}\,dz = \int_{T/\sigma}^\infty \frac{e^{-u^2/2}}{\sqrt{2\pi}}\,du = \begin{cases} Q\left(\dfrac{T}{\sigma}\right), & T > 0 \\ 1 - Q\left(\dfrac{|T|}{\sigma}\right), & T < 0 \end{cases}$$

find the threshold normalized by σ for a probability of false alarm of 10^{-3}. Determine the corresponding probability of detection, given by

$$P_D = 1 - P_{\text{miss}}$$

where

$$P_{\text{miss}} = \int_{-\infty}^T \frac{e^{-(z-A)^2/2\sigma^2}}{\sqrt{2\pi\sigma^2}}\,dz = \int_{(A-T)/\sigma}^\infty \frac{e^{-u^2/2}}{\sqrt{2\pi}}\,du = \begin{cases} Q\left(\dfrac{A-T}{\sigma}\right), & T < A \\ 1 - Q\left(\dfrac{T-A}{\sigma}\right), & T > A \end{cases}$$

and plot versus A/σ. This determines the operating characteristic of a Neyman–Pearson test, for it can be shown that it is a likelihood ratio test like the Bayes' test with the threshold fixed by the allowed false alarm probability.

7-8. Referring to Example 7-6, the game of chance is modified so that a random variable with pdf

$$f_N(n) = \tfrac{1}{2}e^{-|n|}$$

is added to the up-face number before giving it to the contestant.

(a) Design a maximum likelihood test to determine the best guess if the number given to the contestant is 5.2.

(b) Determine an expression for the average probability of making an error (i.e., the probability that any other number is guessed other than the correct one).

7–9 COMPUTER EXERCISES

7-1. Generalize a Monte Carlo simulation of Example 7-1 to include an arbitrary ratio A/σ. Derive an expression that minimizes $C(D)$ with respect to T/σ. Compare the output of your simulation with (7-27) and (7-29) for specific values of A/σ and T/σ.

7-2. Design a Monte Carlo simulation of Example 7-4. Repeatedly generate sets of 10 length measurements (10 cm plus a Gaussian measurement error of variance 0.1) and compare with the confidence interval boundaries to determine the power of the test. Compare with Figure 7-3.

7-3. Design a Monte Carlo simulation of Example 7-6. You can simplify the test to find the minimum of the absolute value as pointed out in the example. Assume that \$1 is awarded for each right guess and \$1 is taken away for each wrong guess. What is the expected winnings over a long period of playing the game?

8 Reliability

8–1 INTRODUCTION

The term *reliability* has many connotations. The time-dependent definition of reliability is the *probability that a system is functioning at time t after being put in service at time zero.* For example, a light bulb ages as it is used, and we can rightly ask the question: what is the probability that it is operating at 500 hours, or 1000 hours? On the other hand, we can consider the probability of a system performing the function for which it was designed. An example of this case is a booster rocket: it either performs its function or it does not. For this aspect of reliability, we use the principles developed in preceding chapters involving probabilities of compound events to find the probability that the component parts of a system will perform their separate functions so that the overall system can perform its function. This is called *system reliability.* We consider these two aspects of reliability in this chapter. In short, reliability is the probability that a system (i.e., an interconnection of several components to perform a desired task) will perform that task for a stated period of time.

One aspect of reliability theory that we will not consider to any great extent is time to repair. That is, for repairable components, the repair time can be considered a random variable and this can be factored into computing the expected time between failures, which is the sum of the *mean time to failure* (MTTF) plus the *mean time to repair* (MTTR).

8–2 TIME-DEPENDENT RELIABILITY

The time to failure of a system is the interval of time between its being put into service and its failure. It is a random variable which we denote by T and that has a cumulative probability distribution function

$$F(t) = P(T \leq t) \tag{8-1}$$

which is the probability that the system fails prior to time t. The reliability, on the other hand, is the probability that the device functions for times t greater than T and is given by

$$R(t) = 1 - F(t) = P(T > t) \tag{8-2}$$

The *mean time to failure* is the expectation of the time to failure. Let the pdf of the time to failure be denoted by $f(t)$; by definition, it is

$$f(t) = \frac{dF(t)}{dt} \tag{8-3}$$

In terms of this pdf, the mean time to failure is

$$E(T) = \int_0^\infty tf(t)\,dt \tag{8-4}$$

where the lower limit is zero because the failure time cannot be less than zero. Substituting (8-3) and making use of (8-2), we may rewrite this as

$$E(T) = \int_0^\infty \frac{dF(t)}{dt}\,dt = -\int_0^\infty t\frac{dR(t)}{dt}\,dt = \int_0^\infty R(t)\,dt \tag{8-5}$$

which follows by integration by parts with $u = t$ and $dv = (dR/dt)\,dt$.

Example 8-1

The failure pdf for a device is given by

$$f(t) = \lambda e^{-\lambda t}, \qquad t \geq 0 \tag{8-6}$$

Find the following:

(a) The failure cdf, $F(t)$

(b) The reliability

(c) The mean time to failure

Solution (a) The failure cdf is the integral of the failure pdf:

$$F(t) = \int_0^t \lambda e^{-\lambda\zeta}\,d\zeta = -e^{-\lambda\zeta}|_0^t = 1 - e^{\lambda t}, \qquad t \geq 0 \tag{8-7}$$

(b) From (8-2), the reliability is

$$R(t) = 1 - F(t) = e^{-\lambda t}, \qquad t \geq 0 \tag{8-8}$$

(c) The expected time to failure, from (8-5), is

$$E(T) = \int_0^\infty e^{-\lambda t}\,dt = \frac{1}{\lambda} \tag{8-9}$$

The probability that a system functioning at time t will fail prior to time T is the conditional cdf

$$F(x|T > t) = \frac{P(T \leq x, T > t)}{P(T > t)} \tag{8-10}$$

If $x < t$, this probability is clearly zero. Using the definition of $F(t)$, we have

$$P(T \leq x, T > t) = F(x) - F(t) \tag{8-11}$$

Substituting (8-11) and (8-2) into (8-10), we have

$$F(x|T > t) = \frac{F(x) - F(t)}{1 - F(t)} = \frac{F(x) - F(t)}{R(t)}, \qquad x > t \tag{8-12}$$

Differentiating with respect to x, we get the conditional pdf

$$f(x|T > t) = \frac{f(x)}{1 - F(t)} = \frac{f(x)}{R(t)}, \qquad x > t \tag{8-13}$$

Consider the product

$$f(x|T > t)dx = P\left[\text{system fails in time interval } (x, x + dx) \atop \text{given it operates past time } t\right] \tag{8-14}$$

That is, the conditional failure pdf times an increment in the independent variable is the probability that the item has survived to time t but will fail in the increment of time (x, $x + dx$). Such information is known in the insurance field as actuarial data.

Example 8-2

Find the conditional failure pdf corresponding to the failure pdf of Example 8-1.

Solution From (8-13), we obtain

$$f(x|T > t) = \frac{f(x)}{1 - F(t)} = \frac{\lambda e^{-\lambda x}}{e^{-\lambda t}} = \lambda e^{-\lambda(x-t)}, \qquad x > t \tag{8-15}$$

For the exponential failure pdf it is seen that the failure pdf conditioned on the failure time being greater that some time t is simply the unconditional failure pdf shifted to $x = t$. This is called the *memoryless property.*

Another concept of use in time-varying reliability theory is that of *failure rate*. Let $n(t)$ be the number of (macroscopically the same) devices operating at time t, and let the number that fail in the interval (t, $t + dt$) be dn. The fraction of those that fail to the total number at time t is

$$F_f = \frac{dn}{n(t)} \tag{8-16}$$

However, this is exactly the same quantity as given in (8-14) with $x = t$, so we equate (8-16) and (8-14) to obtain

$$f(t|T > t)dt = \frac{dn}{n(t)} \tag{8-17a}$$

or, dividing through by dt, we obtain

$$h(t) = f(t|T > t) = \frac{1}{n(t)}\frac{dn}{dt} = \frac{f(t)}{R(t)} \tag{8-17b}$$

which is the failure rate.

Example 8-3

Substituting (8-6) and (8-8) into (8-17b), we see that the failure rate for a device having an exponential failure pdf is simply λ devices per second. Suppose that a certain batch of light bulbs fails at a rate of $\lambda = 0.001$ per year. What is the reliability?

Solution Since $\lambda = 0.001$, we find from (8-8) that

$$R(t) = e^{-\lambda t} = e^{-0.001t}, \qquad t > 0 \tag{8-18}$$

The failure rate being a constant is the exception rather than the rule. Devices generally have a higher failure rate at the beginning of their lifetimes due to imperfections in the manufacturing process that escape inspection. After this initially higher failure rate, sometimes called the infant mortality rate in the case of humans, the normal lifetime of the device takes place and the failure rate stabilizes at some acceptable lower value (otherwise, one would not use it). Finally, after the useful lifetime of the device, the failure rate begins to climb due to wear-out. These three regions in the failure rate versus time of service, called the *break-in, stable,* and *wear-out periods* of the device, are illustrated in Figure 8-1. This is sometimes called the *bathtub curve.*

Differentiating (8-2) and using (8-3), we find that

$$\frac{dR(t)}{dt} = -\frac{dF(t)}{dt} = -f(t) \tag{8-19}$$

Substituting this into (8-17b), we obtain

$$h(t) = -\frac{dR(t)/dt}{R(t)} = -\frac{d}{dt}\ln R(t) \tag{8-20}$$

Integration gives

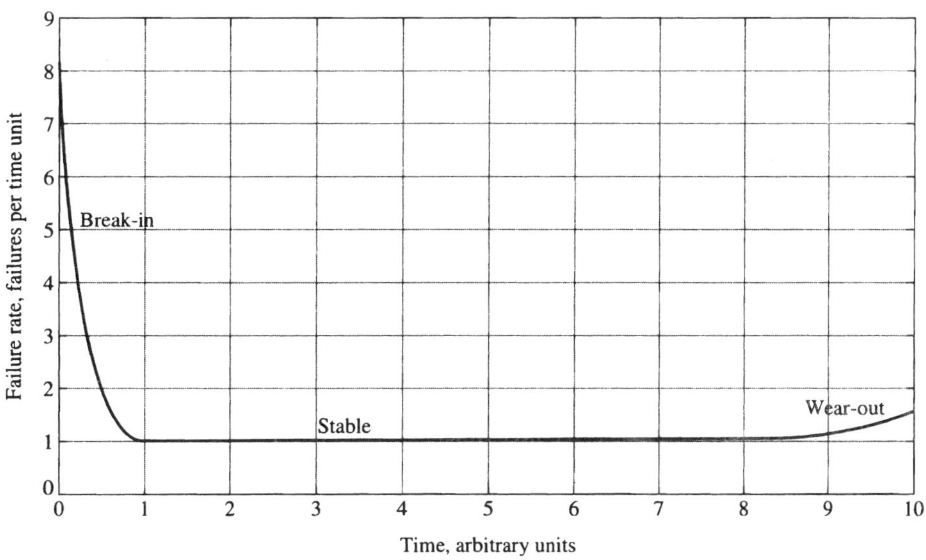

Figure 8-1 Failure rate curve for typical manufactured devices.

$$R(t) = \exp\left[-\int_0^t h(\lambda)\,d\lambda\right] \tag{8-21}$$

Note that if the failure rate is constant, the reliability from (8-21) is

$$R(t) = e^{-kt} \tag{8-22}$$

where k is a proportionality constant.

Example 8-4

Consider a device with a uniform failure pdf, given by

$$f(t) = \begin{cases} \dfrac{1}{T}, & 0 \le t \le T \\ 0, & \text{otherwise} \end{cases} \tag{8-23}$$

Find the reliability and the failure rate.

Solution The failure cdf, by integrating (8-23), is

$$F(t) = \int_0^t f(\tau)\,d\tau = \frac{1}{T}\int_0^t d\tau = \frac{t}{T}, \qquad 0 \le t \le T \tag{8-24}$$

Thus the reliability is

$$R(t) = 1 - \frac{t}{T}, \qquad 0 \le t \le T \tag{8-25}$$

From (8-17b), the failure rate is

$$h(t) = \frac{f(t)}{R(t)} = \frac{1/T}{1 - t/T} = \frac{1}{T - t}, \qquad 0 \le t \le T \tag{8-26}$$

These are plotted in Figure 8-2.

Example 8-5

Given the actuarial data for U.S. males and females in Table 8-1, compute the failure rate for both sexes.

Solution We apply (8-17b) by subtracting successive column entries and dividing by the bottom column entry in each case. For example, the failure rate for boys at age 1 year is

$$\frac{1}{n(t)}\frac{dn}{dt} = \frac{100{,}000 - 81{,}049}{81{,}049} = 0.234$$

Note the presence of the burn-in, stable, and wear-out periods from the data.

8-3 RELIABILITY OF SYSTEMS

A system is an interconnection of devices to perform a desired task. When examining the reliability of a system, we break it down into its component subsystems or devices, to each

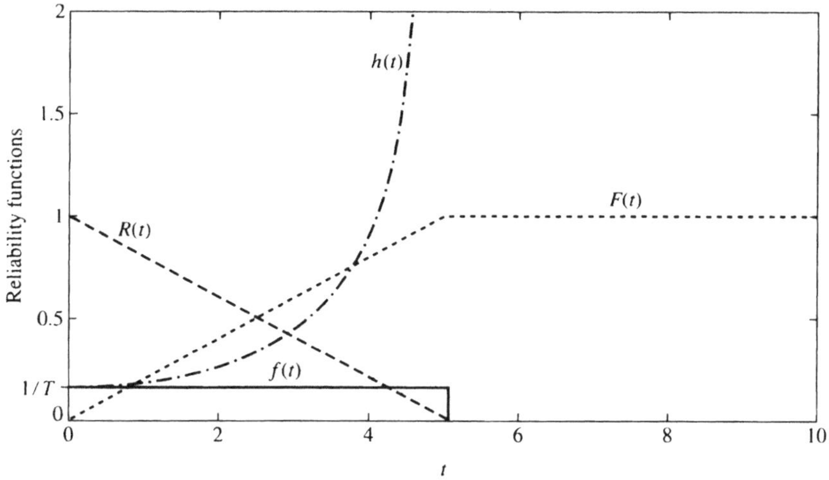

Figure 8-2 Various reliability functions for Example 8-4.

of which we somehow assign a reliability (e.g., inferred from past uses of a given device); by considering the interrelationship of the various subsystems, we are able to calculate the reliability of the overall system. Two simple examples will illustrate such procedures, after which we consider the general theory of reliability of systems. We consider series, parallel, and standby systems.

Example 8-6

The launch of a small satellite requires the following sequence of operations with associated probabilities of success. Each individual action is assumed independent of the others.

1. Ignition of booster rocket, 0.99
2. Bolts holding satellite to main rocket blow, 0.98
3. Gas jets spin satellite for stability, 0.965
4. Booster rocket shutdown at desired final velocity, 0.97

Each of these actions must happen in series, and if a single action fails, the launch fails. What is the probability of mission success, or reliability?

Solution The system success is the intersection of the four separate subsystem successes, or

$$\begin{aligned} P(T_{MS} > t) &= P[(T_1 > t) \cap (T_2 > t) \cap (T_3 > t) \cap (T_4 > t)] \\ &= \prod_{i=1}^{4} P(T_i > t) \qquad (8\text{-}27) \\ &= (0.99)(0.98)(.0965)(0.97) = 0.908 \end{aligned}$$

It is seen that the reliability of a system consisting of a series of subsystems is less than or equal to the reliability of the subsystem with the lowest reliability.

Table 8-1 ACTURIAL DATA FOR U.S. MALES AND FEMALES

Age	Males	Females	Failure Rate (yr) Male	Female
0	100,000	100,000		
1	81,049	83,807	0.234	0.193
2	76,601	79,276	0.058	0.057
3	74,654	77,246	0.026	0.026
4	73,378	75,933	0.017	0.017
5	72,430	74,940	0.014	0.013
10	70,315	72,654	0.006	0.006
15	68,827	70,924	0.004	0.005
20	66,790	68,678	0.006	0.007
25	63,989	65,932	0.009	0.008
30	61,021	62,965	0.010	0.009
35	57,767	59,759	0.011	0.011
40	54,137	56,389	0.013	0.012
45	49,982	52,906	0.017	0.013
50	45,385	49,249	0.020	0.015
55	39,986	44,787	0.027	0.020

Source: Reported in Ansley J. Coale et al., *Regional Model Life Tables and Stable Populations,* 2nd ed. New York: Academic Press, 1983, pp. 37 and 46 (level 10 data).

Example 8-7

Brakes are an important component of any vehicle. Most road vehicles have two independent systems: hydraulic and mechanical. If one fails, the other takes over (actually, only the mechanical takes over if the hydraulic goes bad, but we shall assume that either one may take over in case of failure of the other). Given the following reliabilities for the braking subsystems of a large open-pit mining truck, what is the overall reliability of the braking system for the truck?

1. Hydraulic, 0.98
2. Mechanical, 0.95

Solution For this system, both subsystems must fail for the overall system to fail. Thus

$$
\begin{aligned}
P(T_B \leq t) &= P[(T_1 \leq t) \cap (T_2 \leq t)] \\
&= \prod_{i=1}^{2} P(T_i \leq t) \qquad (8\text{-}28) \\
&= (1 - 0.98)(1 - 0.95) = 0.001
\end{aligned}
$$

which is the overall probability of failure. The probability of success, or reliability, is

$$P(T_B > t) = 1 - P(T_B \leq t) = 1 - 0.001 = 0.999 \qquad (8\text{-}29)$$

Note that the reliability of a parallel system is greater than the reliability of the best subsystem.

From the examples above, we may deduce the following reliability relationships for the two types of systems (i.e., series and parallel).

Series Systems

For a series system, the successful operation of the overall system depends on the successful operation of each of its subsystems. That is,

$$S = S_1 \cap S_2 \cap \cdots \cap S_n \tag{8-30}$$

Thus the probability of success of the overall system, or system reliability, assuming that the subsystems are independent, is

$$\begin{aligned} R_s(t) &= P(T_{\text{series}} > t) = P\big[(T_1 > t) \cap (T_2 > t) \cap \cdots \cap (T_n > t)\big] \\ &= \prod_{i=1}^{n} P(T_i > t) = \prod_{i=1}^{n} R_i(t) \end{aligned} \tag{8-31}$$

We can show that the failure rate for a series system is the sum of the failure rates for each subsystem as follows. Consider the logarithm of (8-31), which is

$$\ln[R_s(t)] = \ln[P(T_{\text{series}} > t)] = \sum_{i=1}^{n} \ln[P(T_i > t)] \tag{8-32}$$

The negative derivative of the logarithm of the reliability is the failure rate according to (8-20). Thus

$$\begin{aligned} h_s(t) &= -\frac{d}{dt}\ln[R_s(t)] \\ &= \sum_{i=1}^{n}\left\{-\frac{d}{dt}\ln[P(T_i > t)]\right\} \\ &= \sum_{i=1}^{n} h_i(t) \end{aligned} \tag{8-33}$$

Example 8-8

Suppose that the time-dependent reliabilities for the subsystems of Example 8-6 are exponential according to

$$\begin{aligned} R_1(t) &= e^{-\lambda_1 t} \\ R_2(t) &= e^{-\lambda_2 t} \\ R_3(t) &= e^{-\lambda_3 t} \\ R_4(t) &= e^{-\lambda_4 t} \end{aligned} \tag{8-34}$$

Find the reliability of the system, the mean time to failure, and the failure rate.

Solution The system reliability, according to (8-31), is

$$R_s(t) = \prod_{i=1}^{4} R_i(t) = e^{-(\lambda_1+\lambda_2+\lambda_3+\lambda_4)t} \tag{8-35}$$

From (8-5), the mean time to failure is

$$E(T_s) = \int_0^\infty R_s(t)\,dt = \int_0^\infty e^{-(\lambda_1+\lambda_2+\lambda_3+\lambda_4)t}\,dt = \frac{1}{\lambda_1+\lambda_2+\lambda_3+\lambda_4} \tag{8-36}$$

For a subsystem with an exponential reliability, the failure rate is the constant in the exponent. From (8-33), therefore, the failure rate of the system is

$$h_s(t) = \lambda_1 + \lambda_2 + \lambda_3 + \lambda_4 \tag{8-37}$$

Parallel Systems

For a parallel system, all subsystems must fail before the system fails. Thus for independent subsystems,

$$F_{\text{paral}} = F_1 \cap F_2 \cap \cdots \cap F_n \tag{8-38}$$

The probability of failure is therefore

$$\begin{aligned} F_p(t) = P(T_{\text{paral}} \le t) &= P\left[(T_1 \le t) \cap (T_2 \le t) \cap \cdots \cap (T_n \le t)\right] \\ &= \prod_{i=1}^{n} P(T_i \le t) \\ &= \prod_{i=1}^{n} F_i(t) \\ &= \prod_{i=1}^{n} \left[1 - R_i(t)\right] \end{aligned} \tag{8-39}$$

The reliability for a parallel system is

$$R_p(t) = 1 - F_p(t) = 1 - \prod_{i=1}^{n} \left[1 - R_i(t)\right] \tag{8-40}$$

Example 8-9

Consider two subsystems in parallel with exponential reliabilities given by

$$\begin{aligned} R_a(t) &= e^{-\lambda_a t} \\ R_b(t) &= e^{-\lambda_b t} \end{aligned} \tag{8-41}$$

Find the reliability of the parallel system, the mean time to failure, and the failure rate.

Solution From (8-40), the reliability of the parallel system is

$$R_p(t) = 1 - (1 - e^{-\lambda_a t})(1 - e^{-\lambda_b t}) = e^{-\lambda_a t} + e^{-\lambda_a t} - e^{-(\lambda_a + \lambda_b)t} \tag{8-42}$$

The mean time to failure, from (8-5), is

$$E(T_p) = \int_0^{\infty} R_p(t)\, dt = \frac{1}{\lambda_a} + \frac{1}{\lambda_b} - \frac{1}{\lambda_a + \lambda_b} \tag{8-43}$$

From (8-20), the failure rate is

$$h_p(t) = -\frac{d}{dt} \ln R_p(t) = \frac{\lambda_a e^{-\lambda_a t} + \lambda_b e^{-\lambda_a t} - (\lambda_a + \lambda_b)e^{-(\lambda_a + \lambda_b)t}}{e^{-\lambda_a t} + e^{-\lambda_a t} - e^{-(\lambda_a + \lambda_b)t}} \tag{8-44}$$

Unlike the series case, there is no simple expression for the failure rate of a parallel system.

Standby Systems

In this case, a subsystem operates until it fails, at which time a standby system is switched in its place. For such an arrangement,

$$T_{sb} = T_1 + T_2 \tag{8-45}$$

where T_1 refers to the primary system and T_2 refers to the standby system. Taking the expectation of both sides, we find the mean time to failure to be

$$E(T_{sb}) = E(T_1) + E(T_2) \tag{8-46}$$

Equation (8-45) says that the time to failure of a standby system is the sum of two random variables: the time of failure to the primary system and that of the secondary system. If these two failure times are statistically independent, the failure pdf of the standby system is the convolution of the two subsystem failure pdf's, or

$$f_{sb}(t) = \int_{-\infty}^{\infty} f_1(t - \eta) f_2(\eta)\, d\eta \tag{8-47}$$

The failure cdf is the integral of the failure pdf, and the reliability is 1 minus the failure cdf.

Example 8-10

Find the mean time to failure, the failure pdf, the failure cdf, and the reliability of a standby system with two subsystems that have exponential failure pdf's given by

$$f_i(t) = \lambda e^{-\lambda t}, \qquad t \geq 0 \tag{8-48}$$

Solution The mean time to failure for each subsystem is $1/\lambda$, so the failure rate of the standby system is

$$E(T_{sb}) = \frac{2}{\lambda} \tag{8-49}$$

The failure pdf is, from (8-47), found to be

$$f_{2,sb}(t) = \int_0^t \lambda^2 e^{-\lambda t} e^{-\lambda(t-\lambda)}\, d\tau = \lambda^2 t e^{-\lambda t}, \qquad t \geq 0 \tag{8-50}$$

Note that this pdf is 0 at $t = 0$ (it is very unlikely that both the system and the backup will fail immediately) and a maximum for $t = 1/\lambda$. Integration gives the failure cdf, which is

$$F_{2,sb}(t) = \int_0^t f_{2,sb}(\tau)\, d\tau = 1 - (1 + \lambda t)e^{-\lambda t}, \qquad t \geq 0 \tag{8-51}$$

Finally, the reliability is

$$R_{2,sb}(t) = 1 - F_{2,sb}(t) = (1 + \lambda t)e^{-\lambda t}, \qquad t \geq 0 \tag{8-52}$$

Integration of the reliability from zero to infinity will give the mean time to failure, as found in (8-49), which is a check of the correctness of this result.

We now generalize the results of Example 8-10 to an arbitrary number n of standby systems. To get an idea for the pattern for the failure pdf, we find it for $n = 3$. In (8-47) we replace the first failure pdf in the integrand with the failure pdf of a single exponentially distributed system [(8-48)] and the second failure pdf with that of two exponentially distributed standby systems as derived in Example 8-10 [(8-50)]. The resulting convolution integral becomes

$$f_{3,sb}(t) = \int_0^t \lambda^3 \tau e^{-\lambda t} e^{-\lambda(t-\tau)}\, d\tau = \frac{\lambda_3}{1 \times 2} t^2 e^{-\lambda t}, \qquad t \geq 0 \tag{8-53a}$$

A similar calculation for four standby systems, which would involve convolving (8-50) with itself, would show that the powers on λ and t increase by 1 and a factor of 3 appears in the denominator in addition to the factors of 1 and 2. The generalization to n standby systems is

$$f_{n,sb}(t) = \frac{\lambda^n}{(n-1)!} t^{n-1} e^{-\lambda t}, \qquad t \geq 0 \tag{8-53b}$$

This is the Erlang pdf given by (3-39) with the parameters of the pdf changed appropriately for the present application. Since it is the result of summing n independent exponentially distributed random variables, it is not surprising that the pdf tends to look more Gaussian as n gets large.

We can get the failure cdf by integration of the pdf. The result for two standby systems is given by (8-51). To establish the pattern for arbitrary n we integrate (8-53a) for three standby systems. The integral for the cdf is

$$F_{3,sb}(t) = \int_0^T f_{3,sb}(\tau)\, d\tau = 1 - \left[1 + \lambda t + \frac{(\lambda t)^2}{2}\right] e^{-\lambda t}, \qquad t \geq 0 \tag{8-54}$$

We hypothesize the following result for n arbitrary:

$$F_{n,sb}(t) = \int_0^t f_{n,sb}(\tau)\, d\tau = 1 - \left[1 + \lambda t + \frac{(\lambda t)^2}{2} + \cdots + \frac{(\lambda t)^{n-1}}{(n-1)!}\right] e^{-\lambda t}, \qquad t \geq 0 \tag{8-55}$$

That this is the correct result can be shown by differentiating (8-55) with respect to t and obtaining (8-53b). The reliability for n standby systems is

$$R_{n,sb}(t) = 1 - F_{n,sb}(t) = \left[1 + \lambda t + \frac{(\lambda t)^2}{2} + \cdots + \frac{(\lambda t)^{n-1}}{(n-1)!}\right] e^{-\lambda t}, \qquad t \geq 0 \tag{8-56}$$

This result is plotted as a function of λt in Figure 8-3 for several values of n.

Example 8-11

You are the pilot, Bernt Balchen, flying on November 28 and 29, 1929 with Admiral Byrd's expedition to reach the South Pole from the base Little America on the Ross Ice Shelf. One particular part in your airplane is particularly troublesome in cold weather. Its reliability obeys an exponential model with $\lambda = 0.08\ \text{h}^{-1}$. The flight going to and coming back is expected to last 25 h, and you are told by the admiral to put in enough replacement parts to ensure that the flight has a 90% probability of success due only to this particular failure. Each part, of course, means that other valuable gear cannot be taken, so you want enough spares to ensure the mission success (with 90% reliability) and no more. How many spares, in addition to the one already installed, should you take with you?

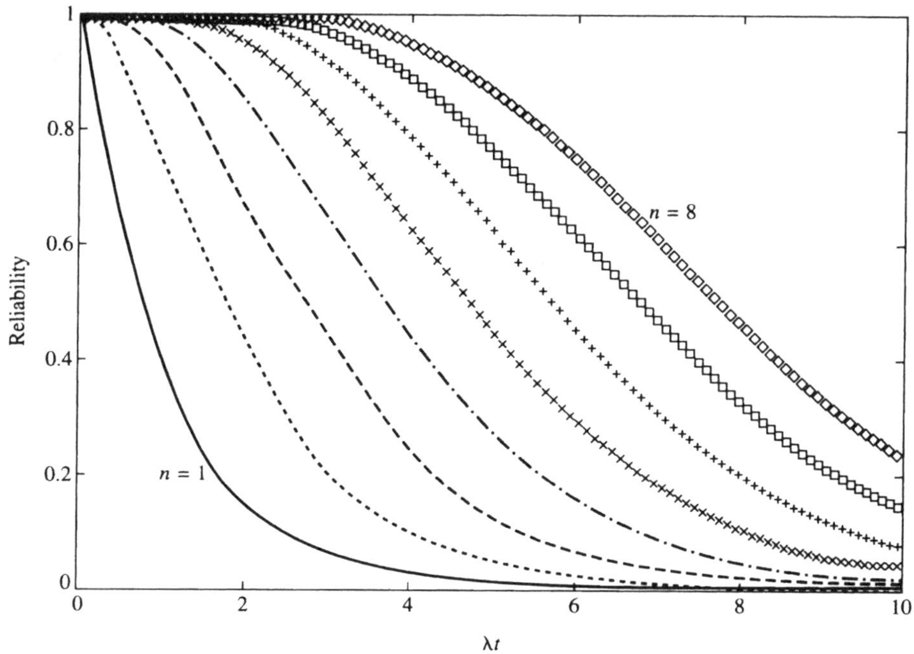

Figure 8-3 Reliability of n standby systems with exponentially distributed failure pdf: $n = 1, 2, 3, 4, 5, 6, 7, 8$.

Solution Given the flight time and value for λ, we compute $\lambda t = (0.08)(25) = 2$. From Figure 8-3 it appears that four parts won't quite do it, but five will be more than enough. Thus you pack four spares in addition to the one installed. Note the mean time to failure for one part is $1/\lambda = 12.5$ h. Five parts appears to be overkill, but we are asking for 90% reliability. A calculation from (8-56) with $\lambda t = 2$ and $n = 5$ shows that the reliability is, in fact, 0.947, but it is only 0.857 for $n = 4$.

8-4 WEIBULL FAILURE MODEL

Consider a failure cdf modeled by

$$F(t) = 1 - e^{-(t/c)^m}, \qquad t \geq 0 \tag{8-57}$$

where m and c are positive constants. The corresponding failure pdf is

$$f(t) = \left(\frac{m}{c}\right)\left(\frac{t}{c}\right)^{m-1} e^{-(t/c)^m}, \qquad t \geq 0 \tag{8-58}$$

Called the *Weibull distribution,* this is used widely in reliability studies. Figure 8-4 shows the Weibull pdf plotted as a function of t/c for several values of m.

Its popularity results from being able to adjust the parameter m to fit a range of failure models. For example, $m = 1$ gives the exponential failure pdf that results in a constant failure rate. If $m > 1$, the failure rate is an increasing function, as will be shown shortly. This is representative of the wear-out failure period. Thus m is referred to as the *shape parameter.*

By straightforward substitution and evaluation, it follows that for $t = c$ the failure cdf (8-57) evaluates to $1 - e^{-1} = 0.632$. That is, the reliability of a system modeled by a Weibull model has reliability 36.8% at time $t = c$. The parameter c is called the *characteristic time.*

From (8-58), the mean time to failure for a Weibull model is

$$E(T) = \int_0^\infty t f_T(t)\, dt = c\Gamma\left(1 + \frac{1}{m}\right) \tag{8-59}$$

where $\Gamma(x)$ is the gamma function defined by (3-35). The variance of the time to failure is

$$\text{Var}(T) = c^2\left[\Gamma\left(1 + \frac{2}{m}\right) - \Gamma^2\left(1 + \frac{1}{m}\right)\right] \tag{8-60}$$

The reliability corresponding to a Weibull failure model, from (8-57), is

$$R(t) = e^{-(t/c)^m}, \qquad t \geq 0 \tag{8-61}$$

and from (8-17b) with (8-61) and (8-58) substituted, the failure rate is

$$h(t) = \frac{m}{c}\left(\frac{t}{c}\right)^{m-1}, \qquad t \geq 0 \tag{8-62}$$

It is shown in Figure 8-5 for several values of m, where it is seen that it does indeed exhibit the features discussed above. The cumulative failure rate is

$$H(t) = \int_0^t h(\lambda)\, d\lambda = \left(\frac{t}{c}\right)^m, \qquad t \geq 0 \tag{8-63}$$

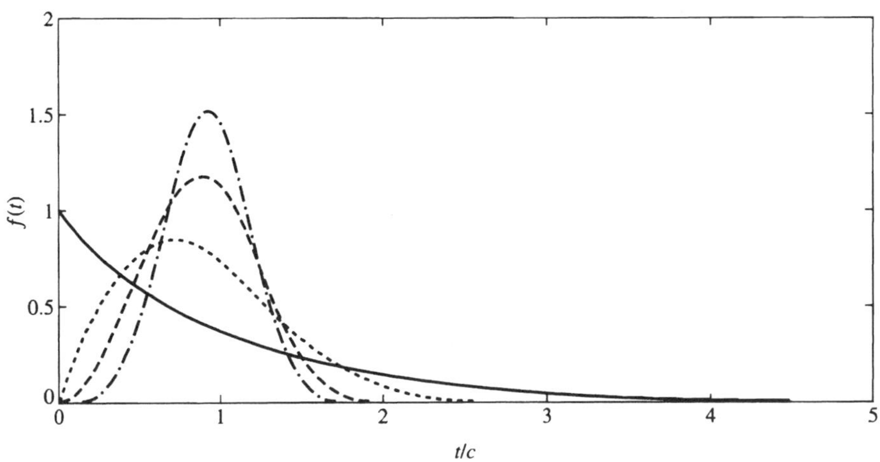

Figure 8-4 Weibull pdf: solid line, $m = 1$; short dashes, $m = 2$; long dashes, $m = 3$; alternating long and short dashes, $m = 4$.

It is of interest to be able to fit the Weibull model to data. To do so, we start with the cdf (8-57). Taking the natural logarithm of $1 - F(t)$, we have

$$-\ln[1 - F(t)] = \left(\frac{t}{c}\right)^m \tag{8-64}$$

Taking the natural logarithm again, we obtain

$$\ln\{-\ln[1 - F(t)]\} = m\ln(t) - m\ln(c) \tag{8-65}$$

Letting the dependent variable be the left-hand side and the independent variable be $\ln(t)$, this is of the form of a straight line where the slope is m and the intercept on the ordinate is $b = -m\ln(c)$.

We can use the following procedure to estimate these parameters. Given a set of data to which we wish to fit a Weibull model, we first construct the cdf histogram, subtract the values from 1, take the negative natural logarithm and the natural logarithm of that, and plot that versus the logarithm of the cdf histogram bin locations. To these points we fit a straight line using regression techniques (see Chapter 5). The slope of this line is the estimate of the parameter m. The ordinate-axis intercept gives $b = -m\ln(c)$, which, by using the value found for m, can be solved for the parameter c.

We illustrate this procedure in Figure 8-6. Two hundred random numbers were generated according to a Weibull probability model with $c = 40$ and $m = 24$. Figure 8-6a shows the pdf histogram, and Figure 8-6b shows the cdf histogram. Figure 8-6c shows $\ln\{-\ln[1 - F(t)]\}$ plotted versus $\ln(t)$, where the t's are the midpoints of the bin values for the histogram. Finally, Figure 8-6d shows a least-mean-squares straight-line fit to these data. The slope of this line is 22.83, which is the estimate for m. The vertical-axis intercept

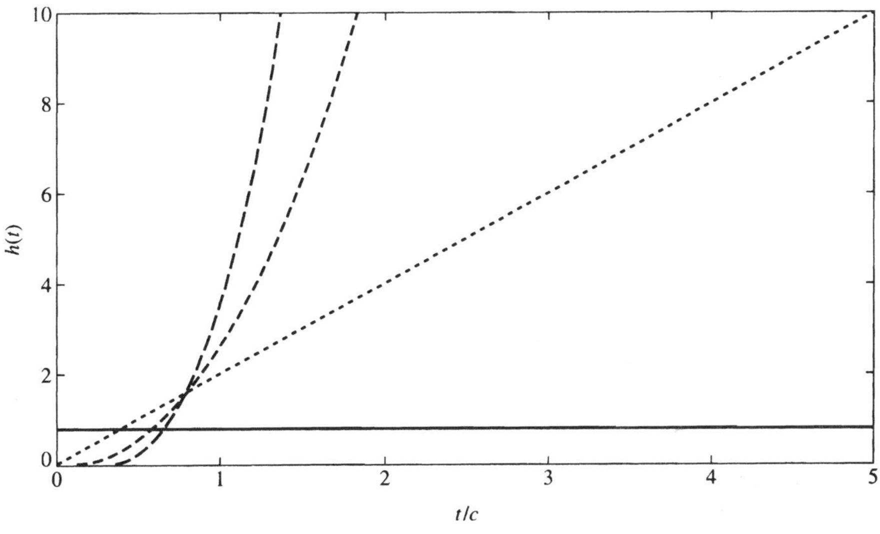

Figure 8-5 Failure rate for Weibull model: solid line, $m = 1$; short dashes, $m = 2$; medium dashes, $m = 3$; long dashes, $m = 4$.

is -83.97 (off the scale to the left), which is an estimate for $-m\ \ln(c)$. Solving for the estimate of c, we obtain

$$\hat{c} = e^{-\hat{b}/\hat{m}} = e^{-(-83.97)/22.83} = 39.54 \tag{8-66}$$

where the circumflexes indicate estimated values. This compares favorably with the actual value of $c = 40$.

8–5 SUMMARY

The topic considered in this chapter is the reliability of systems. The reliability of a system can be defined as the *probability that a system is functioning at time t after being put in service at time zero.* We first considered the time-dependent nature of reliability and then looked at ways to obtain the reliability of a combination of subsystems.

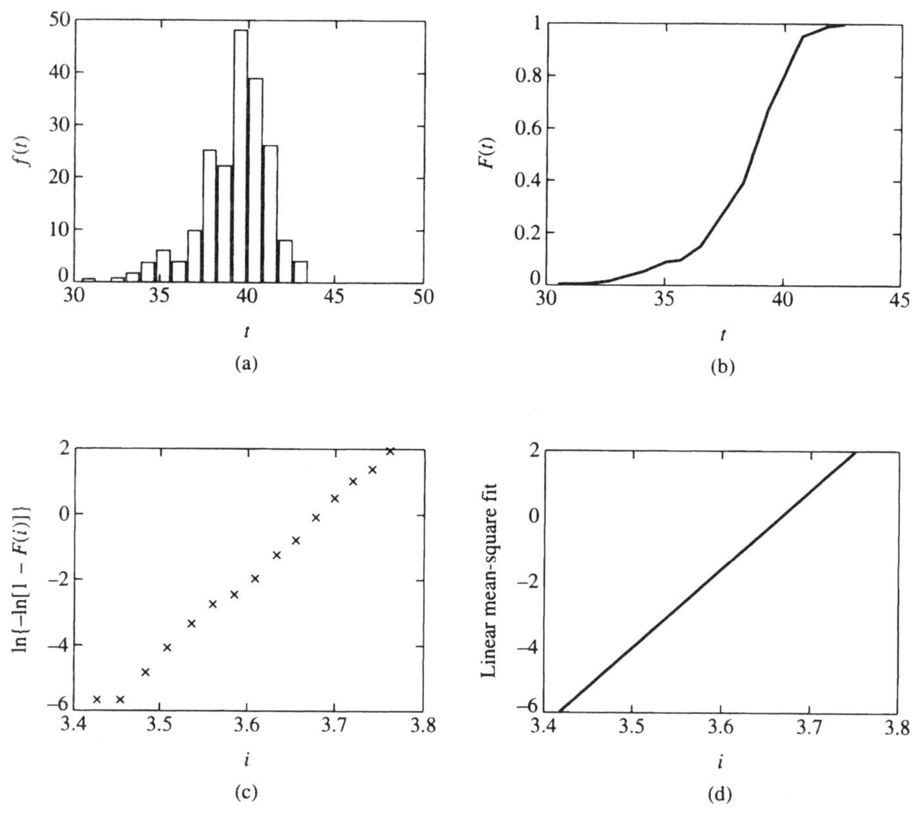

Figure 8-6 Histograms and least-mean-squares straight-line fit pertinent to fitting a Weibull model to data: (a) pdf histogram of data; (b) cdf histogram of data; (c) log-log plot of cdf histogram; (d) straight-line least-squares fit to log-log plot.

1. Let the time to failure of a system by denoted by T. The reliability is defined as

$$R(t) = 1 - F(t) = P(T > t)$$

where $F(t)$ is the cdf of the time to failure.

2. The mean time to failure is defined as

$$E(T) = \int_0^\infty t f(t)\, dt = \int_0^\infty R(t)\, dt$$

where $f(t)$ is the failure pdf, defined as

$$f(t) = \frac{dF(t)}{dt}$$

3. The failure rate of a system can be shown to be

$$h(t) = \frac{1}{n(t)} \frac{dn}{dt} = \frac{f(t)}{R(t)}$$

where $n(t)$ is the number of systems in service at time t. It can also be expressed as

$$h(t) = -\frac{d}{dt} \ln R(t)$$

4. If several subsystems must operate successfully in series for a system to operate successfully, the reliability of the system is

$$R_s(t) = \prod_{i=1}^{n} R_i(t)$$

where $R_i(t)$ is the reliability of the ith subsystem. The failure rate for a system composed of a series of subsystems is

$$h_s(t) = \sum_{i=1}^{n} h_i(t)$$

where $h_i(t)$ is the failure rate of the ith subsystem.

5. A system consisting of several subsystems in parallel has a reliability of

$$R_p(t) = 1 - \prod_{i=1}^{n} \left[1 - R_i(t)\right]$$

where $R_i(t)$ is the reliability of the ith subsystem. There is no simple expression for the failure rate as in the case of a series system.

6. A standby system in which a primary subsystem that goes bad is replaced by a standby subsystem has a mean time to failure of

$$E(T_{sb}) = E(T_1) + E(T_2)$$

where T_i is the time to failure of subsystem i. Their failure pdf is

$$f_{sb}(t) = \int_{-\infty}^{\infty} f_1(t - \eta) f_2(\eta)\, d\eta$$

where $f_i(t)$ is the failure pdf of subsystem i.

7. The Weibull random variable is characterized by the cdf

$$F(t) = 1 - e^{-(t/c)^m}, \qquad t \geq 0$$

and the pdf

$$f(t) = \frac{m}{c}\left(\frac{t}{c}\right)^{m-1} e^{-(t/c)^m}, \qquad t \geq 0$$

When applied to characterizing the failure of a system, the mean time to failure is

$$E(T) = \int_0^\infty t f_T(t)\, dt = \int_0^\infty R(t)\, dt = c\Gamma\left(1 + \frac{1}{m}\right)$$

where $\Gamma(\,\cdot\,)$ is the gamma function and $R(t) = 1 - F(t)$ is the reliability. Its failure rate is

$$h(t) = \frac{m}{c}\left(\frac{t}{c}\right)^{m-1} \qquad t \geq 0$$

It is handy because it can be used to model a wide variety of failure mechanisms through choice of the parameters m and c.

8-6 FURTHER READING

The books by Papoulis (1990, 1993) and Williams (1991) are particularly appropriate reading for the subject material of this chapter.

8-7 PROBLEMS

Section 8-2

8-1. Fill in the steps in the derivation of (8-5).

8-2. A certain brand of automobile battery has an exponential reliability. What value of λ will ensure that 50% of the batteries fail after their 5-year warranty period?

8-3. A light bulb has a constant failure rate of $\lambda = 0.1$ bulb per year.

(a) What is the reliability? What are the dimensions of t in the reliability expression?

(b) What is the mean time to failure? Give its dimensions.

8-4. A certain microchip has a constant failure rate of 2 chips per million per 1000 hours [this is abbreviated either as 2 PPM/K (*parts per million per 1000 hours*) or 2 FIT (*failures in time*)]. What is the mean time to failure in years?

8-5. Plot failure rate curves corresponding to the acturial data given in Table 8-1.

8-6. A certain device has a failure pdf given by

$$f(t) = \begin{cases} \dfrac{2t}{T^2}, & 0 \leq t \leq T \\ 0, & \text{otherwise} \end{cases}$$

(a) Find and plot the corresponding reliability and failure rate curves as a function of t/T.

(b) What is the mean time to failure?

Section 8-3

8-7. Given the combinations of subsystems shown in Figure 8-7. Assume a reliability for each of r and a failure probability for each of $f = 1 - r$. Express the reliability of the overall system as functions of r and f.

8-8. The computer on a certain space mission must have a reliability of 0.999. It is decided to parallel three less reliable computers (triple redundancy). What must be the reliability of each component computer, assuming that the reliabilities of all are the same?

8-9. A certain component must be extremely reliable in a system. Yet the reliability of the component is only 0.95. How many must be paralleled to achieve an overall reliability for this particular part of the system of 0.999?

8-10. An indicator light with a mean time to failure of 100 days must indicate the proper operation of a certain system for 200 days, so two lights are used in standby mode. Using the fact that the mean time to failure is a constant (i.e., exponential failure pdf) find the failure pdf of the two lights in standby configuration. What is the mean time to failure of the standby configuration?

8-11. A circuit board has 20 integrated circuits with a constant failure rate of 5 FITs (see Problem 8-4 for the definition of FIT), 300 resistors with a constant failure rate of 20 FITs, and

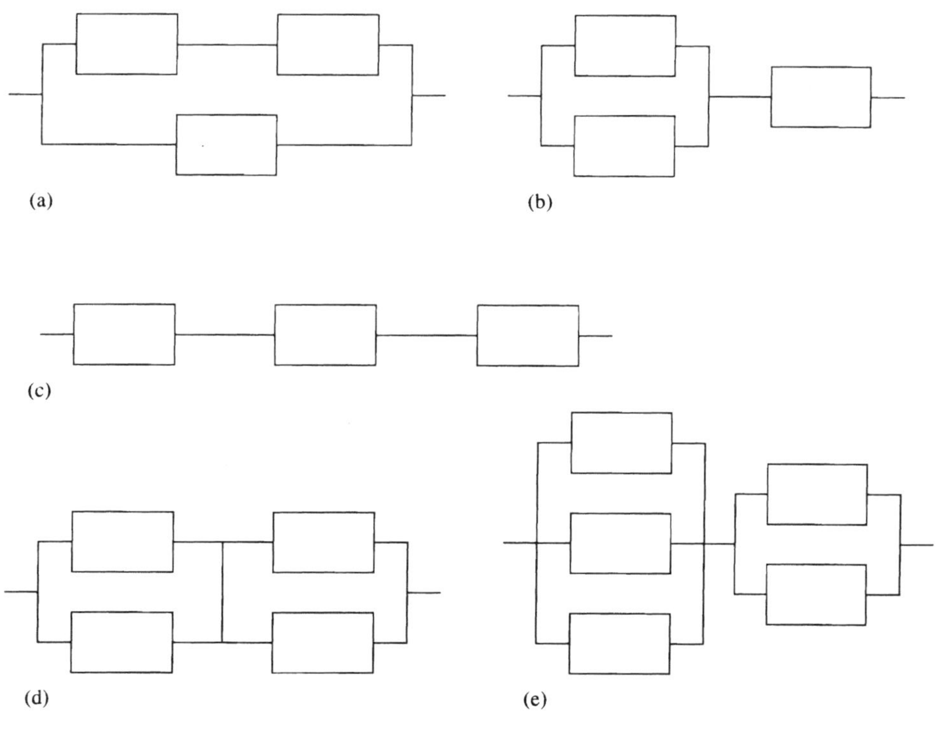

Figure 8-7

10 diodes with a constant failure rate of 10 FITs. All must work for the circuit board to work (implied series connection as far as failure is concerned). At some arbitrary time, the circuit board works properly. What is the probability that it will be working 5000 hours hence. (*Hint:* Note that the constant failure rates indicate an exponential reliability for each component.)

Section 8-5

8-12. A system is modeled as having a reliability distribution that is Weibull with $m = 2$ and $c = 10$ s.

(a) Find the reliability.

(b) What is the mean time to failure?
(*Hint:* The integral $\int_0^\infty e^{-at^2}\,dt = \frac{1}{2}\sqrt{\pi/a}$ will be useful.)

8-8 COMPUTER EXERCISES

8-1. Write a Monte Carlo simulation to evaluate system reliability of series–parallel combinations.

8-2. Write a computer program to show the curves of Figure 8-3. Include a solver to find the number of standby parts required to provide a desired system reliability such as that worked in Example 8-11.

8-3. Write a computer program that generates a block of Weibull distributed numbers, and then fits a Weibull probability law to them as shown in Figure 8-6. Note that this is not realistic in that your program will generate random numbers according to a Weibull distribution and fit a Weibull distribution to them—you will ordinarily be given the former. Test your program by generating several examples of Weibull-distributed data and seeing how close the fits come to the actual data.

9 Introduction to Random Processes

9-1 INTRODUCTION

In Chapter 3, we introduced the idea of a random variable. There, a random variable was defined as a function that maps each point in a sample space to a point on the real line. We were therefore able to work with numbers rather than having to describe the outcomes of our chance experiments in words. For example, when tossing a coin, the outcomes of this chance experiment were described by the numbers 1 (head) and 0 (tail).

In this chapter, we take the concept of a random variable one step further. Now the mapping from the outcomes of our chance experiments will result in functions, usually of time, although most generally they could be functions of space or space and time. Examples are temperature measurements at periodic time instants at your residence, or the fluctuations of air pressure at some point on the surface of an aircraft, voltage fluctuations across a resistor at room temperature, quotes for a certain stock at closing time of the market, and so on. We will consider only examples in this book where the mappings from the sample space result in functions of a single variable, namely time.

For now, we denote this dual dependence on the sample space and time by the notation $X(t, \zeta)$, where the first variable in the parentheses refers to the time dependence and the second variable refers to the dependence on the outcome of the underlying chance experiment. Such a definition, along with the probability measure defined on the sample space, constitutes what is referred to as a *random process*. The second variable, the chance experiment-dependent one, defines what is referred to as an *ensemble of sample functions* of the random process (i.e., for ζ fixed we get a function of time called a sample function).

On the other hand, for each fixed $t = t_k$ from the index set, I, $X(t_k, \zeta)$ is a random variable. Later we suppress the ζ and write simply $X(t)$ for the sample functions.

Random processes[1] can be either discrete time or continuous time, depending on whether the index set I is a countable set (i.e., can be put into one-to-one correspondence with the integers) or continuous (e.g., the real line), respectively. In addition, they can be discrete valued or continuous valued, depending on whether the values the sample functions $X(t, \zeta)$ take on at any specific t are countable or continuous, respectively.

In the next section we consider examples of each type of random process (i.e., discrete time/discrete valued, discrete time/continuous valued, continuous time/discrete valued, and continuous time/continuous valued).

9-2 EXAMPLES OF RANDOM PROCESSES

Example 9-1

As an example of a discrete-time/discrete-valued random process, let the index set I be $\{0, 1, 2, \ldots\}$. At each $t_k \in I$ assume that a fair coin is flipped. If a head is obtained, the amplitude value 1 is assigned, whereas if a tail is obtained, the amplitude value -1 is assigned. Several typical sample functions are shown in Figure 9-1.

Example 9-2

An example of a discrete-time/continuous-valued random process is provided by the recurrence relationship

$$X_{k+1} = X_k + e_k, k = 0, 1, 2, \ldots \tag{9-1}$$

where $\{e_0, e_1, \ldots\}$ is a sequence of random variables that are mutually independent; each member of the sequence is distributed according to the pdf $f_E(e)$. Example sample functions are shown in Figure 9-2 for $f_E(e)$ uniform in the interval $(-1, 1)$.

Example 9-3

An example of a continuous-time/continuous-valued random process is provided by the family of waveforms

$$X(t, \Theta) = A\cos(\omega_0 t + \Theta) \tag{9-2}$$

where A and ω_0 are constants and Θ is a random variable, which for this discussion will be assumed uniformly distributed in the interval $[0, 2\pi)$. An ensemble of sample functions for this random process is illustrated in Figure 9-3. Note that the appearance of a single sample function is not random—it is a periodic signal. However, looking down the ensemble at some time t, we cannot predict what the amplitudes of the sample functions are due to the random starting phases.

Example 9-4

An example of a continuous-time/discrete-valued random process is provided by setting the ordinates of the waveforms shown in Figure 9-3 to 1 if they are greater than zero and to -1 if less than zero. This operation is called *hard limiting*. A hard-limited version of the third sample function of Figure 9-3 is shown in Figure 9-4.

[1]In the mathematical probability literature, random processes are also referred to as *stochastic processes*. In the engineering literature they are sometimes called *random signals*.

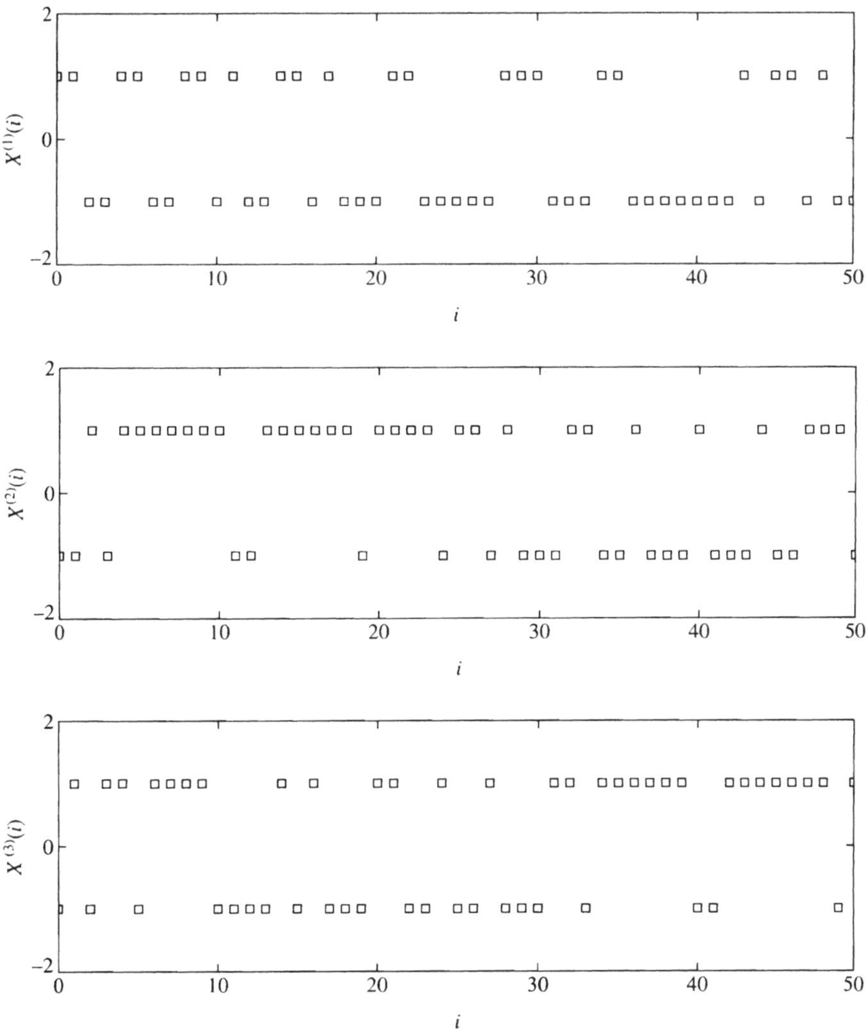

Figure 9-1 Sample functions for a discrete-time/discrete-valued random process.

9–3 STATISTICAL DESCRIPTIONS OF RANDOM PROCESSES

Probability Density Functions

At a given instant in time, a random process can be described by the first-order, or single-variable, pdf of its possible values at that time. Thus if $X(t)$ represents the random process at time t, we denote the first-order pdf, referring to the possible values of the random process at time t, by $f_X(x, t)$. According to (3-14), it can be interpreted as

$$f_X(x, t)\, dx = P[x < X(t) \leq x + dx \text{ at time } t] \tag{9-3}$$

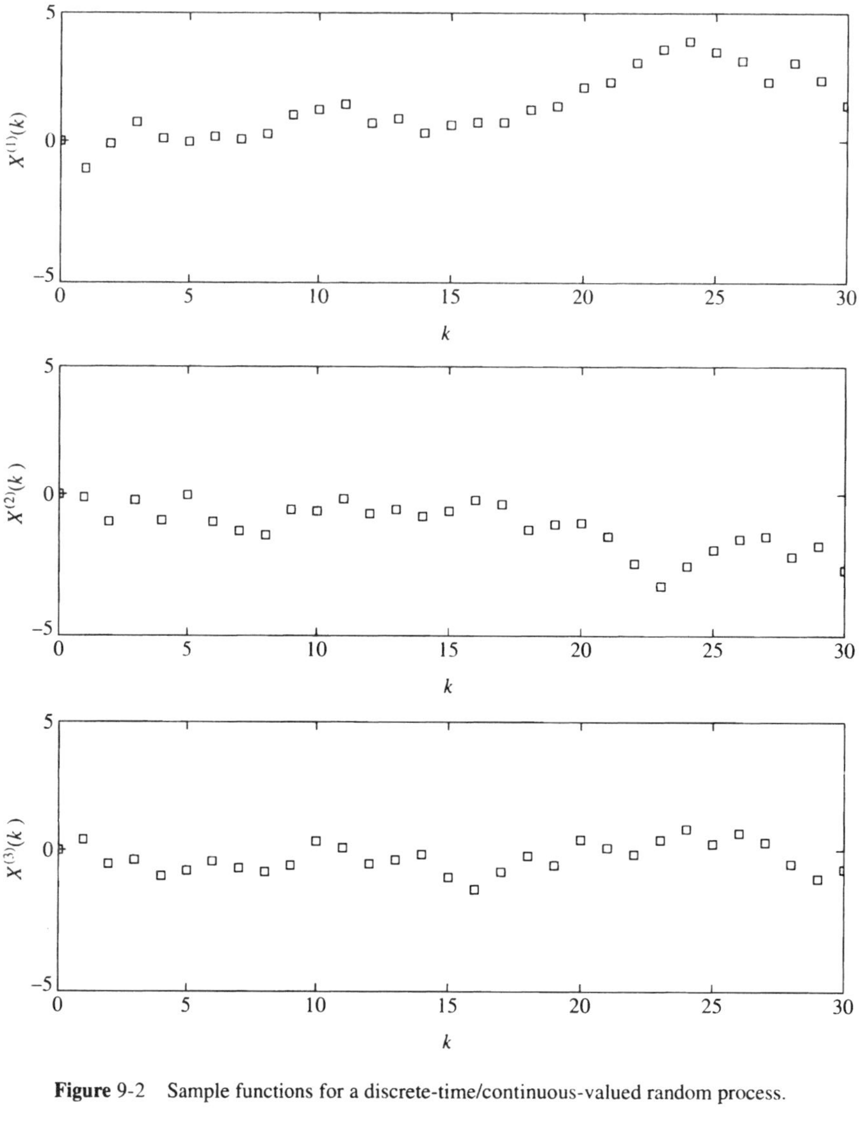

Figure 9-2 Sample functions for a discrete-time/continuous-valued random process.

Similarly, if we are interested in the statistical description of the random process at two time instants, say t_1 and t_2, we use the joint pdf. It can be interpreted as

$$f_{X_1X_2}(x_1, t_1; x_2, t_2)\,dx_1\,dx_2 = P[x_1 < X(t_1) \le x_1 + dx_1 \text{ and } x_2 < X(t_2) \le x_2 + dx_2] \quad (9\text{-}4)$$

A complete statistical description of the random process would consist of the joint pdf of the values of the random process at N arbitrarily chosen time instants. This, of course, requires a tremendous amount of data if we were to determine an empirical distribution as described in Chapter 5. Only in special cases is it possible to obtain enough information to specify the N-fold joint pdf of a random process. Often, we settle for less in the way of a statistical description (e.g., certain averages).

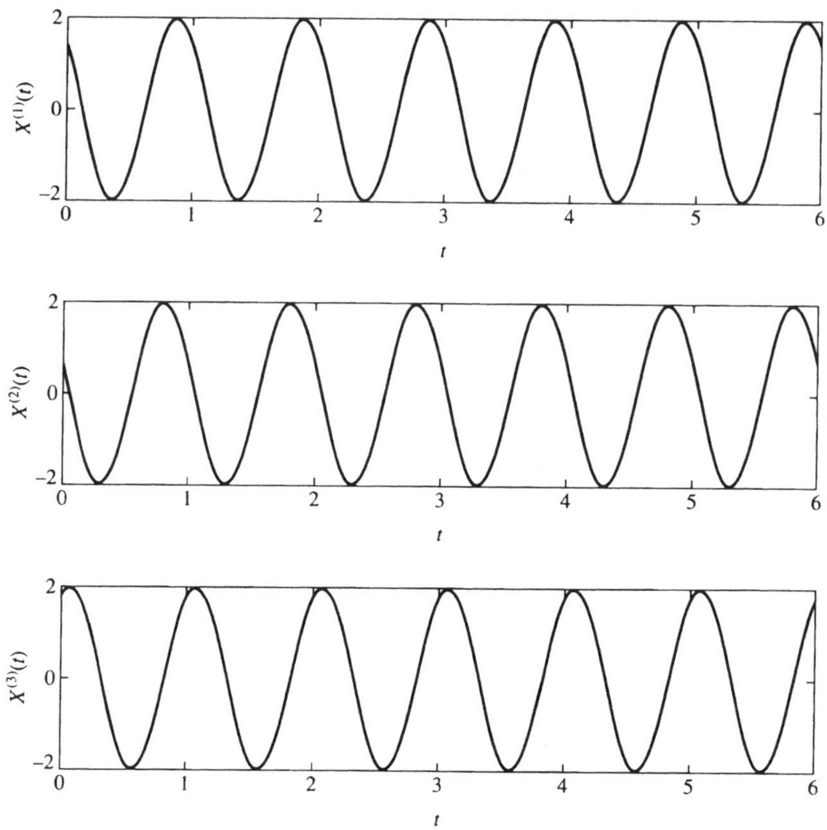

Figure 9-3 Sample functions for a sinusoidal random process with random phase in $(0, 2\pi)$.

Ensemble Averages

One type of average for a random process is obtained by using the pdf's described above to compute them. First-order averages are obtained by using the first-order pdf. For example, if $g[X(t)]$ is some function of the random process at time t, the average of this function of the random process, by (3-82), is

$$E\{g[X(t)]\} = \int_{-\infty}^{\infty} g(x) f_X(x, t)\, dx \tag{9-5}$$

Note that it is, in general, a function of time, although in certain cases it may not be. We can extend this idea to functions of the random process at two time instants, say t_1 and t_2. In this case we would evaluate the integral

$$E\{h[X(t_1), X(t_2)]\} = \int_{-\infty}^{\infty}\int_{-\infty}^{\infty} h(x_1, x_2) f_{X_1X_2}(x_1, t_1; x_2, t_2)\, dx_1\, dx_2 \tag{9-6}$$

where $h(\cdot\,, \cdot)$ is a function of two variables. Note that this average is, in general, a function of both t_1 and t_2.

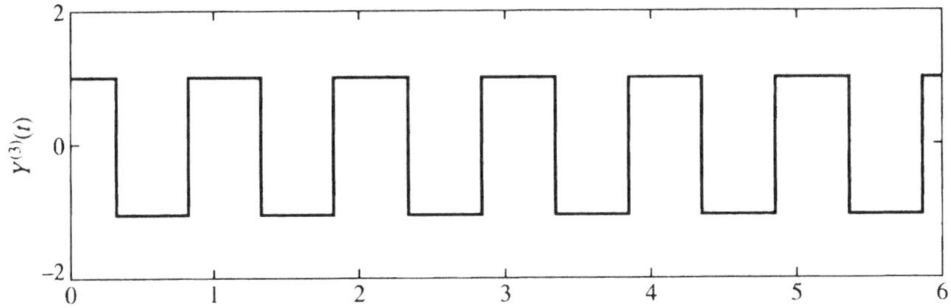

Figure 9-4 Sample function of a continuous-time/discrete-valued random process.

Important single-time averages are the mean and mean square of the random process. These are obtained by letting $g(x)$ be x and x^2, respectively, in (9-5). An important two-time average is the autocorrelation function, which is obtained by letting $h[X(t_1), X(t_2)] = X(t_1)\,X(t_2)$, so that

$$R_X(t_1, t_2) = \int_{-\infty}^{\infty}\int_{-\infty}^{\infty} x_1 x_2 f_{X_1X_2}(x_1, t_1; x_2, t_2)\,dx_1\,dx_2 \tag{9-7}$$

is the definition of the ensemble-average autocorrelation function. These averages will be discussed in more detail shortly. The *autocovariance* is the covariance of the random variables $X(t_1)$ and $X(t_2)$.

Example 9-5

Consider the random process defined by (9-2) and the accompanying discussion. Find its statistical average mean and autocorrelation function.

Solution Using the concept of an average of a function of a random variable, we have

$$E[X(t;\theta)] = E[g(\theta)] = E[A\cos(\omega_0 t + \Theta)] = \int_0^{2\pi} A\cos(\omega_0 t + \theta)\frac{d\theta}{2\pi} = 0 \tag{9-8}$$

which follows by noting the cosine has as much area above the abscissa as below it over a full period. Similarly, the autocorrelation function is computed by using (9-7) with $t_1 = t$ and $t_2 = t + \tau$ (since Θ is the only random variable, we only need to average over it):

$$\begin{aligned} R_X(t, t+\tau) &= E[X(t)X(t+\tau)] = E\{A\cos(\omega_0 t + \Theta)A\cos[\omega_0(t+\tau) + \Theta]\} \\ &= \int_0^{2\pi} A\cos(\omega_0 t + \theta)\,A\cos[\omega_0(t+\tau) + \theta]\frac{d\theta}{2\pi} \\ &= \frac{A^2}{2}\int_0^{2\pi}[\cos(\omega_0\tau) + \cos(\omega_0\tau + \omega_0\tau + 2\theta)]\frac{d\theta}{2\pi} \end{aligned} \tag{9-9}$$

The last integral was obtained by using the trigonometric identity

$$\cos(u)\cos(v) = \tfrac{1}{2}\cos(u+v) + \tfrac{1}{2}\cos(u-v)$$

We can separate the integrals in (9-9) into two integrals to obtain

$$\begin{aligned} R_X(t, t+\tau) = R_x(\tau) &= \frac{A^2}{2}\left[\int_0^{2\pi}\cos(\omega_0\tau)\frac{d\theta}{2\pi} + \int_0^{2\pi}\cos(2\omega_0 t + \omega_0\tau + 2\theta)\frac{d\theta}{2\pi}\right] \\ &= \frac{A^2}{2}\cos(\omega_0\tau) \end{aligned} \tag{9-10}$$

where the last integral over θ is zero because it is the integral of a sinusoid over two periods (if you don't believe this, carry out the integration and substitute the limits). Note that in this case, the autocorrelation function depends only on $\tau = t_2 - t_1$ because the second integral is zero. The mean-square value of $X(t)$ is $R_X(0) = A^2/2$, and its root-mean-square (rms) value is $A/2^{1/2}$.

Strictly Stationary Random Processes

A random process is *strict-sense stationary* if its first-order pdf is independent of time, if its two-time joint pdf depends on the time difference $t_2 - t_1$, and in general, if its N-time joint pdf depends on the time differences $t_2 - t_1, t_3 - t_1, \ldots, t_N - t_1$. That is, the time we pick for the origin makes no difference. A strict-sense stationary random process has a mean and mean square that are independent of time and an autocorrelation function that is dependent only on the time difference $\tau = t_2 - t_1$. Of course, higher-order time averages will be functions of time differences as well because of the time-difference property of the N-fold joint pdf of a strict-sense stationary random process.

Wide-Sense-Stationary Random Processes

If a random process has a mean and mean square that are independent of time, and an autocorrelation function that is dependent only on the time difference $\tau = t_2 - t_1$, but that's all that we can say, it is referred to as *wide-sense stationary*. Higher-order ensemble averages may not be functions of time differences, as they must be for a strict-sense stationary random process. It is apparent that a strict-sense stationary random process is also wide-sense stationary, but it is not necessarily true that a wide-sense-stationary random process is strict-sense stationary. An exception is the case of a Gaussian random process to be discussed later. Wide-sense-stationary Gaussian processes are also strict-sense stationary. The random process considered in Example 9-5 is wide-sense stationary.

Time Averages

For a stationary random process, we can find averages over time. For example, the time-average mean is[2]

$$\langle X(t)\rangle = \lim_{T\to\infty} \frac{1}{2T}\int_{-T}^{T} X(t)\,dt \tag{9-11}$$

The time-average mean-square value is

$$\langle X^2(t)\rangle = \lim_{T\to\infty} \frac{1}{2T}\int_{-T}^{T} [X(t)]^2\,dt \tag{9-12}$$

The time-average autocorrelation function is

$$R_X(\tau) = \langle X(t)X(t+\tau)\rangle = \lim_{T\to\infty} \frac{1}{2T}\int_{-T}^{T} X(t)X(t+\tau)dt \tag{9-13}$$

[2] We use angular brackets to denote time averages. Sometimes an overbar will be used to denote expectations. Thus $\langle X(t)\rangle$ denotes the time average mean and $\overline{X(t)}$ denotes the expectation, or statistical average of $X(t)$.

Since we are averaging over time, it is necessary that the random process be stationary. We could extend this idea to joint averages at more than two time instants. For example, the time average

$$\langle X(t)X(t+\tau_1)X(t+\tau_2)\rangle = \lim_{T\to\infty} \frac{1}{2T}\int_{-T}^{T} X(t)x(t+\tau_1)X(t+\tau_2)\,dt \qquad (9\text{-}14)$$

might be of interest in certain applications.

Ergodic Processes

The question naturally arises whether statistical averages are equal to their corresponding time averages. For example, is the statistical average mean equal to the time-average mean? Or is the statistical average autocorrelation function equal to the time-average autocorrelation function? For a certain class of random processes called *ergodic*, this is indeed the case. *For an ergodic process, all statistical averages are equal to their time-average equivalents.* This is indeed fortunate, for measurement of time averages is what can be implemented practically.

Example 9-6

Consider the random process defined by

$$X(t) = A, \qquad -\infty < t < \infty$$

where A is a random variable, which in general assumes a different value for each sample function. For any time, t, the ensemble average of this random process is simply the mean, m_A, of the random variable A. The time average corresponding to a given sample function is

$$\langle X(t)\rangle = \lim_{T\to\infty} \frac{1}{2T}\int_{-T}^{T} A\,dt = \lim_{T\to\infty} \frac{1}{2T}A(2T) = A \qquad (9\text{-}15)$$

which clearly depends on the sample function chosen and, in fact, is a random variable. Therefore, this random process cannot be ergodic.

One may naturally wonder what the nature of an ergodic process must be. In order that the time average of at function of a sample function be equal to the corresponding ensemble average, it must assume the character of the whole process. Roughly speaking, something new and unexpected must happen as we examine new segments of a sample function of an ergodic random process; yet each new segment must be "typical" of the process as a whole.

Example 9-7

Consider the random process defined by (9-2) and the accompanying discussion. Find its time-average mean and autocorrelation function.

Solution Using the concept of an average of a function of a random variable, we have

$$\langle X(t)\rangle = \lim_{T\to\infty} \frac{1}{2T}\int_{-T}^{T} A\cos(\omega_0 t + \Theta)\,dt = 0 \qquad (9\text{-}16)$$

which follows by noting that the cosine has as much area above the abscissa as below it over symmetrical limits about zero (or integrate it, substitute the limits, and take the limit). Similarly, the time-average autocorrelation function is

$$R_X(\tau) = \langle X(t)X(t + \tau)\rangle$$

$$= \lim_{T\to\infty} \frac{1}{2T}\int_{-T}^{T} A\cos(\omega_0 t + \theta)\, A\cos[\omega_0(t + \tau) + \theta]\, dt$$

$$= \frac{A^2}{2}\lim_{T\to\infty}\frac{1}{2T}\int_{-T}^{T}[\cos(\omega_0\tau) + \cos(2\omega_0 t + \omega_0\tau + 2\theta)]\, dt$$

$$= \frac{A^2}{2}\lim_{T\to\infty}\frac{1}{2T}\left\{\int_{-T}^{T}\cos(\omega_0\tau)dt + \int_{-T}^{T}\cos(2\omega_0 t + \omega_0\tau + 2\theta)\, dt\right\}$$

$$= \frac{A^2}{2}\cos(\omega_0\tau) \tag{9-17}$$

This is the same result as obtained in Example 9-5 for the ensemble average autocorrelation function. Is this process ergodic? We can't say for sure. Even though the statistical average mean equals the time-average mean, and the statistical-average autocorrelation function equals the time-average autocorrelation function, it is necessary for all possible statistical averages to be equal to the corresponding time averages for the random process to be ergodic. Sometimes the term *ergodic in the wide sense* is used to denote random processes like this one where time and ensemble averages are equal up to and including second order (i.e., the mean, variance, and autocorrelation function).

9–4 AUTOCORRELATION FUNCTION PROPERTIES

The autocorrelation function of a random process is an important two-time average. In case the random process is statistically stationary, the autocorrelation function is a function only of the difference of the two time instants, t_1 and t_2, at which this joint average is taken. This time difference is often denoted by the variable $\tau = t_2 - t_1$, and we shall do so here. We list several additional properties of the autocorrelation function of a stationary random process in this section, and then consider a random process for which the autocorrelation function is easily computed. Properties of the autocorrelation function of a stationary random process are as follows:

1. $|R_X(\tau)| \le R_X(0)$; that is, the value of the autocorrelation function at the origin is a relative maximum.
2. $R_X(-\tau) = R_X(\tau)$; that is, the autocorrelation function is an even function.
3. $\lim_{|\tau|\to\infty} R_X(\tau) = \{E[X(t)]\}^2$, provided that the process is ergodic and that this limit exists; that is, the value of the autocorrelation function as its argument approaches infinity is equal to the mean squared.
4. The autocorrelation function of a periodic random process (i.e., a random process whose sample functions are periodic functions of the independent variable) is also periodic.

5. The Fourier transform of the autocorrelation function, referred to as the *power spectral density*, is real and nonnegative.[3]

If, in addition, the random process is ergodic, we can interchange the statistical average and time average autocorrelation functions. We see that for a random process $X(t)$, the time-average autocorrelation function evaluated for $\tau = 0$ gives

$$R_X(0) = \lim_{T\to\infty} \frac{1}{2T}\int_{-T}^{T} X^2(t)\,dt = \langle X^2(t)\rangle \overset{\text{(by ergodicity)}}{=} E[X^2(t)] = R_X(0) \quad (9\text{-}18)$$

This is the average power of the random process, which is also the average power of each sample function.

The proofs of these properties are not difficult. We now briefly sketch how each may be proved. The first property is obtained by considering the nonnegative quantity

$$[X(t) \pm X(t+\tau)]^2 \geq 0 \quad (9\text{-}19)$$

We square this to obtain

$$X^2(t) \pm 2X(t)X(t+\tau) + X^2(t+\tau) \geq 0 \quad (9\text{-}20)$$

The expectation can now be taken term by term to give

$$E[X^2(t)] \pm 2E[X(t)X(t+\tau)] + E[X^2(t+\tau)] \geq 0 \quad (9\text{-}21)$$

For a stationary process, the first and last expectations are equal to $R_X(0)$; the middle expectation is equal to $R_X(\tau)$. Equation (9-21) therefore reduces to

$$R_X(0) \pm 2R_X(\tau) + R_X(0) \geq 0 \quad (9\text{-}22)$$

Rearranging the inequalities (one with the $+$ sign and one with the $-$ sign), we obtain

$$-R_X(0) \leq R_X(\tau) \leq R_X(0) \quad (9\text{-}23)$$

which is equivalent to property 1.

Property 2 can be proved by making the substitution $t = t + \tau$ in the definition of the statistical average autocorrelation function. We have

$$R_X(\tau) = E[X(t)X(t+\tau)] = E[X(t'-\tau)X(t')] = R_X(-\tau) \quad (9\text{-}24)$$

where in the second expectation, we note that for a stationary process it makes no difference what we take as the time origin (i.e., the final average does not depend on the independent variable, no matter what we call it).

Property 3 we justify heuristically by noting that as $|\tau|\to\infty$, the random variables $X(t)$ and $X(t+\tau)$ become statistically independent if the process is not periodic (the process cannot "remember" over an infinite time separation). The expectation of the product of two statistically independent random variables is the product of their expectations.[4] Hence

[3]If the student has not had any Fourier transform theory at this time, this property can be ignored. The Fourier transform of the autocorrelation function is called the *power spectral density* of the process. This is known as the *Weiner–Kinchine theorem*.

[4]Clearly this would not work for the nonergodic process defined in Example 9-6, or for any nonergodic process.

$$\lim_{\tau \to \infty} E[X(t)X(t + \tau)] = E[X(t)]E[X(t + \tau)] = \{E[X(t)]\}^2 \tag{9-25}$$

where we may replace $E[X(t + \tau)]$ by $E[X(t)]$ because of stationarity.

Property 4 follows from the time-average definition of the autocorrelation function (9-13). If each sample function $X(t)$ is periodic, so is the integrand:

$$\text{integrand} = X(t)X(t + \tau)$$

The integration over t does not alter the periodicity, although now the independent variable is τ because t has been integrated over.

Note that properties 1, 2, 4, and 5 are exemplified by the random process of Examples 9-3 and 9-7. For students not yet exposed to Fourier transforms, property 5 may be meaningless. Nevertheless, the Fourier transform (9-17) is

$$S_X(f) = \frac{A^2}{4}[\delta(f - f_0) + \delta(f + f_0)] \tag{9-26}$$

where $f_0 = \omega_0/2\pi$, f is the Fourier transform variable in hertz, and $\delta(f)$ is the unit impulse or delta function. The function $S_X(f)$ is the power spectral density. It gives the density of power of the random process with frequency. When integrated over all frequency, we obtain the total average power of the process. Recalling that integration over a delta function gives unity, we find that integration of (9-26) gives

$$\int_{-\infty}^{\infty} S_X(f)\,df = \frac{A^2}{4}\left[\int_{-\infty}^{\infty} \delta(f - f_0)\,df + \int_{-\infty}^{\infty} \delta(f + f_0)\,df\right] = \frac{A^2}{4}(1 + 1) = \frac{A^2}{2} \tag{9-27}$$

We recognize this as $R_X(0) = E[X^2(t)] = A^2/2$, or the average power of the process.

Example 9-8

A typical toaster uses about 1000 W to toast two slices of bread. Give an expression for the power spectral density of the power it extracts from your household wiring system when toasting.

Solution Although the process of toasting bread is a transient one that begins and ends with the toasting process, we idealize it as steady state during the process of toasting and apply (9-26). Thus the power spectral density for the toasting process is

$$S_{\text{toast}}(f) = 500[\delta(f - 60) + \delta(f + 60)]$$

where $f = 60$ Hz is the nominal frequency of the power. Note that the integral of this power spectral density function over all frequency gives 1000 W.

We close this section with another example of computing the autocorrelation function of a very special random process, the random telegraph wave.

Example 9-9

Consider a random process with sample functions having the properties:

1. The values taken on at any time instant t_0 are either A or $-A$, with each value taken on with equal probability.
2. The number k of switching instants in any time interval T obeys a Poisson distribution:

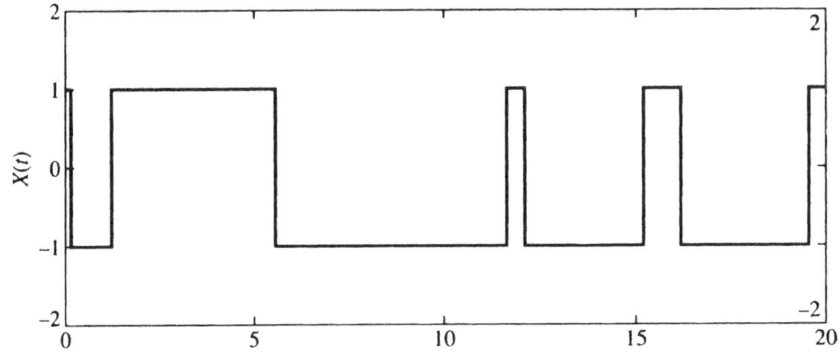

Figure 9-5 Sample function from a random telegraph wave process.

$$P_k = \frac{(\alpha T)^k}{k!} e^{-\alpha T}, \quad k = 0, 1, 2, \ldots \tag{9-28}$$

A typical sample function for this random process is shown in Figure 9-5. Find the autocorrelation function and power spectral density of this random process.

Solution Let τ be any positive time interval. Using the definition of the statistical average autocorrelation function, we have

$$\begin{aligned} R_X(\tau) &= E[X(t)X(t+\tau)] \\ &= A^2 P[X(t) \text{ and } X(t+\tau) \text{ have same sign}] \\ &\quad + (-A^2)P[X(t) \text{ and } X(t+\tau) \text{ have opposite sign}] \\ &= A^2 P[\text{even number of switching instants in } (t, t+\tau)] \\ &\quad - A^2 P[\text{odd number of switching instants in } (t, t+\tau) \end{aligned} \tag{9-29}$$

Using the Poisson distribution, (9-29) can be written as

$$\begin{aligned} R_X(\tau) &= A^2 \sum_{k=0,\, k \text{ even}}^{\infty} \frac{(\alpha\tau)^2}{k!} e^{-\alpha\tau} - A^2 \sum_{k=1,\, k \text{ odd}}^{\infty} \frac{(\alpha\tau)^k}{k!} e^{-\alpha\tau} \\ &= A^2 e^{-\alpha\tau} \sum_{k=0}^{\infty} \frac{(-\alpha\tau)^k}{k!} \\ &= A^2 e^{-\alpha\tau} e^{-\alpha\tau} \\ &= A^2 e^{-2\alpha\tau} \end{aligned} \tag{9-30}$$

A similar result can be derived for τ negative, with the result that the exponent is positive. Thus we can write the autocorrelation function as

$$R_X(\tau) = A^2 e^{-2\alpha|\tau|} \tag{9-31}$$

We have properties 1, 2, 3, and 5 for autocorrelation functions applying to this random process. Thus its average power is

$$P_{\text{tele. wave}} = A^2 \quad \text{W} \tag{9-32}$$

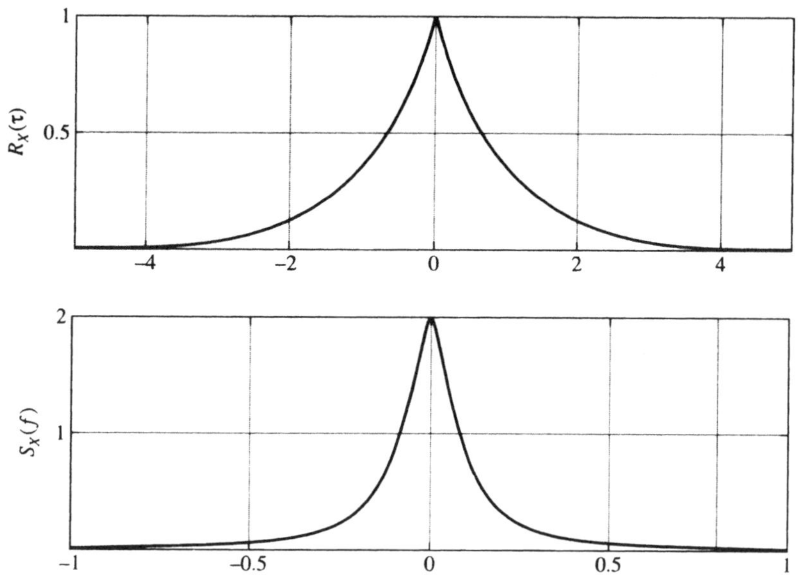

Figure 9-6 Autocorrelation function (a) and power spectral density (b) for a random telegraph signal ($\alpha = 0.5$).

and square of its mean is zero, which implies that the mean is zero (this is obvious from looking at a long-enough segment of the waveform and using ergodicity, which we assume is true). Obviously (9-31) is an even function of τ. For those students with prior exposure, the Fourier transform of (9-31) is

$$S_X(f) = \frac{A^2/\alpha}{1 + (\pi f/\alpha)^2} \tag{9-33}$$

The autocorrelation function and power spectral density of the random telegraph wave process are sketched in Figure 9-6.

When integrated over all frequencies, the power spectral density should give the average power of the process. We may check this by carrying out the integral

$$\begin{aligned} P &= \int_{-\infty}^{\infty} \frac{A^2/\alpha}{1 + (\pi f/\alpha)^2}\, df \\ &= 2\int_0^{\infty} \frac{A^2/\alpha}{1 + (\pi f/\alpha)^2}\, df \\ &= \frac{2A^2}{\pi}\int_0^{\infty} \frac{du}{1 + u^2} = \frac{2A^2}{\pi}\tan^{-1}\Big|_0^{\infty} = A^2 = P_{\text{tele. wave}} \end{aligned} \tag{9-34}$$

where the second integral follows by evenness of the integrand and the change of variables $u = \pi f/\alpha$ was used to get the last integral. Since we get the average power of the telegraph wave first found from (9-32), we gain confidence that the power spectral density is appropriately named.

9-5 CROSS-CORRELATION AND COVARIANCE FUNCTIONS

Joint Random Processes

It is often of interest to consider two related random processes, say $X(t)$ and $Y(t)$. An example is the input and output random processes of a system, such as a filter. To describe fully such joint random processes, as they are called, we would need the joint pdf of the values of each at an arbitrary number N of time instants, $f_{\mathbf{XY}}(x_1, t_1; x_2, t_2; \ldots; x_N, t_N; y_1, t_1; y_2, t_2; \ldots; y_N, t_N)$. If this joint pdf is independent of the time origin, or a function only of the time differences $t_2 - t_1, t_3 - t_1, \ldots, t_N - t_1$, the processes are said to be *jointly stationary in the strict sense*. If their joint moments of second order are either constants (i.e., the one-time averages, such as variances) or functions of time differences (i.e., the two-time averages, such as $E[X(t_1)Y(t_2)]$), the processes are said to be *jointly stationary in the wide sense*.

Cross-Correlation Function

Consider two wide-sense jointly stationary random processes, $X(t)$ and $Y(t)$. Their cross-correlation function is defined as

$$R_{XY}(\tau) = E[X(t)Y(t + \tau)] \tag{9-35}$$

If the cross-correlation function is zero, the processes are called *uncorrelated* or *orthogonal*. For statistically independent random processes, the expectation in (9-35) can be factored into the product of the separate expections over the two random processes:

$$\begin{aligned} R_{XY}(\tau) &= E[X(t)]E[Y(t + \tau)] && \text{statistically independent processes} \\ &= E[X(t)]E[Y(t)] && \text{stationary and independent} \end{aligned} \tag{9-36}$$

The cross-correlation function of two jointly stationary random processes has the following properties:

1. $R_{XY}(-\tau) = R_{YX}(\tau)$
2. $|R_{XY}(\tau)| \leq [R_X(0)R_Y(0)]^{1/2}$
3. $|R_{XY}(\tau)| \leq \frac{1}{2}[R_X(0) + R_Y(0)]$

The proofs of these properties can be carried out in a manner similar to those of the autocorrelation function and will be left to the problems.

Example 9-10

Consider the random processes of Example 9-3 as the input to a filter that simply modifies the amplitude of each sample function to B and shifts the phase by a fixed amount ϕ. Find the input and output autocorrelation functions and the cross-correlation function between input and output. Show that the foregoing properties are satisfied.

Solution From Example 9-5 we have the autocorrelation function of the input, given by (9-10). From this result it is clear that the autocorrelation function of the output is

$$R_Y(\tau) = \frac{B^2}{2}\cos(\omega_0\tau) \tag{9-37}$$

The cross-correlation function is found from

$$
\begin{aligned}
R_{XY}(\tau) &= E[X(t)Y(t+\tau)] = E\{A\cos(\omega_0 t + \theta)B\cos[\omega_0(t+\tau)+\theta+\phi]\} \\
&= \frac{AB}{2}E\{\cos(\omega_0\tau+\phi) + \cos[\omega_0(2t+\tau)+2\theta+\phi]\} \qquad (9\text{-}38)\\
&= \frac{AB}{2}\cos(\omega_0\tau+\phi)
\end{aligned}
$$

A similar derivation shows that

$$R_{YX}(\tau) = \frac{AB}{2}\cos(\omega_0\tau - \phi) \qquad (9\text{-}39)$$

It is readily verified that the properties for the cross-correlation function given above are satisfied. For example, property 1 becomes

$$R_{XY}(-\tau) = \frac{AB}{2}\cos(-\omega_0\tau+\phi) = \frac{AB}{2}\cos(\omega_0\tau-\phi) = R_{YX}(\tau)$$

by the eveness of the cosine. Property 2 becomes

$$|R_{XY}(\tau)| = \frac{AB}{2}|\cos(\omega_0\tau+\phi)| \le \frac{AB}{2}$$

because the absolute value of the cosine is bounded by 1. Property 3 gives exactly the same result as property 2 for this example.

Example 9-11

Express the autocorrelation function of a random process consisting of the sum of two other random processes in terms of their autocorrelation and cross-correlation functions.

Solution The desired autocorrelation function can be written as

$$R_Z(\tau) = E\{[X(t)+Y(t)][X(t+\tau)+Y(t+\tau)]\} \qquad (9\text{-}40)$$

Multiplying out the two sums and taking the expectation term by term, we obtain

$$
\begin{aligned}
R_Z(\tau) &= E\{[X(t)X(t+\tau) + X(t)Y(t+\tau) + Y(t)X(t+\tau) + Y(t)Y(t+\tau)]\} \\
&= R_X(\tau) + R_{XY}(\tau) + R_{YX}(\tau) + R_Y(\tau) \\
&= R_X(\tau) + R_{XY}(\tau) + R_{XY}(-\tau) + R_Y(\tau) \qquad (9\text{-}41)
\end{aligned}
$$

If the processes are orthogonal, (9-41) becomes

$$R_Z(\tau) = R_X(\tau) + R_Y(\tau) \qquad (9\text{-}42)$$

Setting $\tau = 0$ and recalling (9-18), we find that the power in the sum of two random processes is the sum of the powers in the separate random processes only if they are orthogonal.

Note that if $X(t) = Y(t)$, (9-41) becomes $R_Z(\tau) = 4R_X(\tau)$ and the power in $Z(t)$ is four times the power in $X(t)$. This is in stark contrast to (9-42), where the powers in the separate processes simply add. One can understand from this fact the importance of having soldiers not marching in step when crossing a long bridge span; or the greater efficiency of a laser light source, such as a laser diode, with all the emitted light coherent (or very nearly so), as opposed to a noncoherent light source such as a light-emitting diode.

Covariance Function

The covariance function of two jointly wide-sense random processes is found by first subtracting off their means before multiplication and taking the expectation of the product:

$$C_{XY}(\tau) = E\{[X(t) - E(X(t))][Y(t+\tau) - E(Y(t+\tau))]\} \qquad (9\text{-}43)$$

This can be simplified to

$$\begin{aligned} C_{XY}(\tau) &= E[X(t)Y(t+\tau)] - E[X(t)]E[Y(t+\tau)] \\ &= R_{XY}(\tau) - \mu_X\mu_Y \end{aligned} \tag{9-44}$$

where $R_{XY}(\tau)$ is their cross-correlation function, $\mu_X = E[X(t)]$, and $\mu_Y = E[Y(t)]$. Properties of the covariance function can be derived from those of the cross-correlation function. Note that the covariance function of orthogonal random processes is zero.

9–6 GAUSSIAN RANDOM PROCESSES

A random process is called *Gaussian* if its first-order pdf at an arbitrary time t is Gaussian as given by (3-28); if its joint pdf at two arbitrary times, t and $t + \tau$, is Gaussian as given by (4-52); and if its joint pdf at N time instants is the N-fold generalization of (4-52).[5] For a stationary process with zero mean, (4-52) simplifies to

$$f(x_1, x_2; \tau) = \frac{\exp\left\{-\dfrac{x_1^2 - 2r(\tau)x_1x_2 + x_2^2}{2\sigma^2[1 - r^2(\tau)]}\right\}}{2\pi\sigma^2\sqrt{1 - r^2(\tau)}} \tag{9-45}$$

In (9-45), x_1 refers to the possible values of the random process at time t, $X(t)$, and x_2 refers to the possible values of the random process at time $t + \tau$, $X(t + \tau)$. Since the process is stationary, the joint pdf does not depend on both time instants, t and $t + \tau$, but only on their difference, τ. The parameter σ^2 is the variance, which is time independent due to the stationarity of the random process. It is also equal to the mean-square value because the random process has zero mean. The function $r(\tau)$ is the normalized autocorrelation function, also called the *covariance,* given by

$$r(\tau) = \frac{R(\tau)}{\sigma^2} = \frac{R(\tau)}{R(0)} = \frac{E[X(t)X(t+\tau)]}{E[X^2(t)]} \tag{9-46}$$

From (9-46), we see that if a stationary Gaussian random process has zero mean, we can completely specify its twofold joint pdf by knowing the parameter σ^2 and function $r(\tau)$ [or equivalently, $R(\tau)$]. Gaussian processes are very important models for various applications, including noise, measurement errors, and corrupted signals. They have many important properties, perhaps the most important of which is that any linear operation on a Gaussian random process produces another Gaussian random process or Gaussian random variable.

Another important property is that the sum of two or more Gaussian random processes, independent or not, is Gaussian. This can be seen from Example 4-19.

Finally, because of the central limit theorem stated in Chapter 4, the superposition or addition of a large number of random processes that are not Gaussian will tend to a Gaussian random process provided that the first and second moments of the component random processes are finite. Quite often, the latter observation occurs when a non-Gaussian random

[5]This will not be given here, because we will need only the second-order pdf.

process is "filtered" by a fixed, linear system. The latter can be an electronic circuit designed to filter voltages, or a large vehicle such as an airplane. For example, the random pressure fluctuations on the covering of aircraft due to turbulence when heard in the cabin as sound variations are approximately Gaussian because they result from the superposition of a large number of turbulent eddies "filtered" by the aircraft structure.

Given the discussion above, it should be apparent that Gaussian random processes are very important models for analysis and design of systems subject to random variations. We do not have the time or the tools to delve into this fascinating subject in detail in this book. However, an example will illustrate the procedures typically used.

Example 9-12

Consider the integrator-thresholder shown schematically in Figure 9-7. The input is either

$$Y(t) = A + N(t),\ 0 \le t \le T \text{ or } Y(t) = -A + N(t),\ 0 \le t \le T \tag{9-47}$$

with equal probability. The parameters A and T are nonrandom and assumed specified for a particular application. The noise $N(t)$ is a Gaussian random process with zero mean and autocorrelation function

$$R(\tau) = \sigma^2 r(\tau) = \frac{N_0}{2}\delta(\tau) \tag{9-48}$$

where N_0 is a constant called the power spectral density of the noise and $\delta(\tau)$ is the unit impulse, or delta, function.[6] The output of the integrator at the end of T seconds is compared with a threshold set at zero. The object of this comparison is to determine whether the first component of $Y(t)$ is A or $-A$. (This is a very simple example of a signal detection problem that might occur, for example, in a digital communication system.) Determine the following:

(a) The probability of making an error given that A was really present

(b) The probability of making an error given that $-A$ was really present

(c) The average probability of making an error, which should be plotted as a function of A^2T/N_I

Solution The output of the integrator can be written as

$$Z = \pm AT + N_0 \tag{9-49}$$

where N_I is a Gaussian random variable given by

$$N_I = \int_0^T N(t)dt \tag{9-50}$$

It is Gaussian because it is the result of a linear operation on a Gaussian random process (the student will have to take this on faith as a result of the discussion above). We may write down its pdf by finding its mean and variance. Its mean is

$$E(N_I) = E\left[\int_0^T N(t)dt\right] = \int_0^T E[N(t)]dt = 0 \tag{9-51}$$

where the expectation operation and integral can be interchanged because the expectation operation is really another integral [of $N(t)$ times its first-order pdf]. The variance of N_I is

[6]Such a random process, called *white* because all frequencies are present in equal power and therefore resembles white light in this respect.

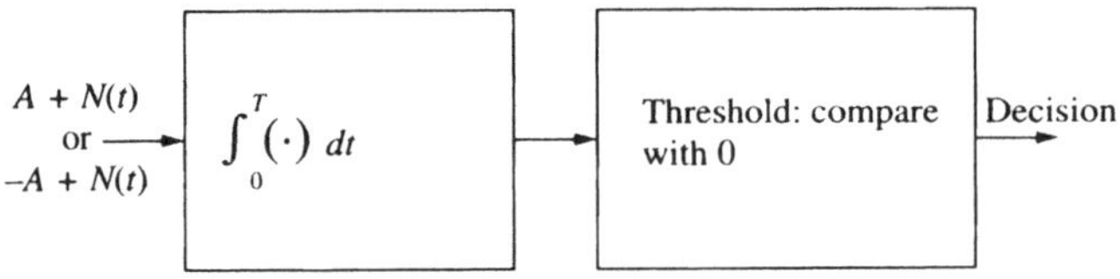

Figure 9-7 Integrator-threshold device for detecting a constant signal in Gaussian noise.

somewhat more complicated to find, but still straightforward. Since the mean of N_0 is zero, its variance and second moment are equal. Thus we calculate

$$\sigma_Z^2 = E(N_I^2) = E\left\{\left[\int_0^T N(t)dt\right]^2\right\} = E\left[\int_0^T\int_0^T N(t)N(\zeta)dtd\zeta\right]$$
$$= \int_0^T\int_0^T E[N(t)N(\zeta)]dtd\zeta = \int_0^T\int_0^T \frac{N_0}{2} r(\zeta - t)dtd\zeta$$
$$= \int_0^T\int_0^T \frac{N_0}{2}\delta(\zeta - t)dtd\zeta = \int_0^T \frac{N_0}{2}dt = \frac{N_0T}{2} \tag{9-52}$$

The step in going from the expectation of a square of integrals to the expectation of an iterated integral allows us to interchange the operations of expectation and integration over time.

(a) The probability of an error given that A was really present is

$$P(E\,|\,A \text{ present}) = P(AT + N_I < 0) = P(N_I < -AT)$$
$$= \int_{-\infty}^{-AT} \frac{e^{-z^2/2\sigma_Z^2}}{\sqrt{2\pi\sigma_Z^2}}dz = \int_{AT}^{\infty} \frac{e^{-z^2/2\sigma_Z^2}}{\sqrt{2\pi\sigma_Z^2}}dz \tag{9-53}$$

where the last integral follows by symmetry of the integrand. The change of variables

$$u = \frac{z}{\sigma_Z} \tag{9-54}$$

allows the integral in (9-53) to be put in the form

$$P(E\,|\,A \text{ present}) = \int_{AT/\sigma_Z}^{\infty} \frac{e^{-u^2/2}}{\sqrt{2\pi}}dz = Q\left(\frac{AT}{\sigma_Z}\right) \tag{9-55}$$

where $Q(\cdot)$ is the Q-function defined in Chapter 3 and tabulated in Appendix C.

(b) By symmetry, the probability of error given $-A$ was transmitted is the same.

(c) The average probability of error is

$$P_E = \tfrac{1}{2}P(E\,|\,A \text{ present}) + \tfrac{1}{2}P(E\,|\,-A \text{ present})$$
$$= Q\left(\frac{AT}{\sigma_Z}\right) = Q\left(\sqrt{\frac{2A^2T}{N_0}}\right) \tag{9-56}$$

where (9-52) has been substituted for σ_Z^2. The quantity A^2T/N_0 in (9-56) is called the signal-to-noise ratio. Figure 9-8 shows the probability of error plotted versus the signal-to-noise ratio, $SNR = A^2T/N_0$.

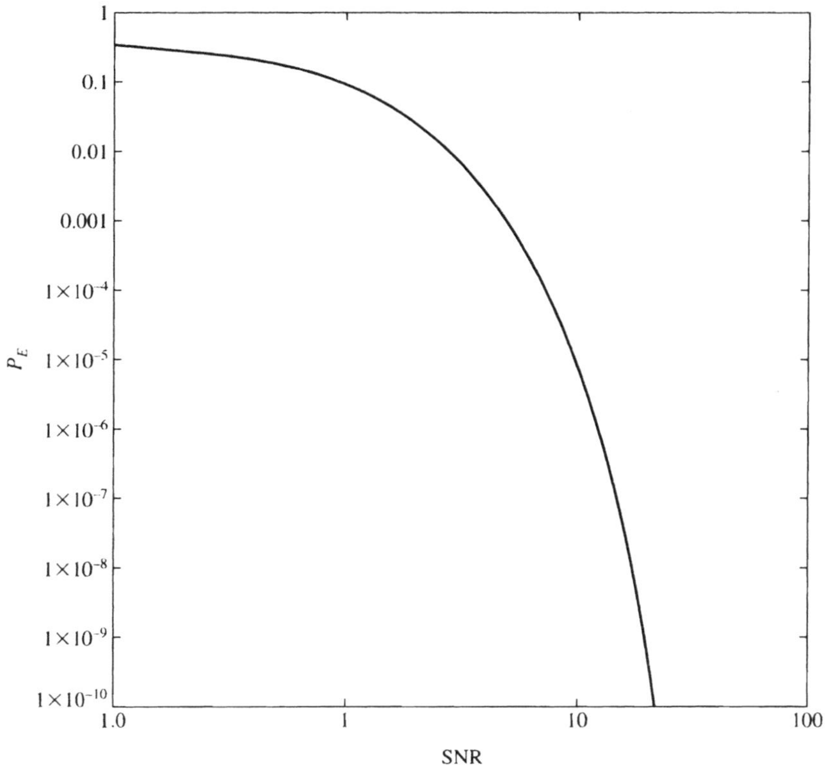

Figure 9-8 Probability of error versus signal-to-noise ratio for Example 9-12.

9–7 DISCRETE-TIME RANDOM PROCESSES: THE AUTOREGRESSIVE PROCESS

As a final example of a random process, we consider a discrete-time process defined by the difference equation

$$X_{k+1} = \alpha X_k + (1 - \alpha)N_k, \quad k = 0, 1, 2, \ldots \tag{9-57}$$

where α is a parameter with value in the interval $[0, 1]$ and the N_k are identically distributed, independent (iid) random variables. For simplicity, they will be assumed be Gaussian with zero means and variances σ_n^2. Although the most general case would be for X_{k+1} to depend on several past X_k's, we will assume dependence on only the immediately preceding one. This is referred to as a *first-order autoregressive random process*. Writing down (9-57) for several values of k to establish a pattern, we have

$$\begin{aligned} X_1 &= \alpha X_0 + (1 - \alpha)N_0 = (1 - \alpha)N_0 \\ X_2 &= \alpha X_1 + (1 - \alpha)N_1 \\ X_3 &= \alpha X_2 + (1 - \alpha)N_2 \\ X_4 &= \alpha X_3 + (1 - \alpha)N_3 \end{aligned} \tag{9-58}$$

where the initial value for X_k at $k = 0$ is assumed to be zero. Substitution of each equation into the immediately following equation gives

$$X_4 = (1 - \alpha)(\alpha^3 N_0 + \alpha^2 N_1 + \alpha N_2 + N_3) \tag{9-59}$$

Using the fact that the N_k's are iid and zero mean, we find the variance of X_4 to be

$$\begin{aligned}\text{Var}(X_4) &= (1 - \alpha)^2[\alpha^6 \text{Var}(N_0) + \alpha^4 \text{Var}(N_1) + \alpha^2 \text{Var}(N_2) + \alpha^0 \text{Var}(N_3)] \\ &= (1 - \alpha)^2(1 + \alpha^2 + \alpha^4 + \alpha^6)\sigma_n^2 \end{aligned} \tag{9-60}$$

Generalizing to arbitrary k from the pattern established in (9-60), we surmise that

$$\begin{aligned}\text{Var}(X_k) &= (1 - \alpha)^2(1 + \alpha^2 + \cdots + \alpha^{2(k-1)})\sigma_n^2 \\ &= (1 - \alpha)^2\frac{1 - \alpha^{2k}}{1 - \alpha^2}\sigma_n^2 = \frac{1 - \alpha}{1 + \alpha}(1 - \alpha^{2k})\sigma_n^2 \end{aligned} \tag{9-61}$$

which is a result that can be proved by induction. In obtaining (9-61), the summation formula for a geometric series was used, in particular,

$$\sum_{n=0}^{k-1} x^n = \frac{1 - x^k}{1 - x} \tag{9-62}$$

The process has a startup period that approaches steady-state behavior as the index k increases. The speed of this approach to steady state depends on the value of α; for α close to zero, this settling-out period is small, whereas for α close to 1, the settling-out period takes longer. A plot of the variance of the X_k's versus k provides a good indication of the duration required for the process to approach steady state. Examples for a sample function and the variance are plotted versus k in Figures 9-9 and 9-10 for the N_k's Gaussian with unit variance for $\alpha = 0.5$ and 0.95, respectively.

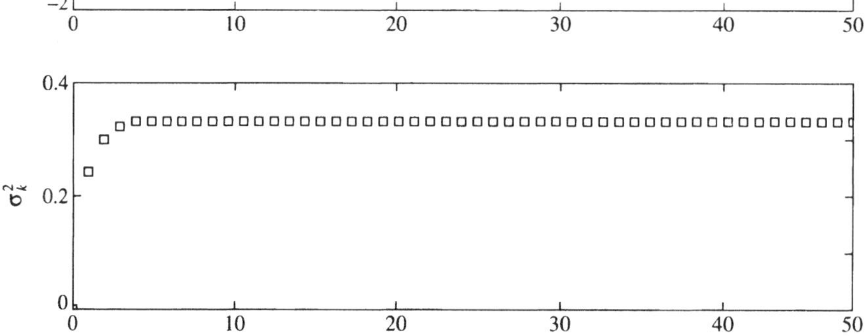

Figure 9-9 Sample function (a) and variance (b) versus the sequence index k for an autoregressive process with $\alpha = 0.5$.

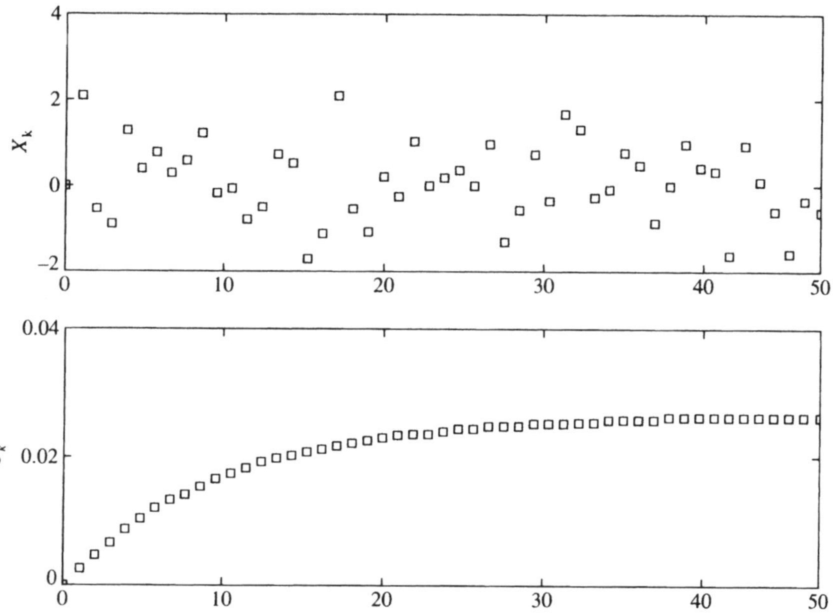

Figure 9-10 Sample function (a) and variance (b) versus sequence index number for an autoregressive process with $\alpha = 0.95$.

As k increases, Figures 9-9 and 9-10 indicate that the process "forgets" about its starting transient and approaches a stationary behavior. We now investigate the autocorrelation function of the process assuming that the transient buildup behavior has died out. Now we assume that the process began back at $k = -\infty$. First, we note that

$$\begin{aligned} R_X(0) = E(X_k^2) &= E\{[\alpha X_k + (1-\alpha)N_k]^2\} \\ &= E[\alpha^2 X_k^2 + 2\alpha(1-\alpha)X_k N_k + (1-\alpha)^2 N_k^2] \\ &= \alpha^2 E(X_k^2) + (1-\alpha)^2\sigma_n^2 = \alpha^2 R_X(0) + (1-\alpha)^2\sigma_n^2 \end{aligned} \tag{9-63}$$

where the fact that N_k is independent of X_k and is zero mean has been used (X_k depends on N_{k-1}, which is independent of N_k). Solving (9-63) for $E(X_k^{\,2}) = R_X(0)$, we obtain

$$R_X(0) = \frac{1-\alpha}{1+\alpha}\sigma_n^2 \tag{9-64}$$

Now consider $R_X(1)$, which can be written as

$$\begin{aligned} R_X(1) &= E(X_k X_{k+1}) \\ &= E\{X_k[\alpha X_k + (1-\alpha)N_k]\} \\ &= \alpha E(X_k^2) + (1-\alpha)E(X_k N_k) \\ &= \alpha R_X(0) \end{aligned} \tag{9-65}$$

where the fact that $E(X_k N_k) = 0$ has again been used. In general, we can show that

$$R_X(M) = \alpha^{|M|}R_X(0) = \alpha^{|M|}\frac{1-\alpha}{1+\alpha}\sigma_n^2 \tag{9-66}$$

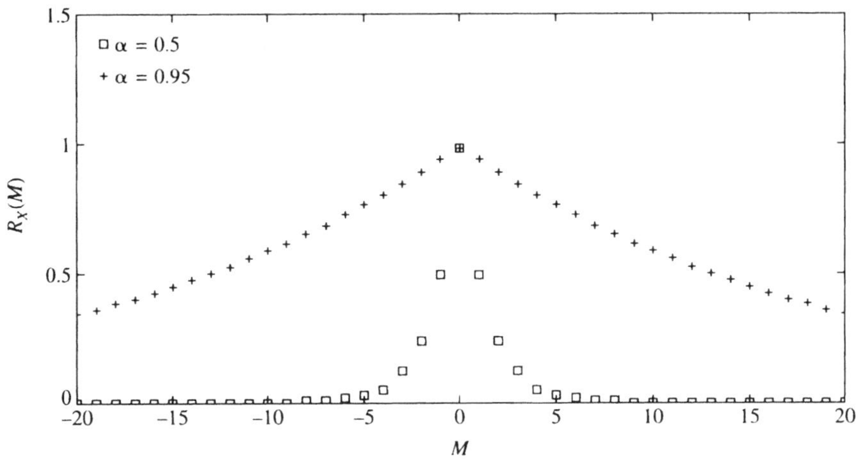

Figure 9-11 Autocorrelation function of an autoregressive process for $\alpha = 0.5$ (squares) and $\alpha = 0.95$ (plus signs) normalized to that maximum value is 1.

Plots of the autocorrelation function are shown in Figure 9-11 for $\alpha = 0.5$ and 0.95, respectively. Since the autocorrelation function shows interdependence between samples of the process separated by the independent variable, we see that the process "forgets" about its past (or is unable to predict its future)—more so as α becomes smaller.

9-8 SUMMARY

In this chapter, an introduction to random processes has been provided. Random processes are useful models for many situations, including random signals and noise. The following are the main topics covered in this chapter.

1. A random process is a mapping of each point of the sample space corresponding to a chance experiment to a function of one or more independent variables. Usually, we are concerned with the case of a single independent variable, which quite often represents time. We denote this dual functional relationship as $X(t, \zeta)$, where t refers to time and ζ refers to the particular outcome of the underlying chance experiment (i.e., point in the sample space).
2. A random process can be pictured by sketches of the possible *sample functions*—that is, the functions of time that result by letting the sample space variable represent different points in the sample space. A simple example is

$$X(t,\Theta) = A\cos(\omega_0 t + \Theta)$$

where A and ω_0 are constants and Θ is a random variable uniformly distributed in the interval $(0, 2\pi)$. (Here Θ corresponds to ζ in the general discussion above.) A specification of the sample functions, including the underlying probability law, is referred to as a *random process,* and the totality of all possible sample functions is called an *ensemble.*

3. For the single independent-variable case, we can have the following situations for a random process:
   (a) Discrete-valued independent variable and discrete–ordinate values
   (b) Discrete-valued independent variable and continuous–ordinate values
   (c) Continuous independent variable and continuous–ordinate values
   (d) Continuous independent variable and discrete–ordinate values

   Examples of these cases are provided by Figures 9-1, 9-2, 9-3, and 9-4, respectively.
4. In general, a random process is described completely by its Nth-order joint pdf at an arbitrary set of N time instants. Usually, we don't need this much information, and settle for the first-order pdf at a single time instant or the joint pdf at two time instants.
5. *Ensemble averages,* or *statistical averages,* of a random process are obtained by using the pdf's discussed in item 4. For example, the mean of a random process, computed as a statistical average, is

$$E[X(t)] = \int_{-\infty}^{\infty} x f_X(x, t)\, dx$$

   and the mean-square value is

$$E[X^2(t)] = \int_{-\infty}^{\infty} x^2 f_X(x, t)\, dx$$

   The autocorrelation function is a two-time average, given by

$$R_X(t, t + \tau) = E[X(t)X(t + \tau)] = \int_{-\infty}^{\infty}\int_{-\infty}^{\infty} x_1 x_2 f_{X_1} f_{X_2}(x_1, t; x_2, t + \tau)\, dx_1\, dx_2$$

6. A random process is *stationary* if its general N-fold joint pdf is a function only of the time differences $t_2 - t_1, t_3 - t_1, \ldots, t_N - t_1$. This means that its first-order pdf is time independent, and moments like the mean and second moment are time independent. The autocorrelation function, defined above, is a function only of the time difference, τ. Indeed, all N-time averages will be functions only of the time differences $t_2 - t_1, t_3 - t_1, \ldots, t_N - t_1$.
7. If a random process is stationary, *time averages* can be defined. Examples are the time-average mean, given by

$$\langle X(t)\rangle = \lim_{T\to\infty}\frac{1}{2T}\int_{-T}^{T} X(t)dt$$

   the time-average second moment, given by

$$\langle X^2(t)\rangle = \lim_{T\to\infty}\frac{1}{2T}\int_{-T}^{T} [X(t)]^2 dt$$

   and the time-average autocorrelation function, given by

$$R_X(\tau) = \langle X(t)X(t + \tau)\rangle = \lim_{T\to\infty}\frac{1}{2T}\int_{-T}^{T} X(t)X(t + \tau)dt$$

   Note that angular brackets are used to denote time averages.
8. An *ergodic process* is one for which statistical averages are equal to the corresponding time averages. For example, the statistical average mean, mean square,

and autocorrelation function are equal to the time average mean, mean square, and autocorrelation function. For a process to be ergodic, all statistical averages must equal their corresponding time averages.

9. Properties of the autocorrelation function of a stationary random process are:
   - **(a)** $|R(\tau)| \leq R(0)$; that is, the value of the autocorrelation function at the origin is a relative maximum.
   - **(b)** $R(-\tau) = R(\tau)$; that is, the autocorrelation function is an even function.
   - **(c)** $\lim_{|\tau|\to\infty} R(\tau) = E^2[X(t)]$, provided that this limit exists; that is, the value of the autocorrelation function as its argument approaches infinity is equal to the mean squared.
   - **(d)** The autocorrelation function of a periodic random process (i.e., a random process whose sample functions are periodic functions of the independent variable) is also periodic.
   - **(e)** The Fourier transform of the autocorrelation function, referred to as the power spectral density, is real and nonnegative.
10. The cross-correlation function of two jointly stationary random processes, $X(t)$ and $Y(t)$, is defined as

$$R_{XY}(\tau) = E[X(t)Y(t+\tau)]$$

It has the following properties:
   - **(a)** $R_{XY}(-\tau) = R_{YX}(\tau)$
   - **(b)** $|R_{XY}(\tau)| \leq [R_{XX}(0)R_{YY}(0)]^{1/2}$
   - **(c)** $|R_{XY}(\tau)| \leq \frac{1}{2}[R_{XX}(0) + R_{YY}(0)]$
11. The covariance function of two jointly stationary random processes is

$$C_{XY}(\tau) = E\{[X(t) - E(X(t))][Y(t+\tau) - E(Y(t+\tau))]\}$$

12. A Gaussian random process is one that has an N-fold joint pdf which is Gaussian. For a stationary Gaussian random process with zero mean, the joint (twofold) pdf is given by

$$f(x, y; \tau) = \frac{\exp\left\{-\dfrac{x^2 - 2r(\tau)xy + y^2}{2\sigma^2[1 - r(\tau)^2]}\right\}}{2\pi\sigma^2\sqrt{1 - r(\tau)^2}}$$

where x refers to the possible values of the random process at time t, $X(t)$, and y refers to the possible values of the random process at time $t + \tau$, $X(t+\tau)$; σ^2 is the variance, and $r(\tau)$ is the covariance, given by

$$r(\tau) = \frac{R(\tau)}{\sigma^2} = \frac{R(\tau)}{R(0)} = \frac{E[X(t)X(t+\tau)]}{E[X^2(t)]}$$

13. A first-order autoregressive process is one obeying the recursive equation

$$X_{k+1} = (1 - \alpha)X_k + \alpha N_k, \; k = 0, 1, 2, \ldots$$

where α is a parameter in the internal $[0, 1]$ and the N_k are identially distributed, independent (iid) random variables. Assuming that it starts at $k = 0$ with $X_0 = 0$, its variance, assuming that the N_k's are zero mean, is

$$\text{Var}(X_k) = \frac{1 - \alpha}{1 + \alpha}(1 - \alpha^{2k})\sigma_n^2$$

A plot of this expression clearly shows that the process has a transient buildup interval that is short for small α and longer the closer α gets to 1. After this initial transient period, the process approaches a stationary process with autocorrelation function

$$R_X(M) = \alpha^{|M|}R_X(0) = \alpha^{|M|}\frac{1 - \alpha}{1 + \alpha}\sigma_n^2$$

9–9 FURTHER READING

The books by Peebles (1987) and Helstrom (1992) have more extensive treatments of random processes than are given here. For a brief treatment oriented toward communication theory applications, see Chapter 5 of Ziemer and Tranter (1995). Schwartz and Shaw (1975) has an interesting treatment of autoregressive processes from the standpoint of signal modeling.

9–10 PROBLEMS

Section 9-2

9-1. A fair die is tossed. Depending on the number of spots on the up face, the following time functions are generated. Plot a representative set of sample functions for each case:

(a) $$X(t,\zeta) = \begin{cases} 3, & \text{3, 5, or 6 spots up} \\ 2, & \text{4 spots up} \\ 1, & \text{1 or 2 spots up} \end{cases}$$

(b) $$X(t,\zeta) = \begin{cases} 2t, & \text{1 or 2 spots up} \\ -2t, & \text{3 or 4 spots up} \\ 0, & \text{5 or 6 spots up} \end{cases}$$

9-2. Classify the random processes of Problem 9-1 according to the following table:

Case	Independent variable	Dependent variable
1	Discrete	Discrete
2	Discrete	Continuous
3	Continuous	Discrete
4	Continuous	Continuous

9-3. A random process is defined only for the time instants . . . , $-2, -1, 0, 1, 2,$ . . . , where the dots (ellipses) indicate that the sequence goes from $-\infty$ to ∞. At each instant, a fair die is tossed, and the value assigned to the random process at this time instant is numerically equal to the number of spots up.

(a) Sketch some typical sample functions for this random process.
(b) Classify it according to the cases given in Problem 9-2.

Section 9-3

9-4. Consider the random process defined by (9-2), but with the pdf for Θ given by

$$f_\Theta(\theta) = \begin{cases} \dfrac{4}{\pi}, & 0 \le \theta \le \dfrac{\pi}{4} \\ 0, & \text{otherwise} \end{cases}$$

(a) Find the ensemble, or statistical, average mean and mean square for this random process.
(b) Find the time-average mean and mean square for a single sample function.
(c) Could this random process be stationary? Could it be Ergodic?

9-5. Find the ensemble average autocorrelation function for the random process defined in Problem 9-4. Note that it is a function of two time instants.

9-6. A random process is defined as follows. In contiguous Δ-second intervals, a coin is tossed; if heads, assign the value A throughout the interval; if tails, assign the value $-A$. The time origin is randomly placed any place in an interval (i.e., never consistently at the beginning or end of an interval.
(a) Sketch some typical sample functions
(b) Show that the autocorrelation function is given by

$$R_X(\tau) = \begin{cases} A^2\left(1 - \dfrac{|\tau|}{\Delta}\right), & -\Delta \le \tau \le \Delta \\ 0, & \text{otherwise} \end{cases}$$

Sketch it.
(c) What is the average power of this random process?

Section 9-4

9-7. Given the following functions of τ, according to the properties of an autocorrelation function given in Section 9-4, which ones are suitable as autocorrelation functions of stationary ergodic random processes? Sketch each. Given reasons for those deemed not suitable.
(a) $R_1(\tau) = e^{-\alpha|\tau|}, \alpha > 0$
(b) $R_2(\tau) = \sin(2\pi\tau)$
(c) $R_3(\tau) = \begin{cases} A, & -\tau_0 \le \tau \le \tau_0 \\ 0, & \text{otherwise} \end{cases}, \quad A \text{ and } \tau_0 > 0$
(*Hint:* The Fourier transform of the last one is

$$S_3(f) = \frac{\sin(2\pi\tau_0 f)}{2\pi\tau_0 f}$$

9-8. Given the autocorrelation function of Problem 9-6:
(a) What is the mean-square value for this random process?
(b) What is the mean?
(c) Is the random process periodic?

9-9. A stationary random process is known to have a first-order pdf that is Gaussian. Its autocorrelation function is given by

$$R_X(\tau) = \begin{cases} 2 - |\tau|, & |\tau| \leq 1 \\ 1, & \text{otherwise} \end{cases}$$

From this information, write down its first-order pdf.

Section 9-5

9-10. Prove the properties for the cross-correlation function for jointly stationary random processes given just below (9-36).

9-11. Two random processes have cross-correlation function

$$R_{XY}(\tau) = \sin\left(3\pi f_0 \tau + \frac{\pi}{3}\right)$$

where f_0 is a positive constant. Their autocorrelation functions are

$$R_X(\tau) = \cos(2\pi f_0 \tau)$$

and

$$R_Y(\tau) = \cos(4\pi f_0 \tau)$$

(a) What is the average power of each?
(b) What is the autocorrelation function of their sum?
(c) What is the average power of their sum?

9-12. Show that (9-41) follows from (9-40).

Section 9-6

9-13. A zero-mean stationary Gaussian random process has average power 4 W and covariance function

$$r(\tau) = \begin{cases} 1 - |\tau|/2, & -2 \leq \tau \leq 2 \text{ s} \\ 0, & \text{otherwise} \end{cases}$$

(a) Write down the joint pdf at two time instants separated by 1s.
(b) Write down the joint pdf at two time instants separated by 3s.

9-14. Rework Example 9-12 for the case where the covariance function is the same as given in Problem 9-13. [*Hint:* By examining areas in the $\zeta - t$ plane, write the next-to-last integral of (9-49) as a single integral that can be integrated.]

9-15. Using the joint pdf (9-45) and the corresponding marginal pdf on $X(t)$ [just set $r(\tau) = 0$ and recognize the joint pdf as the product of two marginals], find the conditional pdf of $Y = X(t + \tau)$ given $X = X(t), f(y \mid x; \tau)$, and show that it is, in fact, given by

$$f(y \mid x;\tau) = \frac{\exp\left\{-\frac{[y - r(\tau)x]^2}{2\sigma^2[1 - r^2(\tau)]}\right\}}{\sqrt{2\pi\sigma^2[1 - r^2(\tau)]}}$$

where $r(\tau)$ is the correlation coefficient. That is, the conditional pdf of the amplitudes of a zero-mean Gaussian random process at time $t + \tau$, given the amplitude at time t, is Gaussian with conditional mean given by

$$E[X(t + \tau) \mid X(t) | = r(\tau)X(t)$$

and conditional variance

$$\text{Var}[X(t+\tau)\,|\,X(t)] = \sigma^2[1 - r^2(\tau)]$$

Section 9-7

9-16. A first-order autoregressive process is used to model speech. Adjacent sample values are to have normalized correlation of no more than 0.6, where by *normalized correlation* we mean (9-66) normalized by its peak value. What should α be set equal to? Plot $R_X(M)$ versus M with its peak value normalized to 1 (i.e., the variance of the process).

9-17. A first-order autoregressive process has $\alpha = 0.3$. Assume that it has reached steady state. If sample values are jointly Gaussian, use (9-45) to write down the joint pdf of adjacent sample values if $\sigma_n^2 = 1$.

9–11 COMPUTER EXERCISES

9-1. Making use of the development of Problem 9-15, design a random number generator that will generate a zero-mean Gaussian random sequence of numbers with a specified variance σ^2 and covariance function $r(\tau)$. Generate a sequence of numbers for the covariance function

$$r(\tau) = \begin{cases} 1 - \dfrac{|\tau|}{\tau_0}, & -\tau_0 \leq \tau \leq \tau_0 \text{ seconds} \\ 0, & \text{otherwise} \end{cases}$$

and plot for various values of $r(\tau)$. Note how the memory affects the plotted sequence.

9-2. Design a sample function generator for a second-order autoregressive process that obeys the recurrence relationship

$$X_{k+1} = \beta_1 X_k + \beta_2 X_{k-1} + \alpha N_k$$

Let $\alpha = 1 - \beta_1 - \beta_2$ and plot several sample functions for various values of β_1 and β_2. Can you derive the autocorrelation function for the stationary process?

10 Random Processes Through Systems[1]

10–1 INTRODUCTION

In this chapter we consider the processing of stationary random processes by systems. For simplicity, we assume that the input has been present from the infinite past. Two classes of systems are considered: zero-memory nonlinear and fixed linear systems.[2] For the former we can compute the first-order output pdf given the first-order input pdf by using the method of transformation of random variables discussed in Chapter 3. We can also find various moments of the output random process if the transfer characteristic is expressible as a series with only a few terms.

In the case of fixed linear systems, various moments of the output process may be found in terms of moments of the input process. The most important of these are the mean and second moments, including mean-square value (or variance) and the autocorrelation function. For the latter it is often easier to find the power spectral density of the output random process and inverse Fourier-transform it to obtain the autocorrelation function if desired. Finding output pdf's is, in general, difficult. The exception to this is if the input to the fixed linear system is Gaussian, in which case the output is Gaussian. Any-order

[1]Depending on the background of the students, the instructor may choose to omit this chapter.

[2]Linear system theory is overviewed in Section A-4. Zero-memory nonlinear systems are illustrated there by example.

joint pdf may then be written down by knowing the mean, variance, and autocorrelation function of the output, although we exhibited only the second-order Gaussian pdf in Chapter 4.

Other concepts discussed in this chapter are those of white noise, noise-equivalent bandwidth of a fixed linear system, thermal noise and calculations involving systems and thermal noise, representation of narrowband noise processes, and optimum fixed linear systems for estimating random processes.

Although much more of this important topic could be covered, this chapter provides an introduction to understanding the effects of systems on random processes and some of the applications of the theory.

10–2 ZERO-MEMORY NONLINEAR SYSTEMS WITH RANDOM WAVEFORMS AS INPUTS

In this section we consider systems with input–output relationships of the form

$$Y(t) = g[X(t)] \tag{10-1}$$

where $g[\cdot]$ is a known function. The output depends on the input at the present time only and not on past or future values of time. Such a nonlinear system is called *zero memory*.

Probability Density Functions

The first-order pdf or cdf of the output, at time t, can be found by using the techniques of Section 3-4, since for t fixed, (10-1) represents the transformation of random variable X to random variable Y. We consider two examples to further illustrate the techniques to be used.

Example 10-1

Obtain the marginal pdf of the output at time t of the dead-zone device with input–output characteristic shown in Figure 10-1 if the input is a stationary Gaussian random process with zero mean and variance σ^2.

Solution From the transfer characteristic of Figure 10-1, it is clear that the following is true:

$$\begin{aligned} F_Y(y;t) &= P[Y(t) \le y] = P[X(t) \le y+c] = F_X(y+c\,;t) \quad y>0 \;\; (y=x-c,\, y>0) \\ F_Y(y;t) &= P[Y(t) \le y] = P[X(t) \le y-c] = F_X(y-c\,;t) \quad y<0 \;\; (y=x+c,\, y>0) \end{aligned} \tag{10-2}$$

Differentiating with respect to y, we obtain

$$\begin{aligned} f_Y(y;t) &= \frac{\partial F_Y(y;t)}{\partial y} = \frac{\partial F_X(y+c;t)}{\partial y} = f_X(y+c;t), \quad y>0 \\ f_Y(y;t) &= \frac{\partial F_Y(y;t)}{\partial y} = \frac{\partial F_X(y-c;t)}{\partial y} = f_X(y-c;t), \quad y<0 \end{aligned} \tag{10-3}$$

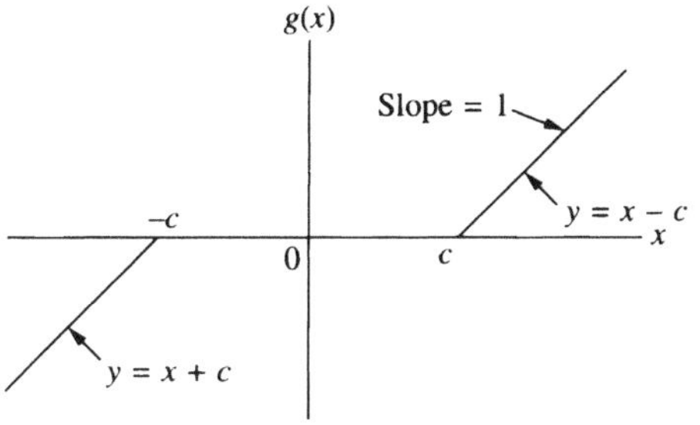

Figure 10-1 Transfer characteristic of a dead-zone device.

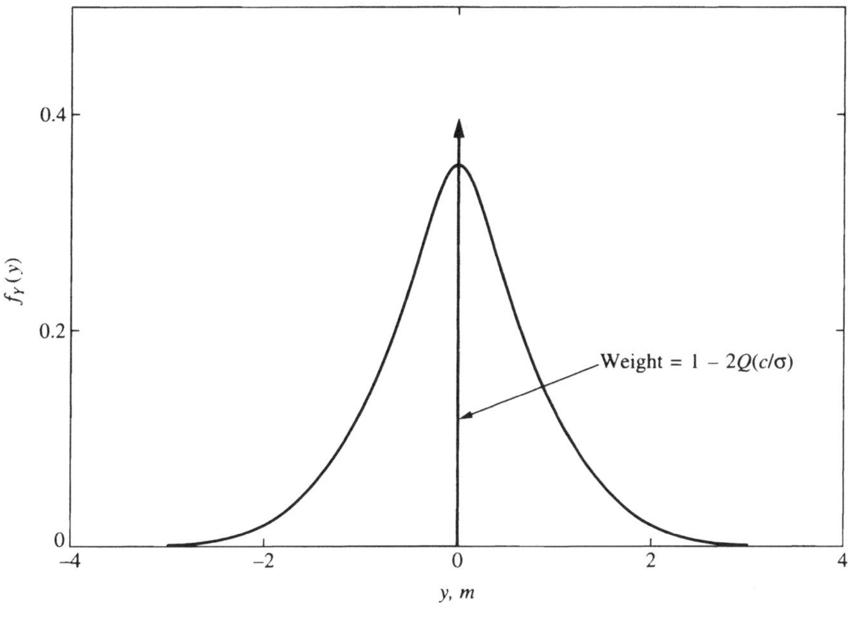

Figure 10-2 Output pdf of a dead-zone device with stationary Gaussian noise as the input.

In addition, there is a finite probability that $Y = 0$, which is given by $P(Y = 0) = P(-c \leq X \leq c) = 1 - 2Q(c/\sigma)$. Substituting the Gaussian pdf on the right-hand side of (10-3), we obtain

$$f_Y(y;t) = \begin{cases} \dfrac{e^{-(y+c)^2/2\sigma^2}}{\sqrt{2\pi\sigma^2}}, & y > 0 \\ \dfrac{e^{-(y-c)^2/2\sigma^2}}{\sqrt{2\pi\sigma^2}}, & y < 0 \\ [1 - 2Q(c/\sigma)]\delta(y), & y = 0 \end{cases} \tag{10-4}$$

The pdf of the output is sketched in Figure 10-2.

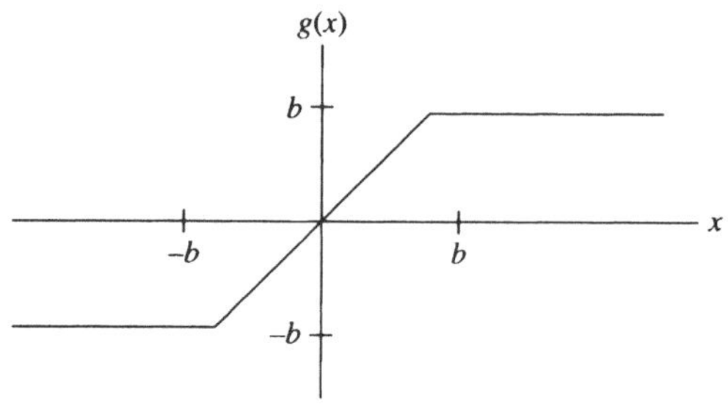

Figure 10-3 Transfer characteristic of a limiter.

Example 10-2

Obtain the marginal pdf at time t of the output of the limiter with transfer characteristic shown in Figure 10-3 if the input is a stationary Gaussian random process with zero mean and variance σ^2.

Solution From the transfer characteristic, we deduce that for $y =$ some arbitrary level:

If	$y \geq b$	then $g(x) \leq y$ for every x;	hence $F_Y(y; t) = 1$;
If	$-b \leq y < b$	then $g(x) \leq y$ for $x \leq y$;	hence $F_Y(y; t) = F_X(y; t)$;
If	$y < -b$	then $g(x) \leq y$ for no x;	hence $F_Y(y; t) = 0$

The pdf takes some care. We note that, in general, there will be a jump in the value of $F_Y(y; t)$ at $y = -b$ and $y = b$. This simply reflects the fact that $Y(t)$ takes on the values b and $-b$ with finite probability. Thus the pdf must have delta functions at these values of y with areas equal the respective probabilities of $Y(t)$ taking on the values of b and $-b$. For the Gaussian pdf assumed in this example, these probabilities are equal due to symmetry and are given by

$$P(Y \leq -b) = P(Y \geq b) = \int_b^\infty \frac{e^{-x^2/2\sigma^2}}{\sqrt{2\pi\sigma^2}}\,dx = \int_{b/\sigma}^\infty \frac{e^{-u^2/2}}{\sqrt{2\pi}}\,du = Q\left(\frac{b}{\sigma}\right) \tag{10-5}$$

Thus the pdf of Y at any time t is

$$\begin{aligned} f_Y(y; t) &= \frac{dF_Y(y; t)}{dy} \\ &= Q\left(\frac{b}{\sigma}\right)\delta(y + b) + \frac{e^{-y^2/2\sigma^2}}{\sqrt{2\pi\sigma^2}}[u(y + b) - u(y - b)] + Q\left(\frac{b}{\sigma}\right)\delta(y - b) \end{aligned} \tag{10-6}$$

where $u(y \pm b)$ is a unit step beginning at $y = -b$ or b, respectively. The pdf of the output is illustrated in Figure 10-4.

Autocorrelation Functions

The autocorrelation function at the output of a zero-memory nonlinear system can be found by expressing the system transfer characteristic as a series. In cases where this series representation has several terms, the method may be unwieldly to carry out, but the principle remains the same. The following two examples illustrate the technique.

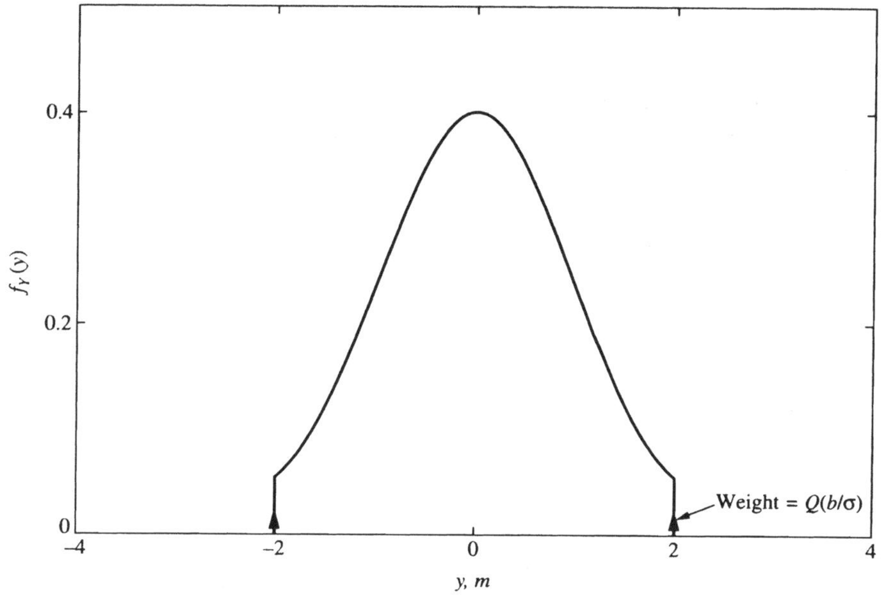

Figure 10-4 Output pdf for Example 10-2 with stationary Gaussian noise at the input; $b = 2$.

Example 10-3

Consider a square-law device with transfer characteristic

$$Y(t) = X^2(t) \tag{10-7}$$

and an input that is a wide-sense-stationary Gaussian random process of zero mean and with autocorrelation function $R_X(\tau)$. Find the mean and autocorrelation function of the output.

Solution The mean is found from

$$E[Y(t)] = E[X^2(t)] = R_X(0) = \text{Var}[X(t)] \tag{10-8}$$

where the last equation follows because $\{X(t)\}$ has zero mean by assumption. The autocorrelation function follows from

$$\begin{aligned} R_Y(\tau) &= E[Y(t)Y(t+\tau)] = E[X^2(t)\,X^2(t+\tau)] \\ &= E[X(t)X(t)X(t+\tau)X(t+\tau)] \end{aligned} \tag{10-9}$$

It can be shown that for a set of four jointly Gaussian random variables Z_1, Z_2, Z_3, and Z_4 with zero means,

$$E(Z_1Z_2Z_3Z_4) = E(Z_1Z_2)\,E[Z_3Z_4) + E(Z_1Z_3)\,E(Z_2Z_4) + E(Z_1Z_4)\,E(Z_2Z_3) \tag{10-10}$$

This can be applied directly to the last equation of (10-9) by letting $Z_1 = X(t)$, $Z_2 = X(t)$, $Z_3 = X(t+\tau)$, and $Z_4 = X(t+\tau)$, to obtain

$$R_Y(\tau) = R_X^2(0) + 2R_X^2(\tau) \tag{10-11}$$

where $R_X(\tau) = E[X(t)\,X(t+\tau)]$ is the autocorrelation function of the input. The power spectral density of the output is the Fourier transform of the output autocorrelation function and can be written as

$$S_X(f) = R_X^2(0)\delta(f) + 2S_X(f) * S_X(f) \tag{10-12}$$

where $S_X(f)$ is the power spectral density of the input, or the Fourier transform of $R_X(\tau)$, and the asterisk denotes convolution.

Figure 10-5 shows a plot of the input and output autcorrelation functions and power spectral densities for the case of the input having a triangular autocorrelation function.

Example 10-4

Reconsider Example 10-3 for the case of the input consisting of a cosinusoidal signal plus wide-sense-stationary Gaussian noise of zero mean[3]:

$$X(t) = s(t) + n(t) = A\cos(\omega_0 t + \theta) + n(t) \tag{10-13}$$

Assume that A and ω_0 are constants and that θ is a random phase uniformly distributed in $(0, 2\pi)$ which is independent of $\{n(t)\}$.

Solution Making use of the fact that $n(t)$ and $s(t) = A\cos(\omega_0 t + \theta)$ both have zero means, it follows that

$$\begin{aligned} R_Y(\tau) &= E\{[s(t) + n(t)]^2[s(t+\tau) + n(t+\tau)]^2\} \\ &= E[s^2(t)s^2(t+\tau)] + E[s^2(t)n^2(t+\tau)] \\ &\quad + 4E[s(t)s(t+\tau)]E[n(t)n(t+\tau)] \\ &\quad + E[n^2(t)s^2(t+\tau)] + E[n^2(t)n^2(t+\tau)] \\ &= E[s^2(t)s^2(t+\tau)] + E[s^2(t)]E[n^2(t+\tau)] \\ &\quad + 4R_s(\tau)R_n(\tau) + E[n^2(t)]E[s^2(t+\tau)] + E[n^2(t)n^2(t+\tau)] \\ &= E[s^2(t)s^2(t+\tau)] + 2R_s(0)R_n(0) + 4R_s(\tau)R_n(\tau) + R_n^2(0) + 2R_n^2(\tau) \end{aligned} \tag{10-14}$$

where the statistical independence of $s(t)$ and $n(t)$ has been used along with stationarity. Also, the definitions of the autocorrelation functions of $s(t)$ and $n(t)$ have been used along with the result obtained in (10-11) to simplify $E[n^2(t)n^2(t+\tau)]$. Note that due to the interaction of signal and noise in the square-law device, the output consists of three types of terms: signal × signal, signal × noise, and noise × noise. By substituting $s(t)$ in the first expectation in the last equation and using trigonometric identities,[4] we obtain

$$\begin{aligned} E[s^2(t)s^2(t+\tau)] &= E\left\{\frac{A^2}{2}[1 + \cos 2(\omega_0 t + \theta)]\frac{A^2}{2}[1 + \cos 2(\omega_0(t+\tau) + \theta)]\right\} \\ &= \frac{A^4}{4} + \frac{A^4}{8}\cos(2\omega_0\tau) \end{aligned} \tag{10-15}$$

where the fact that the expectation of a sum is the sum of the expectations and the fact that $E\{\cos[2\omega_0 t + \theta)]\} = 0$ have been used.

The power spectral density is the Fourier transform of (10-14). Using Fourier transform theorems, we obtain

[3]We depart from our usual practice of using uppercase letters for random variables (in this case a random process at a particular time) because we are already using an uppercase S for spectral density.

[4]In particular, $\cos^2\theta = \frac{1}{2} + \frac{1}{2}\cos(2\theta)$ and $\cos\theta\cos\phi = \frac{1}{2}\cos(\theta + \phi) + \frac{1}{2}\cos(\theta - \phi)$.

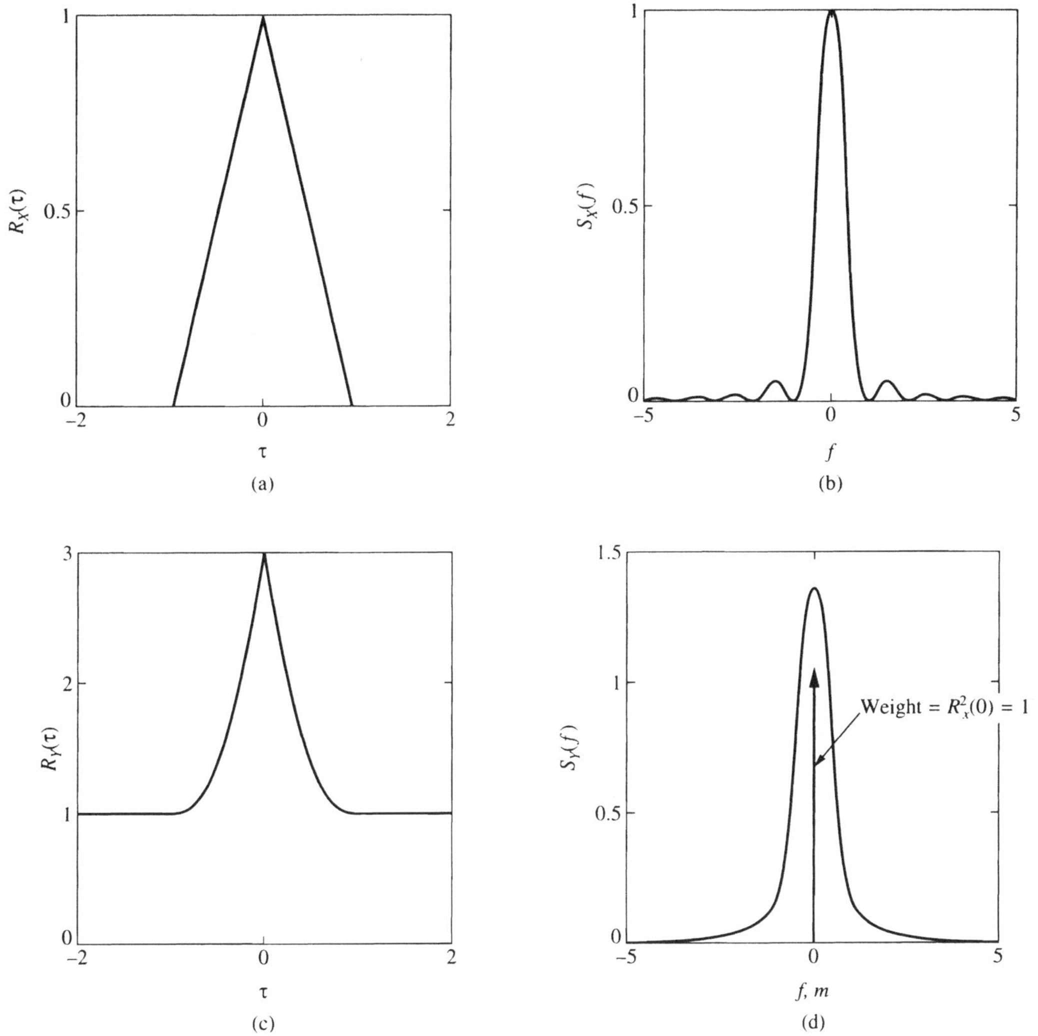

Figure 10-5 Assumed autocorrelation function and power spectral density of the input [(a) and (b), respectively], and corresponding ouput autocorrelation function and power spectral density [(c) and (d), respectively] for a square-law device with stationary Gaussian noise at its input.

$$\begin{aligned} S_Y(f) &= \frac{A^4}{16}[\delta(f+2f_0)+\delta(f-2f_0)] \\ &\quad + \left[\frac{A^4}{4} + 2R_s(0)R_n(0) + R_n^2(0)\right]\delta(f) + 4S_s(f) * S_n(f) + 2S_n(f) * S_n(f) \end{aligned} \tag{10-16}$$

where $f_0 = \omega_0/2\pi$. The first term in brackets is due to interaction of signal with signal, the second set of terms in brackets is due to the dc components produced by the squaring action, the third term is due to interaction of signal with noise, and the last term is due to interaction of noise with itself. We can simplify the third term further by noting that [recall (9-26)]

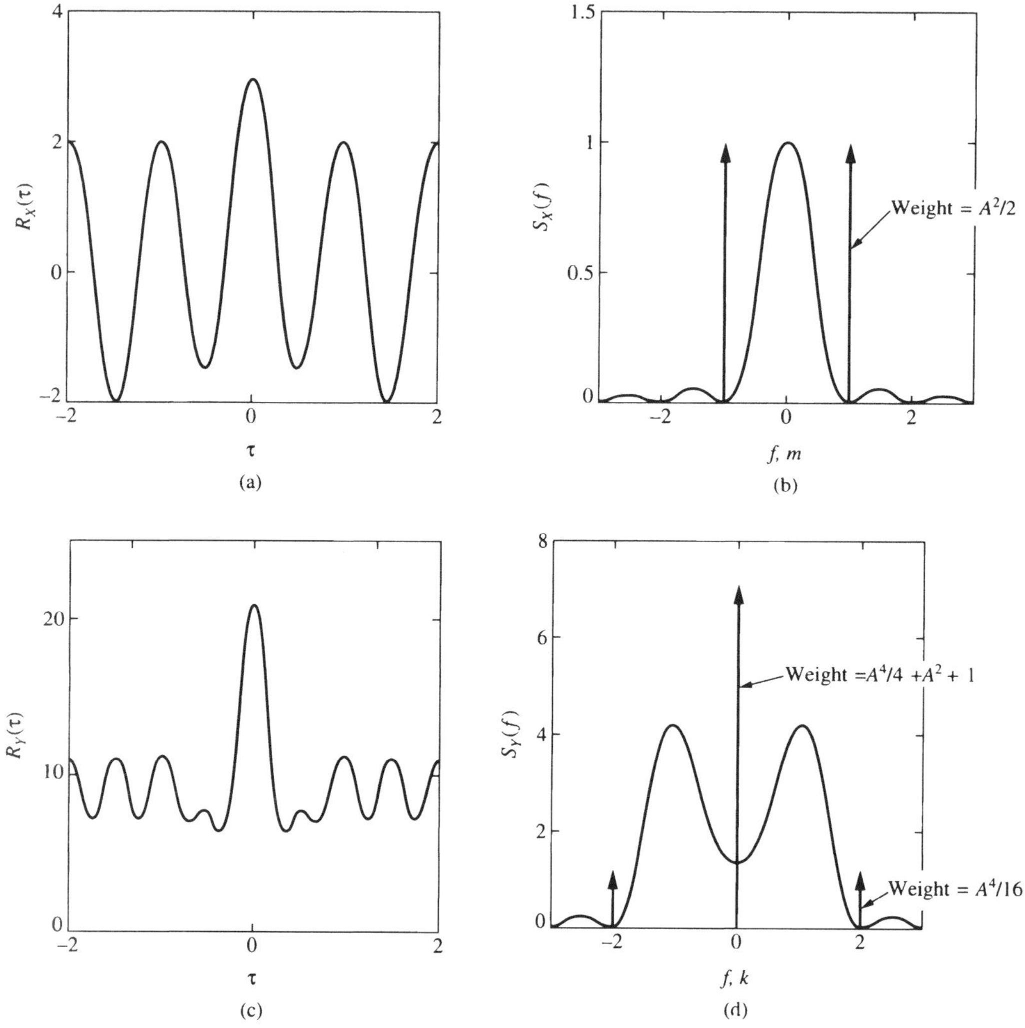

Figure 10-6 Input and output autocorrelation functions and power spectral densities of a square-law device for an input cosinusoid (amplitude 2 and frequency 1 Hz) of random phase plus stationary Gaussian noise with triangular autocorrelation function of unit width and height.

$$4S_s(f) * S_n(f) = 4\left[\frac{A^2}{4}\delta(f - f_0) + \frac{A^2}{4}\delta(f + f_0)\right] * S_n(f) = A^2[S_n(f - f_0) + S_n(f + f_0)] \tag{10-17}$$

Thus the simplified form for the power spectral density of the output due to cosinusoidal signal plus Gaussian noise at the input is

$$\begin{aligned} S_Y(f) = {} & \frac{A^4}{16}[\delta(f + 2f_0) + \delta(f - 2f_0)] + \left[\frac{A^4}{4} + 2R_s(0)R_n(0) + R_n^2(0)\right]\delta(f) \\ & + A^2[S_n(f - f_0) + S_n(f + f_0)] + 2S_n(f) * S_n(f) \end{aligned} \tag{10-18}$$

A plot of the input and output autocorrelation functions and spectra are shown in Figure 10-6 for the case where the input signal component is a cosinusoid of amplitude 2 and frequency 1 Hz and the input noise of autocorrelation function is triangular of width 1 and height 1.

10–3 FIXED LINEAR SYSTEMS WITH STATIONARY RANDOM PROCESSES AS INPUTS

Joint Input–Output Random Processes

The key relationship for characterizing input–output random processes for fixed linear systems is the superposition integral given by (A-25) in Appendix A. In the simplest applications of this integral to studying the effect of a system on a random process, we assume that the input process is wide-sense stationary and has been present at the input forever. Practically speaking, we assume that the input process has been present long enough so that any starting transients for the system have long since died out. Thus let the input random process to a fixed linear system with impulse response $h(t)$ be $\{X(t)\}$ and let the output random process be $\{Y(t)\}$. Then for each sample function of the input, the superposition integral defines a corresponding sample function of the output:

$$Y^{(i)}(t) = \int_{-\infty}^{\infty} h(\tau)\, X^{(i)}(t-\tau)\, d\tau = \int_{-\infty}^{\infty} h(t-\lambda) X^{(i)}(\lambda)\, d\lambda \qquad (10\text{-}19)$$

The superscript (i) emphasizes that (10-19) holds for each sample function of the input and output processes. For simplicity of notation, we will drop these superscripts in future equations.

Output Mean of a Fixed Linear System with Random Input

Since the input is wide-sense stationary and is assumed to have been present for a sufficiently long time that any transients in the output have died out, it is reasonable to assume that the output process is stationary also. The mean of the output is

$$\begin{aligned} m_Y = E[Y(t)] &= E\left[\int_{-\infty}^{\infty} h(t-\lambda) X(\lambda)\, d\lambda\right] \\ &= \int_{-\infty}^{\infty} h(t-\lambda) E[X(\lambda)]\, d\lambda \\ &= \int_{-\infty}^{\infty} h(t-\lambda) m_X\, d\lambda \\ &= m_X \int_{-\infty}^{\infty} h(\zeta)\, d\zeta \end{aligned} \qquad (10\text{-}20)$$

where the expectation can be taken inside the superposition integral because it is really another integral. The last equation follows by noting that the mean of $\{X(t)\}$, m_X, is a constant for a stationary process and the change of variables $\zeta = t - \lambda$ has been made in the last integral.

The transfer function of a fixed linear system is the Fourier transform of the impulse response:

$$H(f) = \int_{-\infty}^{\infty} h(\zeta)e^{-j2\pi f\zeta}\, d\zeta \tag{10-21}$$

Setting $f = 0$ on both sides of this relationship, we have that $H(0) = \int_{-\infty}^{\infty} h(\zeta)\, d\zeta$. Thus (10-20) may also be written as

$$m_Y = H(0)m_x \tag{10-22}$$

The transfer function of the system evaluated at zero frequency gives the *dc gain* of the system.

In Chapter 9 an ergodic random process was stated as having the property that all statistical averages, or expectations, are equal to their time-average equivalents. Thus, if the input and output processes are jointly ergodic for the linear system under consideration, (10-22) states that the dc level at the output of the system is equal to the dc level of the input multiplied by the dc gain of the system.

Example 10-5

Consider an RC lowpass filter with circuit diagram shown in Figure 10-7 with the input preceded by a square-law device as considered in Example 10-4. The input to the square-law device is a stationary random process with zero mean and a variance of 1. What is the dc level at the output of the lowpass RC filter?

Solution Using ac sinusoidal steady-state analysis and voltage division, the ratio of output to input phasors for the RC lowpass filter, which is also the transfer function, is

$$\frac{V_{\text{out}}(j2\pi f)}{V_{\text{in}}(j2\pi f)} = H(f) = \frac{R_2/(1 + j2\pi R_2C)}{R_1 + R_2/(1 + j2\pi R_2C)} = \frac{R_2}{R_1 + R_2 + j2\pi f R_1R_2C} \tag{10-23}$$

The value of the transfer function at $f = 0$ is $R_2/(R_1 + R_2)$, and the expectation, or dc level, of the input is $E[X^2(t)] = \text{var}[W(t)] = 1$, which follows because the mean of the stationary random process at the input of the square-law device is zero. Thus the mean, or dc level, of the output of the RC lowpass filter is

$$m_Y = H(0)m_X = \frac{R_2}{R_1 + R_2} \times 1 = \frac{R_2}{R_1 + R_2} \tag{10-24}$$

Input–Output Correlation Functions

We now consider the relationship between input and output autocorrelation functions for a fixed linear system with stationary random input. First, consider the cross-correlation function of the output with the input. We want the expectation

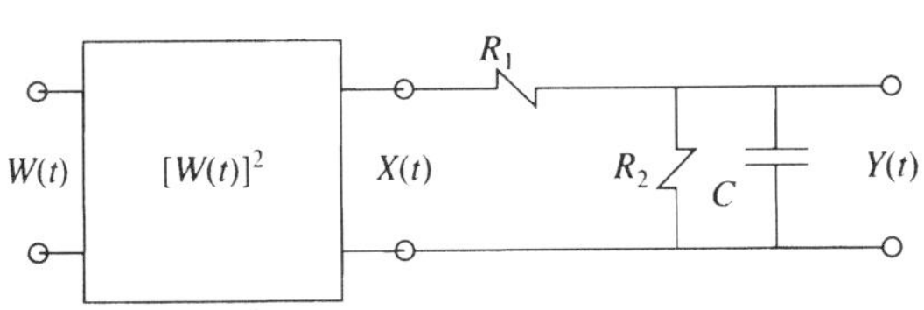

Figure 10-7 Circuit diagram for Example 10-5

$$R_{XY}(\tau) = E[X(t)Y(t + \tau)] \tag{10-25}$$

Rewrite the first integral of (10-19) with another variable of integration, say ζ, so that we can add τ to t in the argument of $Y(t)$ and to $X(t - \zeta)$ inside the integral. Then take the expectation of both sides to get

$$\begin{aligned} R_{XY}(\tau) &= E[X(t)Y(t + \tau)] \\ &= E\left[\int_{-\infty}^{\infty} h(\zeta)X(t)X(t + \tau - \zeta)\, d\zeta\right] \\ &= \int_{-\infty}^{\infty} h(\zeta)E[X(t)X(t + \tau - \zeta)]\, d\zeta \\ &= \int_{-\infty}^{\infty} h(\zeta)R_X(\tau - \zeta)\, d\zeta \\ &= h(\tau) * R_X(\tau) \end{aligned} \tag{10-26}$$

where the last relationship holds by definition of the superposition integral [see (A-25)]. Using the convolution theorem of Fourier transforms, the Fourier transform of (10-26) becomes

$$S_{XY}(f) = H(f)S_X(f) \tag{10-27}$$

where $S_{XY}(f)$ is the Fourier transform of $R_{XY}(\tau)$ and is called the cross-power spectral density of the random processes $\{X(t)\}$ and $\{Y(t)\}$.

A similar development to that used to get (10-26) results in the relationship

$$R_{YX}(\tau) = E[Y(t)X(t + \tau)] = h(-\tau) * R_X(\tau) \tag{10-28}$$

Finally, the autocorrelation function of the output is

$$\begin{aligned} R_Y(\tau) &= E[Y(t)Y(t + \tau)] \\ &= E\left[\int_{-\infty}^{\infty} h(\lambda)X(t - \lambda)\, d\lambda \int_{-\infty}^{\infty} h(\zeta)X(t + \tau - \zeta)\, d\zeta\right] \\ &= \int_{-\infty}^{\infty}\int_{-\infty}^{\infty} h(\lambda)h(\zeta)E[X(t - \lambda)X(t + \tau - \zeta)]\, d\zeta d\lambda \\ &= \int_{-\infty}^{\infty}\int_{-\infty}^{\infty} h(\lambda)h(\zeta)R_X(\tau + \lambda - \zeta)\, d\zeta d\lambda \\ &= h(\tau) * R_{YX}(\tau) \\ &= h(-\tau) * R_{XY}(\tau) \end{aligned} \tag{10-29}$$

Taking the Fourier transform of the last convolution in (10-29) and using (10-27), we obtain

$$S_Y(f) = H^*(f)H(f)S_X(f) = |H(f)|^2 S_X(f) \tag{10-30}$$

where for $h(\tau)$ real we have used the fact that

$$\begin{aligned}\mathscr{F}[h(-\tau)] &= \int_{-\infty}^{\infty} h(-\tau)e^{-j2\pi f\tau}\,d\tau \\ &= \int_{-\infty}^{\infty} h(\lambda)e^{j2\pi f\lambda}\,d\lambda \quad (\lambda = -\tau) \\ &= \left[\int_{-\infty}^{\infty} h(\lambda)e^{-j\pi f\lambda}\,d\lambda\right]^* \\ &= H^*(f)\end{aligned} \tag{10-31}$$

As before, the superscript * denotes a complex conjugate. Equation (10-30) is a very important relationship that can be used for many applications. It basically contains the same information as (10-29) but is much easier to apply, due to the double integration required for (10-29).

Example 10-6

The input to the RC lowpass filter (i.e., the combination of the two resistors and capacitor) of Example 10-5 is stationary noise with autocorrelation function[5]

$$R_X(\tau) = \frac{N_0}{2}\delta(\tau) \tag{10-32}$$

and power spectral density

$$S_X(f) = \frac{N_0}{2}, \quad -\infty < f < \infty \tag{10-33}$$

Find (a) the power spectral density of the output random processes, (b) the autocorrelation function of the output random process, and (c) the average power of the output random process.

Solution (a) We use (10-30) to obtain the power spectral density. From (10-23) and (10-33), it is

$$\begin{aligned}S_Y(f) &= \frac{N_0}{2}\left|\frac{R_2}{R_1 + R_2 + j2\pi f R_1R_2C}\right|^2 = \frac{N_0R_2^2/2}{(R_1 + R_2)^2 + (2\pi R_1R_2Cf)^2} \\ &= \frac{N_0/2}{(R_1C)^2\{[(R_1 + R_2)/R_1R_2]^2 + (2\pi f)^2\}}\end{aligned} \tag{10-34}$$

(b) The autocorrelation function of the output is the inverse Fourier transform of $S_Y(f)$. We use the Fourier transform pair

[5]This type of random process, called white noise, is discussed further in Section 10-4.

$$Ae^{-\alpha\tau}u(\tau) \longleftrightarrow \frac{A}{\alpha + j2\pi f} \tag{10-35}$$

and partial fraction expansion to obtain

$$\begin{aligned}R_Y(\tau) &= \frac{N_0}{2R_1^2C^2}\mathscr{F}^{-1}\left\{\frac{1}{(R_1+R_2)/(R_1R_2C)+j2\pi f}\,\frac{1}{(R_1+R_2)/(R_1R_2C)-j2\pi f}\right]\\ &= \frac{N_0}{2R_1^2C^2}\mathscr{F}^{-1}\left[\frac{A}{(R_1+R_2)/(R_1R_2C)+j2\pi f}+\frac{B}{(R_1+R_2)/(R_1R_2C)-j2\pi f}\right]\\ &= \frac{N_0}{2R_1^2C^2}\left[A\exp\left(-\frac{R_1+R_2}{R_1R_2C}\tau\right)u(\tau)+B\exp\left(\frac{R_1+R_2}{R_1R_2C}\tau\right)u(-\tau)\right]\end{aligned} \tag{10-36}$$

The constants A and B are found to be

$$A = B = \frac{R_1R_2C}{2(R_1+R_2)} \tag{10-37}$$

[To check this, simply substitute (10-37) into the middle equation of (10-36) and recombine to show that the first line of (10-36) results], which results in the following expression for the output autocorrelation function:

$$\begin{aligned}R_Y(\tau) &= \frac{(R_2/R_1)N_0}{4(R_1+R_2)C}\left[\exp\left(-\frac{R_1+R_2}{R_1R_2C}\tau\right)u(\tau)+\exp\left(\frac{R_1+R_2}{R_1R_2C}\tau\right)u(-\tau)\right]\\ &= \frac{R_2N_0}{4(R_1+R_2)R_1C}\exp\left(-\frac{R_1+R_2}{R_1R_2C}|\tau|\right)\end{aligned} \tag{10-38}$$

(c) The average power o f the output random process is

$$E[Y^2(t)] = R_Y(0) = \frac{R_2N_0}{4(R_1+R_2)R_1C} \tag{10-39}$$

The same result could have been obtained by integrating the output power spectral density. The output autocorrelation function and power spectral density are plotted in Figure 10-8a and b, respectively.

Example 10-7

Obtain the cross-correlation function of the input and output for the RC filter of Figure 10-7.

Solution Applying (10-27), we have

$$S_{XY}(f) = \frac{N_0R_2/2}{R_1+R_2+j2\pi fR_1R_2C} = \frac{N_0/(2R_1C)}{(R_1+R_2)/(R_1R_2C)+j2\pi f} \tag{10-40}$$

where (10-23) and (10-33) have been used. Using the Fourier transform pair (10-35), we obtain

$$R_{XY}(\tau) = \frac{N_0}{2R_1C}\exp\left(-\frac{R_1+R_2}{R_1R_2C}\tau\right)u(\tau) \tag{10-41}$$

The cross-correlation function of input and output is plotted in Figure 10-8c.

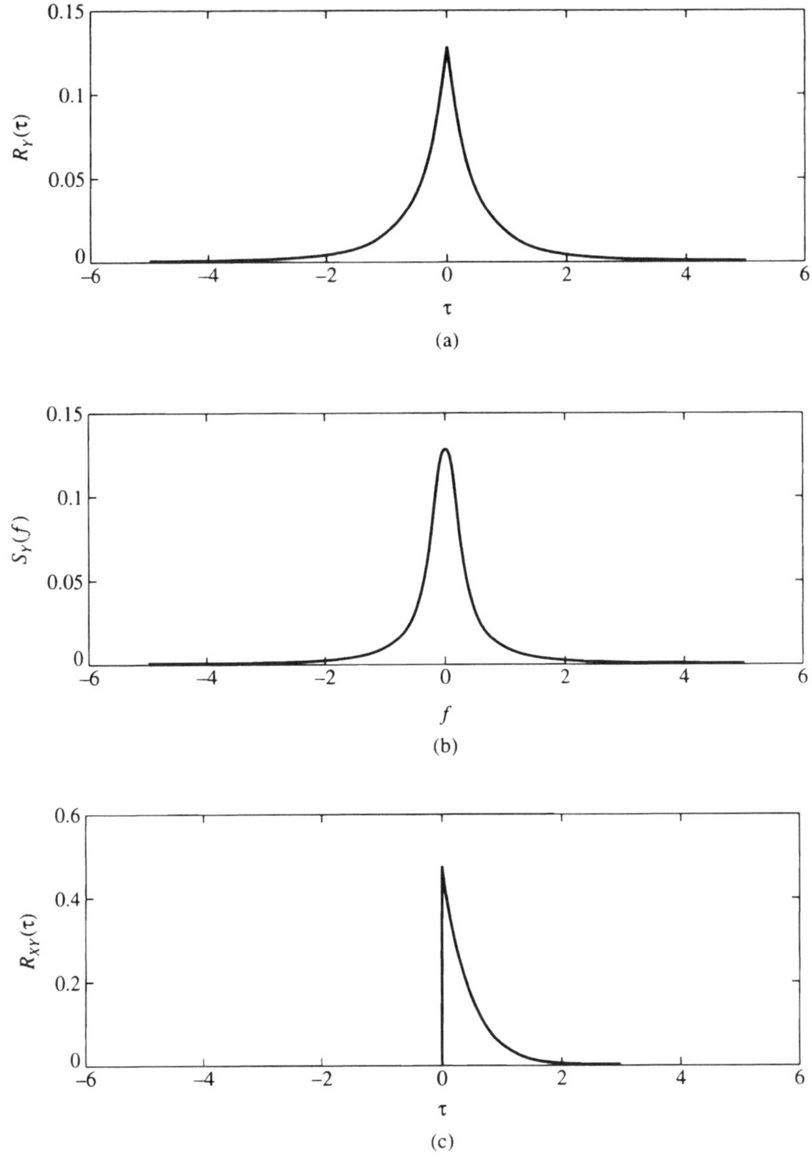

Figure 10-8 (a) Output autocorrelation function and (b) Output power spectral density for Example 10-6 (c) Cross-correlation function of input with output for Example 10-7.

Output Probability Density Function

In general, it is difficult, if not impossible, to obtain the probability density function of the output of a fixed linear system in response to a random process at its input with arbitrary probability density function. The one exception to this statement is if the input random process is stationary and Gaussian. If it has been applied for a very long time at the input, the output random process is stationary and Gaussian. This can be justified as follows. As shown in Chapter 4, the sum of two Gaussian random variables, independent or not, is a Gaussian random variable (see Example 4-19). This can be extended to any finite number of Gaussian random variables. The superposition integral (10-19) is the limit of a sum, although not a finite sum, if the input has been applied forever and the impulse response is semi-infinite in time extent. We will not dwell on this detail, but simply point out that if the input is Gaussian and we take a finite-sum approximation to (10-29), the output is Gaussian. This being the case, all that we need to write down the first-order pdf of the output at any time t are the mean and variance of the output. We can obtain these by means of (10-22) and (10-29) (set $\tau = 0$) or (10-30) (integrate over all f). An example will illustrate the technique.

Example 10-8

Obtain expressions for the first-order pdf and the joint pdf at time separations $\tau = 1$ and 0.1 s for the output of the *RC* lowpass filter of Figure 10-7 with white Gaussian noise at its input if the parameter values are $N_0 = 2$ W/Hz, $R_1 = R_2 = 100$ K Ω, and $C = 1\mu$F.

Solution Since the mean of the input is zero (there is no impulse in the power spectral density at $f = 0$), so is the mean of the output according to (10-22). Thus the variance of the output is given by (10-39) with the parameter values given above substituted:

$$\begin{aligned} \text{var}[Y(t)] &= E\{[Y(t) - m_Y]^2\} \\ &= E[Y^2(t)] = R_Y(0) \\ &= \frac{R_2 N_0}{4(R_1 + R_2)R_1 C} \\ &= \frac{2(100{,}000)}{4(200{,}000)(100{,}000)(10^{-6})} \\ &= 2.5 \end{aligned} \tag{10-42}$$

Thus from (3-28), we have

$$f_Y(y;t) = \frac{e^{-y^2/[2(2.5)]}}{\sqrt{2(2.5)\pi}} = \frac{e^{-y^2/5}}{\sqrt{5\pi}} \tag{10-43}$$

To obtain the joint pdf at times t and $t + 1$ seconds, we apply (4-55). Because the mean is zero at both times (the output process is stationary), we have from the last equation of (4-53) that

$$r = \frac{R_Y(\tau = 1\text{ s})}{R_Y(0)} \tag{10-44}$$

Using (10-38) in (10-44), we obtain

$$r = \exp\left[-\frac{200{,}000}{(100{,}000)(100{,}000)(10^{-6})}(1)\right] = \exp(-20) \approx 0 \tag{10-45}$$

Thus the samples of the output process taken at times t and $t + 1$ seconds are approximately independent (recall that uncorrelated Gaussian random variables are independent), and the joint pdf of the output at these two times is

$$f_{Y_1Y_2}(y_1, y_2; t, t+1) = \frac{e^{-y_1^2/5}}{\sqrt{5\pi}}\frac{e^{-y_2^2/5}}{\sqrt{5\pi}} = \frac{e^{-(y_1^2+y_2^2)/5}}{5\pi} \tag{10-46}$$

For $\tau = 0.1$ s we have

$$r = \exp\left[-\frac{(200{,}000)(0.1)}{(100{,}000)(100{,}000)(10^{-6})}\right] = \exp(-2) = 0.135 \tag{10-47}$$

Substituting into (4-55), we find the joint pdf of the amplitudes at times t and $t + 0.1$ to be

$$\begin{aligned} f_{Y_1Y_2}(y_1, y_2; t, t+\tau) &= \frac{\exp\left[-\dfrac{y_1^2 - 2(0.135)y_1y_2 + y_2^2}{2(2.5)(1 - 0.135^2)}\right]}{2\pi(2.5)\sqrt{1 - 0.135^2}} \\ &= 0.064\exp[-0.204)(y_1^2 - 0.27y_1y_2 + y_2^2)] \end{aligned} \tag{10-48}$$

10–4 WHITE NOISE

A useful model for noise in system analysis is a stationary process having a constant power spectral density for all frequencies (we had this situation in Example 10-8 for the input process but did not expand on it there). Since the average power of a stationary random process is given by the integral over all frequencies of the power spectral density, it is apparent that white noise has infinite power and is therefore a physical impossibility. Nevertheless, it is often a useful model, as we found in Example 10-8, and we give several examples in this section where this is the case. A case in point is the noise generated by electrical components with a resistive part (i.e., resistors, diodes, transistors, etc.) not at absolute zero temperature. The current in such an electrical component consists of the sum total of the currents represented by the motion of charge carriers (either electrons or holes—the absence of an electron), and for temperatures above absolute zero, this charge motion has a chaotic component that manifests itself macroscopically as a white noise random process.

Note that by specifying the power spectral density of the white noise process as a constant for all frequencies, we have implicitly specified that the mean is zero because a nonzero mean would manifest itself as an impulse at zero frequency; to have finite power at a specific frequency, in this case $f = 0$, requires that a finite area be obtained in integrating through that frequency which requires an impulse. Also, by specifying only the power spectral density, we have said nothing about the amplitude probability density function of the white noise process. Often, the additional specification of the process being Gaussian is added.

Autocorrelation Function of a White Noise Process

Let the power spectral density of a white noise random process be represented as

$$S_{wn}(f) = \frac{N_0}{2}, \quad -\infty < f < \infty \quad (f \text{ in hertz}) \tag{10-49}$$

The constant N_0 is sometimes referred to as the *one-sided power spectral density of the white noise process*. Now the Fourier transform of a delta function is the constant 1, which follows by the sifting property of the delta function:

$$\mathcal{F}[\delta(\tau)] = \int_{-\infty}^{\infty} \delta(\tau) e^{-j2\pi f} d\tau = 1$$

Thus it follows directly that the autocorrelation function of white noise corresponding to the power spectral density (10-49) is the inverse Fourier transform of the constant $N_0/2$, or

$$R_{wn}(\tau) = \frac{N_0}{2} \delta(\tau) \tag{10-50}$$

Since the delta function is zero everywhere except where its argument is zero, it follows that samples of a white noise process are uncorrelated no matter how close together the sampling times t_1 and t_2 are taken, where $\tau = |t_2 - t_1|$ (other than zero separation, of course). If, in addition, the white noise process is Gaussian, the samples of a white noise process are statistically independent, no matter how close together they are taken (assuming that they aren't taken at the same instant of time, of course).

Cross-Correlation Function of Input with Output for a Fixed Linear System with White Noise Input

Recall (10-27), and assume white noise at the input of a fixed linear system. Then (10-27), which gives the cross-power spectral density of input with output of the system, becomes

$$S_{XY}(f) = \frac{N_0}{2} H(f) \tag{10-51}$$

where $H(f)$ is the transfer function of the system (the Fourier transform of its impulse response). The inverse Fourier transform of the cross-power spectral density is the cross-correlation function, $R_{XY}(\tau)$, of input with output. Inverse Fourier transforming (10-51), we obtain

$$R_{XY}(\tau) = \frac{N_0}{2} h(\tau) \quad \text{or} \quad h(\tau) = \frac{2}{N_0} R_{XY}(\tau) \tag{10-52}$$

That is, to determine the impulse response of an unknown fixed linear system, we can drive it with white noise and measure the cross-correlation function of input with output. The result is proportional to the impulse response of the system. We might rightly ask how to generate a white noise process when it is physically impossible to do so (infinite power). The answer is that we need only generate a process with a power spectrum that is flat over the band of frequencies passed (or approximately so) by the system. As far as the system is concerned, it will appear to be white.

Note also that the amplitude distribution of the white (or pseudowhite) process can be anything. This measurement technique uses only the delta-correlated nature of the white noise process. In particular, the input process can take on the amplitudes ± 1 and the shortest time interval between the $+1$ and -1 values (or vice versa) need only be short compared with the time constant(s) of the system. There are simple algorithms for generating such processes; the resulting waveforms are called pseudonoise (PN) or pseudorandom (PR). We will not go into the theory of such PN waveforms now but simply point out that PN waveforms are widely used in many applications.

Example 10-9

A fixed linear system is driven by a PN source with $N_0 = 1$ W/Hz. The cross-correlation function between input and output is measured and found to be

$$R_{XY}(\tau) = 2e^{-\tau/10}u(\tau)$$

What is the impulse response of the system? The transfer function?

Solution From (10-52) we have

$$h(\tau) = 4e^{-\tau/10}u(\tau)$$

Using the Fourier transform pair

$$e^{-\alpha\tau}u(\tau) \longleftrightarrow \frac{1}{\alpha + j2\pi f}$$

we find the transfer function to be

$$H(f) = \frac{4}{1/10 + j2\pi f} = \frac{40}{1 + j20\pi f}$$

Noise-Equivalent Bandwidth

It is often convenient to characterize a fixed linear system with white noise at its input by a *noise-equivalent bandwidth*. This is the bandwidth of a fictitious system with transfer function that is constant over the frequency extent of the noise-equivalent bandwidth and zero otherwise but has the same noise power at its output as the actual system. For simplicity, we assume a system that has maximum magnitude transfer function at zero frequency. Let its transfer function be denoted as $H(f)$, and let the white noise at its input have two-sided power spectral density $N_0/2$. From (10-30), the power spectral density of the noise at its output is

$$S_Y(f) = \frac{N_0}{2}|H(f)|^2 \tag{10-53}$$

and the noise power at the output is

$$P_{no} = \int_{-\infty}^{\infty} S_Y(f)df = \frac{N_0}{2}\int_{-\infty}^{\infty}|H(f)|^2 df = N_0\int_0^{\infty}|H(f)|^2 df \tag{10-54}$$

where the last equation follows because $|H(f)|^2$ is an even function since it is a magnitude [the square root of the sum of the squares of the real and imaginary parts of $H(f)$]. The fictitious filter is assumed to have the same maximum response in its passband as the actual filter, say H_0 for $-B_N \le f \le B_N$. For $f < -B_N$ or $f > B_N$ the transfer function of the fictitious filter is zero. This is illustrated in Figure 10-9. Thus, with noise of spectral density $N_0/2$ at its input, the noise power at its output is

$$P_{nof} = \int_{-B_N}^{B_N}\frac{N_0}{2}H_0^2\, df = N_0 H_0^2 B_N \tag{10-55}$$

We want the noise powers at the outputs of the actual filter and fictitious filter equal, which defines the noise-equivalent bandwidth. Thus

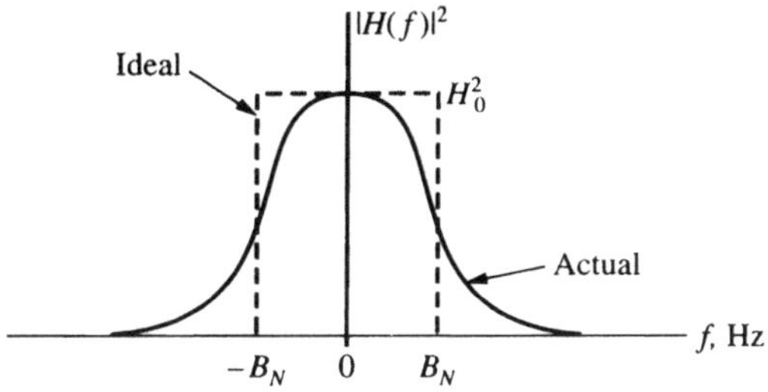

Figure 10-9 Actual magnitude-squared transfer function of a filter and the corresponding ideal magnitude-squared transfer function for computing noise equivalent bandwidth.

$$N_0 \int_0^\infty |H(f)|^2 \, df = N_0 H_0^2 B_N \tag{10-56}$$

or

$$B_N = \frac{1}{H_0^2} \int_0^\infty |H(f)|^2 \, df \tag{10-57}$$

is the noise-equivalent bandwidth.

Example 10-10

Find the noise-equivalent bandwidth for the filter of Example 10-9.

Solution From the last equation in Example 10-9, $H_0 = 40$ and

$$B_N = \frac{1}{40^2} \int_0^\infty \frac{40^2}{1 + (20\pi f)^2} \, df = \frac{1}{20\pi} \int_0^\infty \frac{dx}{1 + x^2} = \frac{1}{20\pi} \tan^{-1} x \Big|_0^\infty = \tfrac{1}{40} \text{ Hz}$$

Thermal Noise Processes

Thermal noise (sometimes called *Johnson noise*) is generated at the terminals of a resistance, say of value R, at Kelvin temperature T and has power spectral density given by[6]

$$S_{\text{thermal}}(f) = 2kTR \quad \text{W/Hz} \quad -\infty < f < \infty \tag{10-58}$$

where $k = 1.38 \times 10^{-23}$ J/K is Boltzmann's constant. For any single-sided bandwidth, B, the average power contained in that bandwidth is

$$P_n = 4kTRB \quad \text{W} \tag{10-59}$$

where it must be remembered that integration over both positive and negative frequencies of the power spectral density (10-58) is required to get total average power. We can view a noisy resistor as a noiseless resistor in series with a voltage source of rms voltage

$$v_n = \sqrt{4kTRB} \quad \text{V rms} \tag{10-60}$$

[6]This is an approximation with the frequency-dependent expression given by

$$S_{\text{qm}}(f) = \frac{2hfR}{\exp(hf/kT) - 1}$$

where h is Planck's constant (6.6254×10^{-34} J-s). For frequencies below infrared (approximately 10^{12} Hz), this expression is well approximated as a constant given by (10-58).

where we get power in this case by squaring the voltage and *not* dividing by the resistance. To apply this to a circuit composed of several resistors, the other fact needed is that noise voltages produced by separate resistors are uncorrelated.

If resistances, capacitances, and inductors are present in a network, we can still apply (10-59) except that now the "resistance" is the real part of the equivalent impedance of the network. This is called *Nyquist's formula*.

Example 10-11

Find the average noise power in a 2-MHz bandwidth at the terminals of the resistive circuit shown in Figure 10-10a. Assume a temperature of 290 K. Do this two ways: (a) by replacing each resistor with an equivalent-noise voltage source, and (b) by applying Nyquist's theorem.

Solution (a) In Figure 10-10b, c, and d, noise-equivalent circuits are shown for each resistor considered noisy in turn and replaced by a noiseless resistor in series with a voltage source with rms voltage given by (10-60). From Figure 10-10b, resistor 1 produces an rms noise voltage at the output of

$$v_{o1} = \frac{R_2}{R_1 + R_2} v_{n1} = \frac{R_2}{R_1 + R_2} \sqrt{4kTR_1B} = 2.83 \times 10^{-7} \text{ V rms}$$

found by using voltage division. Similarly, from Figure 10-10c, resistor 2 produces an rms noise voltage at the output of

$$v_{o2} = \frac{R_1}{R_1 + R_2} v_{n2} = \frac{R_1}{R_1 + R_2} \sqrt{4kTR_2B} = 2.83 \times 10^{-7} \text{ V rms}$$

Finally, from Figure 10-10d, since the full voltage of resistor 3 appears across the output, we have

$$v_{o3} = v_{n3} = \sqrt{4kTR_3B} = 4.00 \times 10^{-7} \text{ V rms}$$

Since the noise powers at the outputs are uncorrelated, the rms voltages add as the square root of the sum of the squares, and the total rms noise voltage at the output is

$$v_0 = \sqrt{v_{o1}^2 + v_{o2}^2 + v_{o3}^2} = 5.66 \times 10^{-7} \text{ V rms} = 0.566 \ \mu\text{V rms}$$

(b) To apply Nyquist's theorem, we find the equivalent resistance of the circuit (two 10-Ω resistors in parallel in series with a 5-Ω resistor $= 10\ \Omega$) and apply (10-60) to obtain

$$V_o = \sqrt{4(1.38 \times 10^{-23})(290)(10)(2 \times 10^6)} = 5.66 \times 10^{-7} \text{ V rms}$$

which is the same result as found by using superposition.

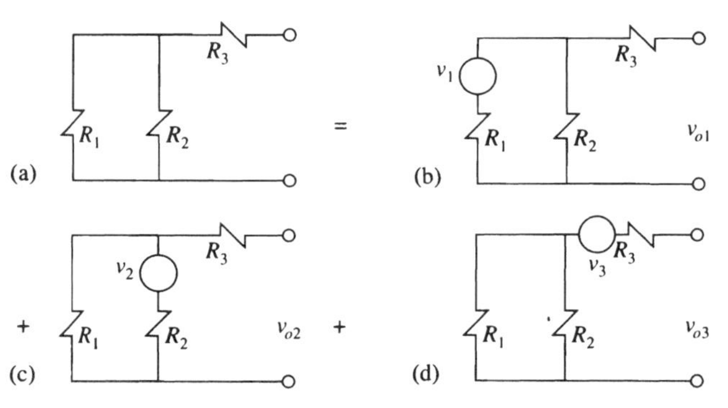

Figure 10-10 Noisy resistive circuit and three circuits for applying superposition taking one resistor at a time.

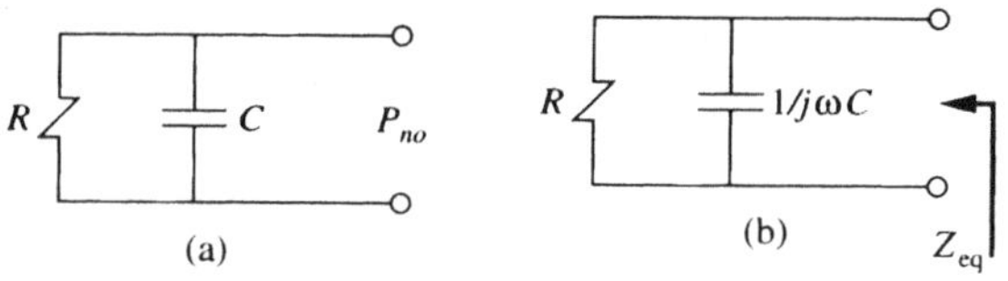

Figure 10-11 Circuit for Example 10-12.

Example 10-12

Find the average noise power at the output of the *RC* circuit shown in Figure 10-11a (considering all frequencies from 0 to ∞).

Solution The equivalent impedance looking back into the circuit, using ac sinusoidal steady-state analysis, is

$$Z_{eq}(\omega) = \frac{R/j\omega C}{R + 1/j\omega C} = \frac{R}{1 + j\omega RC}$$

The real part of this expression is

$$R_{eq}(f) = \frac{R}{1 + (2\pi fRC)^2}$$

where $f = \omega/2\pi$ in hertz has been substituted. The noise spectral density at the output, from (10-58), is

$$S_o(f) = 2kTR_{eq}(f) = \frac{2kTR}{1 + (2\pi fRC)^2}$$

The total noise power at the output is the integral of the spectral density over all frequencies:

$$P_{no} = \int_{-\infty}^{\infty} S_o(f)\,df = \int_0^{\infty} \frac{4kTR}{1 + (2\pi RCf)^2}\,df$$

$$= \frac{2kT}{\pi C}\int_0^{\infty} \frac{dx}{1 + x^2} \quad (x = 2\pi RCf)$$

$$= \frac{2kT}{\pi C}\tan^{-1} x \Big|_0^{\infty} = \frac{kT}{C}$$

where the integral can be doubled and the integration carried out from 0 to ∞ because the integrand is even. Surprisingly, the result is independent of *R*.

Noise Figure and Friis's Formula

Electronic systems can often be modeled as cascades of two-terminal devices, or subsystems, as shown in Figure 10-12. Signal plus noise is present in the input of the first subsystem, that gets passed through each succeeding subsystem, which adds its own noise. Thus the noise appearing at the output of the last subsystem has power greater than that appearing at the input to

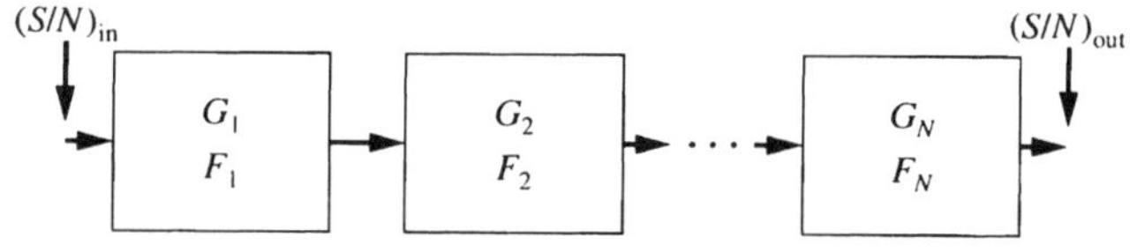

Figure 10-12 Cascade of subsystems making up a system.

the first subsystem unless all subsystems are noiseless, which is an idealization that does not happen in practice. The fact that noise is added by the kth subsystem is characterized by a parameter called a *noise figure, F_k*, which can be defined to be the ratio of signal power to noise power (called the *signal-to-noise ratio*) at a subsystem's input to signal-to-noise ratio at its output:

$$F_k = \frac{(S/N)_{\text{in},\,k}}{(S/N)_{\text{out},\,k}} \tag{10-61}$$

where $(S/N)_{\text{in},\,k}$ is the signal-to-noise power ratio (hereafter simply referred to as the signal-to-noise ratio) at the input and $(S/N)_{\text{out},\,k}$ is the signal-to-noise ratio at the output. Since each subsystem adds its own noise, except for an ideal system, $(S/N)_{\text{out},\,k} < (S/N)_{\text{in},\,k}$ and $F_k > 1$ for practical systems. Usually, for electronic devices, the noise figure is given by the manufacturer in decibels, which is

$$F_{\text{dB}} = 10 \log_{10} F_{\text{ratio}} \tag{10-62}$$

The question remains: How can the noise figure of a cascade of subsystems be found given the noise figure of each subsystem? This is accomplished by applying Friis's formula,[7] which is

$$F_{\text{overall}} = F_1 + \frac{F_2 - 1}{G_1} + \frac{F_3 - 1}{G_1 G_2} + \cdots + \frac{F_N - 1}{G_1 G_2 \cdots G_{N-1}} \tag{10-63}$$

where G_k is the power gain, or ratio of output to input signal power, for the kth subsystem.

Example 10-13

A cascade of three subsystems make up an overall system. The power gains and noise figures for each subsystem are given in the table below. What is the noise figure of the overall system? What is the power gain of the overall system?

Subsystem	G_k (dB)	F_k (dB)
1	10	3
2	−6	6
3	40	5

Solution We must first convert all gains and noise figures to ratios using the inverse of (10-62):

$$G_{k,\,\text{ratio}} = 10^{G_{k,dB}\ 10} \quad \text{and} \quad F_{k,\,\text{ratio}} = 10^{F_{k,dB}\ 10} \tag{10-64}$$

The results are given in the following table

Subsystem	G_k (ratio)	F_k (ratio)
1	10	$1.995 \approx 2$
2	0.25	$3.981 \approx 4$
3	10^4	3.162

[7]See Ziemer and Tranter (1995, Appendix A), for a deviation of Friis's formula.

From Friis's formula, we compute the overall noise figure to be

$$F_{\text{ratio}} = 2 + \frac{4 - 1}{10} + \frac{3.162 - 1}{10 \times 0.25} = 3.15, \quad F_{\text{dB}} = 10 \log_{10}(3.15) = 4.99 \text{ dB} \qquad (10\text{-}65)$$

Note that the result does not depend on the gain of the last subsystem.

The overall gain is the product of the separate subsystem gains expressed as ratios or, in dB, the sum of the subsystem gains:

$$G_{\text{dB}} = G_{1,\text{dB}} + G_{2,\text{dB}} + G_{3,\text{dB}} = 10 - 6 + 40 = 44 \text{ dB}$$

or

$$G_{\text{ratio}} = 10^{4.4} = 2.512 \times 10^4$$

Example 10-14

Given the same system as in Example 10-13, except that now the gain of the first stage is to be chosen so that the overall noise figure is 4 dB, what should the gain be?

Solution Since 4 dB is a ratio of 2.512, we have the equation

$$2.512 = 2 + \frac{4 - 1}{G_1} + \frac{3.162 - 1}{G_1 \times 0.25} \qquad (10\text{-}66)$$

or, when solved for G_1, we obtain $G_{1,\text{ratio}} = 22.75$ or $G_{1,\text{dB}} = 13.57$ dB.

10–5 NARROWBAND RANDOM PROCESSES

In many applications, particularly the analysis and design of communication systems, narrowband bandpass filters[8] are used in the front end of the systems under consideration. In the case of communication systems, this would be an amplifier near the front end of the receiver, for example. The noise added at the input stages, where the signal amplitude is small and therefore has power of the same order of magnitude as the noise generated there, is white. Once it passes through the narrow bandpass filter near the input[9] it is what is referred to as a *narrowband random process* or *narrowband noise*. A convenient representation for such noise is

$$n(t) = n_c(t) \cos(\omega_0 t + \theta) - n_s(t) \sin(\omega_0 t + \theta) \qquad (10\text{-}67)$$

where $n_c(t)$ and $n_s(t)$ are referred to as the quadrature noise components, $\omega_0 = 2\pi f_0$, and θ is an arbitrary phase angle. Equation (10-67) can be rewritten as

$$n(t) = R(t) \cos[\omega_0 t + \phi(t) + \theta] \qquad (10\text{-}68)$$

where

$$R(t) = \sqrt{n_c^2(t) + n_s^2(t)} \qquad (10\text{-}69)$$

[8]By *narrowband* it is meant that the bandwidth of the frequency response function of the filter is much less than the center frequency, usually by at least a factor of 10.

[9]Depending on the structure of the receiver, this could either be an input amplifier, often referred to as the RF (radio-frequency) amplifier, or an amplifier [called the IF (intermediate frequency)] amplifier after a stage of amplification that is not narrowband and a mixer.

and

$$\phi(t) = \tan^{-1}\frac{n_s(t)}{n_c(t)} \tag{10-70}$$

The following can be shown about the random processes $n_c(t)$, $n_s(t)$, $R(t)$, and $\phi(t)$:[10]

1. $n_c(t)$, $n_s(t)$, $R(t)$, and $\phi(t)$ are lowpass, or slowly varying with respect to $n(t)$.
2. $n_c(t)$ and $n_s(t)$ have zero mean if $n(t)$ does and the same variance as $n(t)$.
3. $n_c(t)$ and $n_s(t)$ are uncorrelated *at the same time instant.*
4. If $S(f)$ is the power spectral density of $n(t)$, then the power spectral densities of the quadrature components are

$$S_{n_c}(f) = S_{n_s}(f) = \text{Lp}[S_n(f - f_0) + S_n(f + f_0)] \tag{10-71}$$

   where Lp[·] indicates the low-frequency part of the spectrum inside the brackets.
5. The cross power spectral density of $n_c(t)$ and $n_s(t)$ is

$$S_{n_c n_s}(f) = j\text{Lp}[S_n(f - f_0) - S_n(f + f_0)] \tag{10-72}$$

   Thus $n_c(t)$ and $n_s(t)$ are uncorrelated (i.e., for any time separation τ) if

$$\text{Lp}[S_n(f - f_0) - S_n(f + f_0)] = 0 \tag{10-73}$$

   This property says that $n_c(t)$ and $n_s(t)$ are uncorrelated if the power spectral density of $n(t)$ is symmetrical about $f = f_0, f > 0$.
6. If $n(t)$ is Gaussian so are $n_c(t)$ and $n_s(t)$.
7. If $n(t)$ is Gaussian and property 5 holds, $n_c(t)$ and $n_s(t)$ are statistically independent.
8. If $n(t)$ is Gaussian, $R(t)$ is Rayleigh and $\phi(t)$ is uniformly distributed in $(0, 2\pi)$.

Example 10-15

Consider a Gaussian bandpass random process with the power spectral density shown in Figure 10-13. Consider two cases for f_0: 7 Hz and 5 Hz. Find the following for each case.

(a) The means of $n_c(t)$ and $n_s(t)$
(b) The variances of $n_c(t)$ and $n_s(t)$
(c) The cross-spectral density of $n_c(t)$ and $n_s(t)$
(d) The cross-correlation function of $n_c(t)$ and $n_s(t)$
(e) The joint pdf of $n_c(t)$ and $n_s(t)$ at the same time instant
(f) The pdf of $R(t)$

Solution

(a) Since the power spectral density of $n(t)$ does not have an impulse at zero frequency, the mean of $n(t)$ is zero and, therefore, so are the means of $n_c(t)$ and $n_s(t)$. This holds for both $f_0 = 5$ and 7 Hz.

(b) Because the mean of $n(t)$ is zero, its variance and mean-square value are equal and can be found as the integral of its power spectral density over all frequency, which is 24 W by inspection of Figure 10-13a. This holds for both $f_0 = 5$ and 7 Hz.

[10]See Ziemer and Tranter (1995, Chap. 5) for proofs.

(c) For $f_0 = 7$ Hz, the cross-spectral density is zero. For $f_0 = 5$ Hz, the cross-spectral density of $n_c(t)$ and $n_s(t)$ is given by (10-72) and is shown shaded in Figure 10-13d multiplied by $-j$ to make the plot real.

(d) For $f_0 = 7$ Hz, the cross-correlation function is zero. For $f_0 = 5$ Hz, the cross-correlation function of $n_c(t)$, and $n_s(t)$ is the inverse Fourier transform of the shaded portion of Figure 10-13d without the $-j$ and is given by

$$R_{n_c n_s}(\tau) = \mathcal{F}^{-1}\{-2j\Pi[0.25(f-3)] + 2j\Pi[0.25(f+3)]\}$$

$$= 2j[-4\,\text{sinc}(4\tau)e^{j6\pi\tau} - 4\,\text{sinc}(4\tau)e^{-j6\pi\tau}]$$

$$= 16\,\text{sinc}(4\tau)\sin(6\pi\tau)$$

where $\Pi(x) = 1$, $|x| \leq 1/2$, and 0 otherwise and where Fourier transform pairs and theorems from Appendix A have been used.

(e) $n_c(t)$ and $n_s(t)$ at the same time instant are uncorrelated (but not uncorrelated for any time difference for $f_0 = 5$ Hz) and Gaussian and hence independent, so using the means and variances found in parts (a) and (b), we have

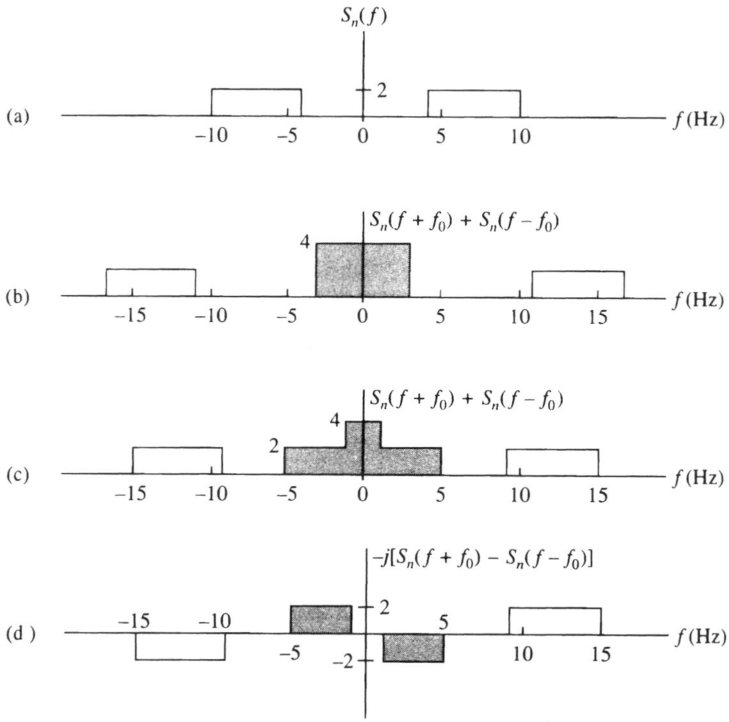

Figure 10-13 Spectra pertinent to Example 10-15: (a) narrowband noise spectrum; (b) sum spectrum for computing spectra of in-phase and quadrature components for $f_0 = 7$ hertz; (c) sum spectrum for computing spectra of in-phase and quadrature components for $f_0 = 5$Hz; (d) difference spectra for computing cross-correlation function between in-phase and quadrature components for $f_0 = 5$ Hz.

$$f_{n_c n_s}(x, y) = \frac{e^{-x^2/48}}{\sqrt{48\pi}} \frac{e^{-y^2/48}}{\sqrt{48\pi}} = \frac{e^{-(x^2+y^2)/48}}{48\pi}$$

This holds for both $f_0 = 5$ and 7 Hz (note that the independent variables in the joint pdf can be anything).

(f) Since $n_c(t)$ and $n_s(t)$ are uncorrelated and Gaussian, $R(t)$ is Rayleigh with pdf

$$f_R(r) = \frac{re^{-r^2/48}}{48\pi}, \quad r \geq 0$$

10-6 LINEAR MEAN-SQUARED ERROR ESTIMATION

In Chapter 6, we considered parameter estimation. Another type of estimation of interest is estimation of a sample function of a random process. This problem can be stated as follows: given two wide-sense-stationary random processes $W(t)$ and $X(t)$, where $W(t)$ is related to a desired signal $S(t)$ through a linear transformation [e.g., $W(t) = dS(t)/dt$] and $X(t)$ is statistically related to $W(t)$ [e.g., $X(t) = W(t) + N(t)$, where $N(t)$ is a stationary process representing noise which is independent of $W(t)$], it is desired to find an estimate for $W(t)$ throughout some range of the independent variable, say $a \leq t \leq b$. In particular, for simplicity, assume that the estimate is to be obtained by a fixed linear system; that is, the estimate $\hat{W}(t)$ for $W(t)$, from the desired observed signal $X(t)$, is to be of the form

$$\hat{W}(t) = \int_{-\infty}^{\infty} h(\lambda)X(t - \lambda)\, d\lambda \tag{10-74}$$

where $h(t)$ is an impulse response for the filter to be found. The goodness criterion to be used in obtaining $h(t)$ is to minimize the mean-squared error between the desired waveform $W(t)$ and the estimate $\hat{W}(t)$:

$$e = E\{[W(t) - \hat{W}(t)]^2\} = \text{minimum} \tag{10-75}$$

or

$$e = E\left\{\left[W(t) - \int_{-\infty}^{\infty} h(\lambda)X(t - \lambda)\, d\lambda\right]^2\right\} = \text{minimum} \tag{10-76}$$

In Section 6-7, it was shown that a general principle applicable to such minimization problems is the *orthogonality principle,* which states that the expectation of the product of the error and the observed data is zero (i.e., are statistically orthogonal). In the case at hand, this orthogonality of error and data takes the form

$$e = E\left\{\left[W(t) - \int_{-\infty}^{\infty} h(\lambda)X(t - \lambda)\, d\lambda\right]X(u)\right\} = 0, \quad a \leq u \leq b \tag{10-77}$$

where the limitation on u indicates the extent over which the estimate is to be based. We will consider the simplest case here, namely that the random process $\{X(t)\}$ is available for all time, or that $a = -\infty$ and $b = \infty$. When we take the expectation of each term in (10-77), interchange the order of expectation and integration, and use the stationarity property of the random process involved, we obtain the relationship

$$R_{XW}(t-u) = \int_{-\infty}^{\infty} R_X(t-\lambda-u)h_0(\lambda)\,d\lambda \tag{10-78}$$

where the subscript on $h_o(t)$ denotes that it is the optimum impulse response,

$$R_{XW}(\tau) = E[X(t)W(t+\tau)] \tag{10-79}$$

is the cross-correlation function of the random processes $\{X(t)\}$ and $\{W(t)\}$, and $R_X(\tau)$ is the autocorrelation function of the random process $\{X(t)\}$. Letting $\tau = t - u$ in (10-78), we obtain

$$R_{XW}(\tau) = \int_{-\infty}^{\infty} R_X(\tau-\lambda)h_0(\lambda)\,d\lambda \tag{10-80}$$

Equation (10-80) is recognized as a convolution integral. Thus we may apply the convolution theorem of Fourier transforms and get

$$S_{XW}(f) = S_X(f)H_0(f) \quad \text{or} \quad H_0(f) = \frac{S_{XW}(f)}{S_X(f)} \tag{10-81}$$

where $S_X(f)$ is the power spectral density of $\{X(t)\}$ and $S_{XW}(f)$ is the cross-power spectral density of $\{X(t)\}$ and $\{W(t)\}$.

The minimum mean-squared error is found by expanding (10-76) and taking the expectation term by term to get

$$e_{\min} = R_W(0) - \int_{-\infty}^{\infty} h_0(\lambda)R_{XY}(\lambda)\,d\lambda \tag{10-82}$$

Fourier transform techniques which we will not go through here allow this to be put into the form

$$e_{\min} = \int_{-\infty}^{\infty} S_W(f)\,df - \int_{-\infty}^{\infty} H_0^*(f)S_{XW}(f)\,df = \int_{-\infty}^{\infty}\left[S_W(f) - \frac{|S_{XW}(f)|^2}{S_X(f)}\right]df \tag{10-83}$$

Example 10-16

Let $X(t) = S(t) + N(t)$, where $S(t)$ and $N(t)$ are uncorrelated stationary random signal and noise, respectively. Let $W(t) = S(t)$. Assume a doubly infinite observation interval so that (10-81) holds. Because $S(t)$ and $N(t)$ are uncorrelated,

$$R_{XW}(\tau) = R_S(\tau) \quad \text{and} \quad R_X(\tau) = R_S(\tau) + R_N(\tau) \tag{10-84}$$

where $R_S(\tau)$ and $R_N(\tau)$ are the autocorrelation functions of the signal and noise, respectively, and $S_S(f)$ and $S_N(f)$ are power spectral densities. From (10-81) we have

$$H_0(f) = \frac{S_S(f)}{S_S(f) + S_N(f)} \tag{10-85}$$

For example, suppose that the noise is white with power spectral density $N_0/2$ and the signal has autocorrelation function $R_S(\tau) = B\exp(-\alpha|\tau|$, where B and α are positive constants. Then

$$S_S(f) = \frac{2B\alpha}{\alpha^2 + (2\pi f)^2} \tag{10-86}$$

and the transfer function of the optimum filter is

$$H_0(f) = \frac{4B\alpha/N_0}{4B\alpha/N_0 + \alpha^2 + (2\pi f^2)} \tag{10-87}$$

This corresponds to the impulse response

$$h_0(t) = Ae^{-|t|/\tau} \tag{10-88}$$

where

$$\frac{1}{\tau} = \sqrt{\frac{4B\alpha}{N_0} + \alpha^2} \quad \text{and} \quad A = \frac{1}{N_0\sqrt{1/\alpha BN_0 + 1/4B^2}} \tag{10-89}$$

Clearly, the optimum filter is noncausal (see Appendix A). We can approximate an optimum causal filter by delaying the double-sided exponential impulse response (10-88) by a fixed amount, say t_0, and setting the part to the left of $t = 0$ to zero.

10–7 SUMMARY

This chapter has provided an introduction to the effect of systems on random processes. A system relates each sample function at its input to a sample function at its output through the input–output relationships $\mathcal{H}[\,\cdot\,]$ as $Y^{(i)}(t)=\mathcal{H}[X^{(i)}(t)]$ where $X^{(i)}(t)$ denotes the ith sample function of the input random process and $Y^{(i)}(t)$ denotes the corresponding ith sample function of the output random process. Basic system theory concepts are reviewed in Appendix A. Depending on the type of system and the statistics of the random process, varying degrees of description of the output random process from the system are possible. The following are the main points presented in this chapter.

1. A zero-memory system is one for which the input–output relationship is of the form

$$Y(t) = g[X(t)]$$

where $g[\,\cdot\,]$ is a known function. The output depends on the input at the present time only and not on past or future values of time. For such systems, the first-order pdf of the output can be obtained by transformation of random variables as first discussed in Section 3-4 because at a given time, we have a functional relationship between two random variables. This was done in this chapter for some cases additional to those considered in Chapter 3, including a dead-zone device and limiter. In addition, it is often possible to relate low-order averages of the output-to-input averages. These included the mean, variance, and autocorrelation function. This was done in the text for the case where the input–output relationship is expressible as a series, although techniques exist for cases where this is not the case, such as hard limiters.

2. In the case of a fixed linear system, input and output sample functions are related by the superposition integral:

$$Y^{(i)}(t) = \int_{-\infty}^{\infty} h(\tau)X^{(i)}(t-\tau)\,d\tau = \int_{-\infty}^{\infty} h(t-\lambda)X^{(i)}(\lambda)\,d\lambda$$

where $h(\tau)$ is the impulse response of the system. For simplicity, we assume that the input random process is stationary and has been present since $t = -\infty$, so that transients have died out. Therefore, the output random process is stationary. The means of the input and output random processes are related by

$$m_Y = m_X \int_{-\infty}^{\infty} h(\tau)\,d\tau$$

or

$$m_Y = H(0)m_x$$

where $H(0)$ is the transfer function of the system

$$H(f) = \int_{-\infty}^{\infty} h(\zeta)e^{-j2\pi f\zeta}d\zeta$$

evaluated at zero frequency. This is called the *dc gain* of the system.

3. The cross-correlation function of the input with output is given by

$$R_{XY}(\tau) = h(\tau) * R_X(\tau)$$

and the Fourier transform of $R_{XY}(\tau)$, or the cross-power spectral density of input with output, is

$$S_{XY}(f) = H(f)S_X(f)$$

4. The autocorrelation function of the output is related to the autocorrelation function of the input by

$$\begin{aligned} R_Y(\tau) &= E[Y(t)Y(t+\tau)] \\ &= \int_{-\infty}^{\infty} h(\lambda)h(\zeta)R_X(\tau+\lambda-\zeta)\,d\zeta \\ &= h(\tau) * R_{YX}(\tau) \\ &= h(-\tau) * R_{XY}(\tau) \end{aligned}$$

5. The power spectral density of the output of a filter with random process at its input is related to the power spectral density $S_X(f)$ of the input by

$$S_Y(f) = H^*(f)H(f)S_X(f) = |H(f)|^2 S_X(f)$$

where $H(f)$ is the transfer function of the filter.

6. For a filter with a stationary Gaussian random process at its input, the output random process is also Gaussian. Thus one can write down the pdf of the output by finding its mean, variance, and autocorrelation function, or equivalently, its power spectral density, and applying (4-55).

7. White noise has a constant power spectral density for all frequencies; that is,

$$S_{wn}(f) = \frac{N_0}{2}, \qquad -\infty < f < \infty \qquad (f \text{ in hertz})$$

By inverse Fourier transformation, the autocorrelation function of white noise is

$$R_{wn}(\tau) = \frac{N_0}{2}\delta(\tau)$$

This says that for a white noise process, successive sample values are uncorrelated no matter how close together they are taken, except at the same instant, of course. It also says that white noise has infinite power and therefore is a physical impossibility.

8. The cross-correlation function of output with input for a fixed linear system with white noise at its input is given by

$$R_{XY}(\tau) = \frac{N_0}{2}h(\tau) \qquad \text{or} \qquad h(\tau) = \frac{2}{N_0}R_{XY}(\tau)$$

which provides a way to measure the impulse response of the system.

9. The noise-equivalent bandwidth of a fixed linear system is the bandwidth of a fictitious system with transfer function that is constant over the frequency extent of the noise-equivalent bandwidth and zero otherwise, with the same noise power at its output as the actual system. In terms of the transfer function, $H(f)$, of a system that has a passband centered at dc, it is given by

$$B_N = \frac{1}{H_0^2}\int_0^\infty |H(f)|^2\,df$$

where $H_0 = H(f)|_{f=0}$

10. Thermal noise is noise generated by resistors at temperatures above absolute zero. The two-sided power spectral density of thermal noise generated by a resistor of resistance R is

$$S_{\text{thermal}}(f) = 2kTR \qquad \text{W/Hz} \qquad -\infty < f < \infty$$

where $k = 1.38 \times 10^{-23}$ J/K is Boltzmann's constant. Thus thermal noise can be approximated as a white noise process, although the actual power spectral density goes to zero for frequencies in the infared region.

11. A noisy resistor can be modeled as a noiseless resistor in series with a rms noise voltage of value

$$v_n = \sqrt{4kTRB} \qquad \text{V rms}$$

To find the output noise power or rms noise voltage from a combination of resistors, this model can be used for each resistor and the principle of superposition applied together with the fact that separate contributions to the output voltage from each resistor are uncorrelated with the rest.

12. If resistances, capacitances, and inductors are present in a network, we can still apply the model above except that now the "resistance" is the real part of the equivalent impedance of the network. This is called *Nyquist's formula*.

13. For a cascade of noisy subsystems, noise is added by each subsystem as the signal and noise pass through it. The noise added by the kth subsystem is characterized by a parameter called the *noise figure,* F_k, which is defined to be the ratio of signal power to noise power (called the *signal-to-noise ratio*) at a subsystem's input to signal-to-noise ratio at its output:

$$F_k = \frac{(S/N)_{\text{in},\,k}}{(S/N)_{\text{out},\,k}}$$

where $(S/N)_{\text{in},\,k}$ is the input signal-to-noise power ratio and $(S/N)_{\text{out},\,k}$ is the output signal-to-noise power ratio.

14. The overall noise figure of a cascade of subsystems is found from Friis's formula, which is

$$F_{\text{overall}} = F_1 + \frac{F_2 - 1}{G_1} + \frac{F_3 - 1}{G_1 F_2} + \cdots + \frac{F_N - 1}{G_1 G_2 \cdots G_{N-1}}$$

where F_k is the noise figure and G_k is the power gain, or ratio of output to input signal power, for the kth subsystem.

15. A narrowband random process is one whose power spectral density is localized in a narrow bandwidth around a center frequency. It may be represented as

$$n(t) = n_c(t)\cos(\omega_0 + \theta) - n_s(t)\sin(\omega_0 t + \theta)$$

where $n_c(t)$ and $n_s(t)$ are referred to as the quadrature noise components, $\omega_0 = 2\pi f_0$ is the center frequency in rad/s, and θ is an arbitrary phase angle. This can be rewritten in envelope and phase form as

$$n(t) = R(t)\cos[\omega_0 t + \phi(t) + \theta]$$

where

$$R(t) = \sqrt{n_c^2(t) + n_s^2(t)}$$

and

$$\phi(t) = \tan^{-1}\frac{n_s(t)}{n_c(t)}$$

The following can be shown about the random processes $n_c(t)$, $n_s(t)$, $R(t)$, and $\phi(t)$:

(a) $n_c(t)$, $n_s(t)$, $R(t)$, and $\phi(t)$ are lowpass, or slowly varying with respect to $n(t)$.

(b) $n_c(t)$ and $n_s(t)$ have zero mean if $n(t)$ does and the same variance as $n(t)$.

(c) $n_c(t)$ and $n_s(t)$ are uncorrelated *at the same time instant.*

(d) If $S(f)$ is the power spectral density of $n(t)$, the power spectral densities of the quadrature components are

$$S_{n_c}(f) = s_{n_s}(f) = \text{Lp}[S_n(f - f_0) + S_n(f + f_0)]$$

where Lp[·] means the lowpass part of [·].

(e) The cross-power spectral density of $n_c(t)$ and $n_s(t)$ is

$$S_{n_c n_s}(f) = j\,\text{Lp}[S_n(f - f_0) - S_n(f + f_0)]$$

Thus $n_c(t)$ and $n_s(t)$ are uncorrelated (i.e., for any time separation τ) if

$$\text{Lp}(S_n(f - f_0) - S_n(f + f_0)] = 0$$

This property says that $n_c(t)$ and $n_s(t)$ are uncorrelated if the power spectral density of $n(t)$ is symmetrical about $f = f_0, f > 0$.

16. In linear mean-squared-error estimation of a functional of a random process, we seek a fixed linear system, or filter, whose output is the estimate and for which the mean-squared error

$$e = E\left\{\left[W(t) - \int_{-\infty}^{\infty} h(\lambda)X(t-\lambda)\,d\lambda\right]^2\right\}$$

is minimized to give the optimum impulse response. The expression above assumes a doubly infinite observation interval, with the input being present from the infinite past. In this case, the transfer function of the optimum filter is provided by

$$H_0(f) = \frac{S_X(f)}{S_{XW}(f)}$$

where $S_X(f)$ is the power spectral density of the observed process and $S_{XW}(f)$ is the cross-power spectral density between of the random process related to the desired signal and the observed process. The minimum mean-square error is given in the case of a doubly infinite observation interval by

$$e_{\min} = \int_{-\infty}^{\infty} S_W(f)\,df - \int_{-\infty}^{\infty} H_0^*(f)S_{XW}(f)\,df = \int_{-\infty}^{\infty}\left[S_W(f) - \frac{|S_{XW}(f)|^2}{S_X(f)}\right]df$$

10-8 FURTHER READING

In addition to the references cited in Section 9-9, see Chapter 5 and Appendix A of Ziemer and Tranter (1995). For more on linear mean-squared-error extimation, see Papoulis (1991, Chap. 14).

10-9 PROBLEMS

SECTION 10-2

10-1. Referring to Example 10-1 and the dead-zone device, what should the ratio c/σ be so that the dc (constant) component and the fluctuating component at the output have equal probabilities?

10-2. Reconsider Example 10-2 with the hard-limiter characteristic shown in Figure 10-14.

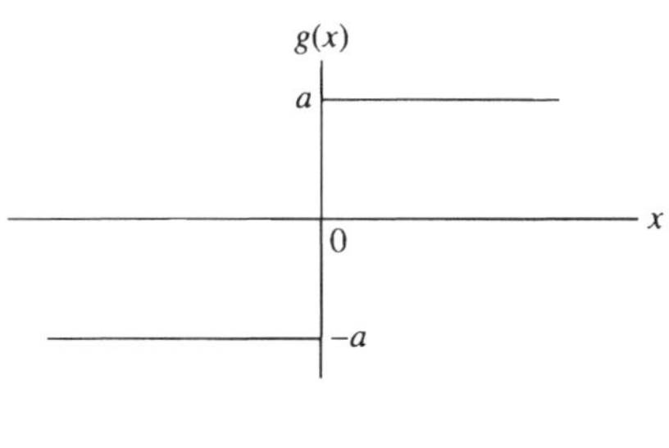

Figure 10-14

Assume stationary Gaussian noise at the input of zero mean and variance σ^2. Sketch a typical sample function at the output. Write down the output pdf. This can be done by inspection.

10-3. Reconsider Example 10-4. Noting that $R_S(0) = A^2/2$ and $R_n(0) = P_n$ is the input noise power, what should the noise power be in terms of the signal amplitude so that the power in the dc component at the output equals four times the power in the sinusoidal component at $\pm 2f_0$?

Section 10-3

10-4. A second-order lowpass Butterworth filter has transfer function

$$H(f) = \frac{1}{1 + j\sqrt{2}\,(f/f_3) - (f^2/f_3^2)}$$

where f_3 is the frequency at which the magnitude of $H(f)$ is 0.707 of its magnitude at $f = 0$. A stationary random process is present at the input with power spectral density

$$S_X(f) = \frac{10}{4 + f^2} + 25\delta(f)$$

If $f_3^2 = 2$, find the following:
(a) The mean of the input
(b) The dc gain of the filter
(c) The mean of the output
(d) The power spectral density of the output
(e) The cross-power spectral density of input and output

10-5. Derive (10-39) by integrating the power spectral density (10-34) of $\{Y(t)\}$ over all frequencies and show that the result is the same as (10-39).

10-6. A system is defined by the input–output relationship

$$Y(t) = X(t) + X(t - T)$$

where T is a constant.
(a) Show that the output autocorrelation function is

$$R_Y(\tau) = 2R_X(\tau) + R_X(\tau + T) + R_X(\tau - T)$$

where $R_X(\tau)$ is the autocorrelation function of the input.
(b) Show that the power spectral density of the output is given by

$$S_Y(f) = 4S_X(f)\cos^2(\pi f T)$$

where $S_X(f)$ is the power spectral density of the input.

10-7. Given a fixed linear system with impulse response

$$h(t) = Ae^{-\alpha t}u(t)$$

and a stationary random process at its input with power spectral density

$$S_X(f) = \frac{B}{\beta^2 + (2\pi f)^2}$$

where A, B, α, and β are positive constants, obtain:
(a) The power spectral density of the output random process
(b) The autocorrelation function of the output random process
(c) The cross-power spectral density of input with output

(d) The cross-correlation function of input with output

10-8. Referring to Example 10-8, suppose that all parameter values are the same as in the example except that $R_1 = 10{,}000\ \Omega$. Find the first- and second-order pdf's of the output random process.

Section 10-4

10-9. Suppose that white noise of double-sided power spectral density $N_0/2$ is passed through a filter with transfer function

$$H(f) = \begin{cases} 1, & -B \leq f \leq B \\ 0, & \text{otherwise} \end{cases}$$

Find the autocorrelation function of the output random process.

10-10. A fixed linear system is driven by white noise with $N_0 = 2$ W/Hz. The cross-correlation function between input and output is measured and found to be

$$R_{XY}(\tau) = 5e^{-\tau/10\pi}u(\tau)$$

Find the transfer function of the filter.

10-11. Find the noise-equivalent bandwidth of a filter with impulse response

$$h(t) = 5e^{-20t}u(t)$$

10-12. Find the noise-equivalent bandwidths of the filters with transfer function magnitudes shown in Figure 10-15.

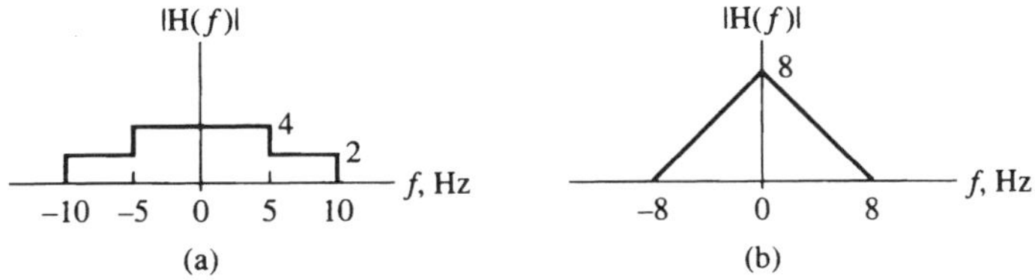

Figure 10-15

10-13. Obtain the rms noise voltages at the outputs of the circuits shown in Figure 10-16 in a bandwidth of 100,000 Hz and for a temperature of 330 K.

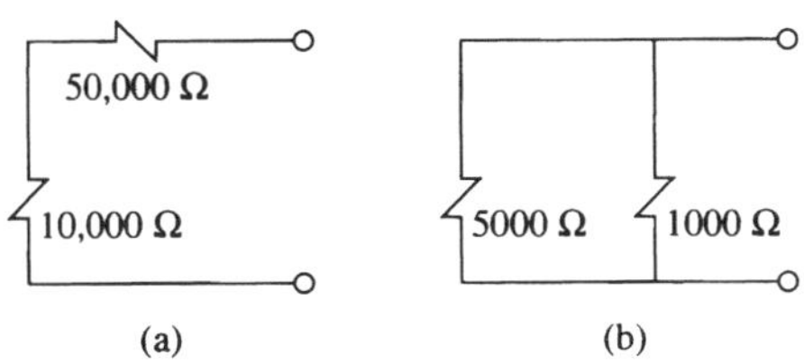

Figure 10-16

10-14. Given a cascade of four subsystems with the power gains and noise figures given in the table below. Find the noise figure of the overall system in dB. What is the overall gain in dB?

Subsystem	G_k (dB)	F_k (dB)
1	15	2.5
2	−5	5
3	20	4
4	60	7

10-15. Given the data of Problem 10-14, what should the gain of the first stage be to give an overall noise figure of 3 dB?

Section 10-5

10-16. Referring to Example 10-15, suppose that $f_0 = 8$ Hz. Redraw Figure 10-13 for this case.

Section 10-6

10-17. Find an expression for the minimum mean-squared error for Example 10-16 in terms of the parameters given in the example (i.e., B, N_0, and α).

10–10 COMPUTER EXERCISES

10-1. Do a computer simulation to determine the first-order pdf at the output of a zero-memory device with Gaussian-distributed noise at the input which is independent sample to sample plus a sinusoid. Experiment with different device characteristics, noise sample variances, σ^2, and sinusoid amplitudes, A. Specifically, for a limiter as shown in Figure 10-3, vary b and $A^2/2\sigma^2$.

10-2. Simulate a narrowband Gaussian noise process by generating a sequence of independent Gaussian noise samples and using a MATLAB filter function, such as "butter", to design a narrowband filter. Study the effect of the bandwidth on the envelope and phase.

A

Some Concepts and Formulas from Linear System Theory

A–1 SPECIAL FUNCTIONS

A few concepts from linear system theory are used in the main body of this book. It is our purpose in this appendix to summarize them for review and handy reference. In this section we define and give some important properties of two special functions used in the text. Normally, these functions would be defined with t as the independent variable, suggesting time. We use the independent variable x in keeping with the context of random variables.

The *unit step function* is defined as

$$u(x) = \begin{cases} 1, & x \geq 0 \\ 0, & x < 0 \end{cases} \tag{A-1}$$

It is useful in turning functions on and off. For example, given a function defined for all x, say $g(x)$, we can select a portion of that function, say the interval $(x_0, x_0 + X)$, by writing

$$\tilde{g}(x) = g(x)u(x - x_0)\,u(x_0 + X - x) \tag{A-2}$$

The first step function "turns on" $g(x)$ at $x = x_0$ and the second step function "turns off" $g(x)$ at $x = x_0 + X$.

The next function to be considered is the *unit impulse function* or *Dirac delta function*, $\delta(x)$. It is *defined* in terms of the integral

$$\int_{-\infty}^{\infty} g(x)\delta(x)\,dx = g(0) \tag{A-3}$$

where $g(x)$ is any test function that is continuous at $x = 0$. A change of variables and redefinition of $g(x)$ results in the *sifting property,*

$$\int_{-\infty}^{\infty} g(x)\delta(x - x_0)\,dx = g(x_0) \tag{A-4}$$

where $g(x)$ is now required to be continuous at $x = x_0$. By considering the special cases $g(x) = 1, x_1 < x < x_2$ and 0 otherwise, the two properties

$$\int_{x_1}^{x_2} \delta(x - x_0)\,dx = 1, \qquad x_1 < x < x_2 \tag{A-5a}$$

and

$$\delta(x - x_0) = 0, \qquad x \neq x_0 \tag{A-5b}$$

are obtained, which provides an alternative definition of the delta function. Based on these two equations, it is helpful sometimes to visualize the impulse function as a family of very narrow, tall functions each with area 1. As some parameter goes to zero (or infinity), the members of this family of functions become narrower and narrower and taller and taller until, in the limit, we have an infinitely narrow, infinitely tall function with area 1.

Other important properties of the delta function that can be proved from (A-3) and properties of the Rieman integral are:

1. $\delta(a\,x) = \dfrac{1}{|a|}\delta(x)$

2. $\delta(-x) = \delta(x)$

3. $g(x)\delta(x - x_0) = g(x_0)\delta(x - x_0)$

4. $\displaystyle\int_{x_1}^{x_2} g(x)\delta^{(n)}(x - x_0)\,dx = (-1)^n g^{(n)}(x_0), \qquad x_1 < x_0 < x_2$

where the superscript (n) denotes the nth derivative, and $g(x)$ and its first n derivatives are assumed continuous at $x = x_0$.

5. $\delta(x) = \dfrac{du(x)}{dx} \qquad$ or $\qquad u(x) = \displaystyle\int_{-\infty}^{x} \delta(\lambda)\,d\lambda$

A–2 CONVOLUTION

Convolution was used in connection with finding the pdf of the sum of two independent random variables. The pdf of the sum random variable was the convolution of the pdf's of the random variables in the sum. It is defined as the operation

$$s(x) = \int_{-\infty}^{\infty} g(\lambda)h(x - \lambda)\,d\lambda = \int_{-\infty}^{\infty} g(x - \lambda)h(\lambda)\,d\lambda \tag{A-6}$$

The second integral can be obtained from the first by a straightforward change of variables. The operation of convolution is seen to involve several operations. These are:

1. Reversal or reflection of one of the functions about $\lambda = 0$.
2. Shifting the reversed function. Taking the first integral above, it is sometimes helpful to write $h(x - \lambda) = h[-(\lambda - x)]$, which is more indicative of the shifting operation after reversal (replacing the argument by its negative).
3. Multiplication of the reversed, shifted function by the other function.
4. Integration of the product for all possible shifts of the reversed function to get the convolution.

Perhaps the most challenging part of carrying out a convolution is that of determining limits on the integration for functions that are zero over certain ranges of their independent variables.

Example A-1

Consider the convolution of a square pulse with a decaying exponential starting at $x = 0$. We define the functions to be

$$g(x) = \begin{cases} 2e^{-3x}, & x \geq 0 \\ 0, & x < 0 \end{cases} \tag{A-7}$$

and

$$h(x) = \begin{cases} 1, & 0 \leq x \leq 2 \\ 0, & x < 0, x > 2 \end{cases} \tag{A-8}$$

We first rewrite them as functions of λ, reverse one of them, say $g(x)$, multiply them, and integrate the product. We must consider three different cases for the overlap. These are illustrated in Figure A-1.

The first case is that of no overlap (Figure A-1a). We assume that $h(x)$ is the function that is reversed and shifted. Replacing x by $\lambda - x$ in (A-8), we obtain

$$h(x - \lambda) = \begin{cases} 1, & x - 2 \leq \lambda \leq x \\ 0, & \lambda < x - 2, \ \lambda > x \end{cases} \tag{A-9}$$

If $x < 0$, the two factors in the integrand of the convolution integral do not overlap and the integration gives 0 since the product of the separate factors is zero.

The second case is that of partial overlap, as shown in Figure A-1b. The limits of integration are now from $\lambda = 0$ (the beginning of the decaying exponential) to $\lambda = x$ (the end of the rectangular pulse). The convolution integral is

$$s(x) = \int_0^x 2e^{-3\lambda} d\lambda = -\tfrac{2}{3}e^{-3\lambda}\Big|_0^x = \tfrac{2}{3}(1 - e^{-3x}), \quad 0 \leq x \leq 2 \tag{A-10}$$

The final case is that of full overlap of the rectangular pulse with the decaying exponential. The factors in the integrand are illustrated in Figure A-1c. From this figure it is apparent that the convolution is

$$s(x) = \int_{x-2}^x 2e^{-3\lambda} d\lambda = -\tfrac{2}{3}e^{-3\lambda}\Big|_{x-2}^x = \tfrac{2}{3}(e^{-3(x-2)} - e^{-3x}), \quad x > 2 \tag{A-11}$$

A plot of the total result for $s(x)$ is shown in Figure A-2.

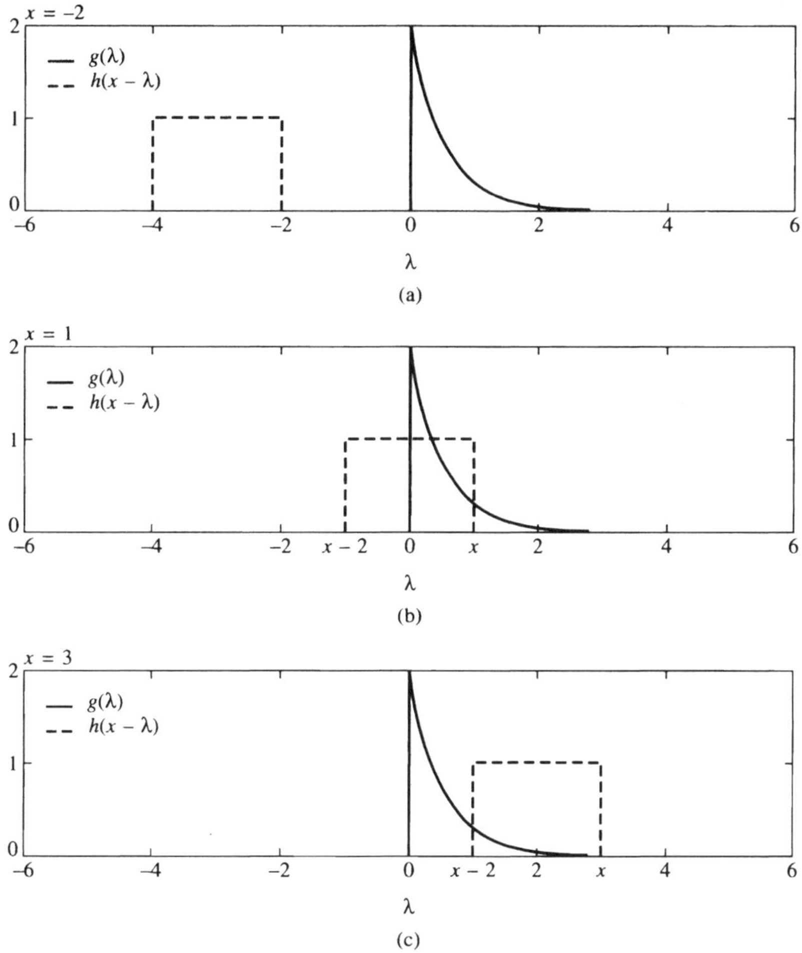

Figure A-1 Factors in the integrand of a convolution of a rectangular pulse with a decaying exponential.

A–3 FOURIER TRANSFORM

The Fourier transform of a function $g(x)$ is provided by the integral

$$\tilde{G}(j\omega) = G(f) = \int_{-\infty}^{\infty} g(x)e^{-j\omega x}dx = \int_{-\infty}^{\infty} g(x)e^{-j2\pi fx}dx \tag{A-12}$$

The inverse Fourier transform gives $g(x)$ from the Fourier transform according to the integral

$$g(x) = \frac{1}{2\pi}\int_{-\infty}^{\infty} \tilde{G}(j\omega)e^{j\omega x}d\omega = \int_{-\infty}^{\infty} G(f)e^{j2\pi fx}df \tag{A-13}$$

It is beyond the scope of this appendix to show that (A-12) and (A-13) are inverses of each other. Note the similarity of the Fourier transform and its inverse when written in

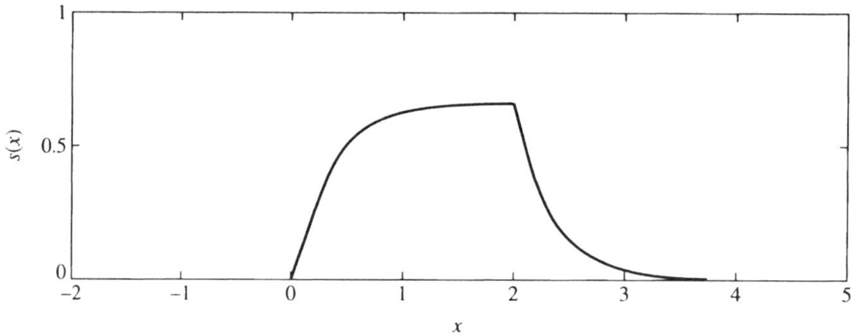

Figure A-2 Plot of the result of convolving a rectangular pulse with a decaying exponential.

terms of $f = \omega/2\pi$; the integrals are identical in form except for the minus sign in the exponent of (A-12).

The Fourier transform is useful in spectral analysis of signals and in linear system analysis. The former application makes use of the fact that the Fourier transform provides a resolution of a signal into its frequency components. That is, a signal can be viewed as being resolved into components at a particular frequency f_0 given by $G(f_0)$ and $G(-f_0)$ as the elemental component

$$\begin{aligned}\Delta g(x) &= G(-f_0)e^{-j2\pi f_0 x}\,\Delta f + G(f_0)e^{2\pi f_0 x}\Delta f \\ &= 2|G(f_0)|\cos\left[2\pi f_0 x + \underline{/G(f_0)}\right]\Delta f\end{aligned} \tag{A-14}$$

where the last equation follows from assuming that $g(x)$ is real [in this case $G(f) = G^*(f)$]. Equation (A-14) shows that the function $g(x)$ can be viewed as built up of a superposition of infinitesimal frequency components via the incremental component (A-14) and the integral in (A-13).

Several elementary Fourier transform pairs and theorems can be proved. It is through these elementary pairs and theorems that the Fourier transforms of more complex functions

Table A-1 SHORT TABLE OF FOURIER TRANSFORM PAIRS (α, τ, AND f_0 ARE CONSTANTS)

Function	Transform
1. $\Pi(x)$	$\mathrm{sinc}(x)$
2. $\Lambda(x)$	$\mathrm{sinc}^2(x)$
3. $\exp(-\alpha x)\ u(x),\ \alpha > 0$	$1/(\alpha + j2\pi f)$
4. $\exp(-\alpha\lvert x\rvert),\ \alpha > 0$	$2\alpha/[\alpha^2 + (2\pi f)^2]$
5. $\exp[-\pi(x/\tau)^2]$	$\tau\exp[-\pi(f\tau)^2]$
6. $\exp(j2\pi f_0 x)$	$\delta(f-f_0)$
7. $\delta(x-x_0)$	$\exp(-j2\pi f x_0)$
8. $\cos(2\pi f_0 x)$	$0.5\delta(f-f_0) + 0.5\delta(f + f_0)$
9. $\sin(2\ \pi f_0 x)$	$-\ 0.5j\delta(f-f_0) + 0.5j\delta(f + f_0)$

Table A-2 FOURIER TRANSFORM THEOREMS (a, a_1, a_2, x_0, AND f_0 ARE CONSTANTS; $g(x)$ is real)

Function	Fourier transform
1. $a_1 g_1(x) + a_2 g_2(x)$	$a_1 G_1(f) + a_2 G_2(f)$
2. $g(x - x_0)$	$G(f) \exp(-j2\pi f x_0)$
3. $g(ax)$	$\lvert a \rvert^{-1} G(f/a)$
4. $G(x)$	$g(-f)$
5. $g(x)\exp(j2\pi f_0 x)$	$G(f - f_0)$
6. $g(x)\cos(2\pi f_0 x)$	$0.5G(f - f_0) + 0.5G(f + f_0)$
7. $d^n g(x)/dx^n$	$(j2\pi f)^n G(f)$ if no dc component
8. $\int_{-\infty}^{\infty} g(\lambda) h(x - \lambda) d\lambda$	$G(f)H(f)$
9. $g(x)h(x)$	$\int_{-\infty}^{\infty} G(\lambda) H(x - \lambda) d\lambda$

can be found. An abbreviated table of Fourier transform pairs is given in Table A-1, and several Fourier transform theorems are summarized in Table A-2. Some special notations for functions are used in Table A-1. These are:

1. The unit width pulse:

$$\Pi(x) = \begin{cases} 1, & \lvert x \rvert \le \frac{1}{2} \\ 0, & \text{otherwise} \end{cases} \tag{A-15}$$

2. The unit triangle:

$$\Lambda(x) = \begin{cases} 1 - \lvert x \rvert, & \lvert x \rvert \le 1 \\ 0, & \text{otherwise} \end{cases} \tag{A-16}$$

3. The sinc function:

$$\text{sinc}(x) = \frac{\sin(\pi x)}{\pi x} \tag{A-17}$$

Example A-2

As an example, suppose that we want the Fourier transform of the signal shown in Figure A-3. It can be written in terms of the triangle and unit width pulse functions defined above as

$$s(x) = \Lambda\left(\frac{x}{2}\right) + \Pi(x) \tag{A-18}$$

Applying superposition, the Fourier transform of $s(x)$ is the sum of the Fourier transforms of each separate function in (A-18). Using the table of Fourier transforms given and the scale-change theorem, we find that

$$S(f) = 2\text{sinc}^2(2f) + \text{sinc}(f) \tag{A-19}$$

A sketch of this Fourier transform is provided in Figure A-4.

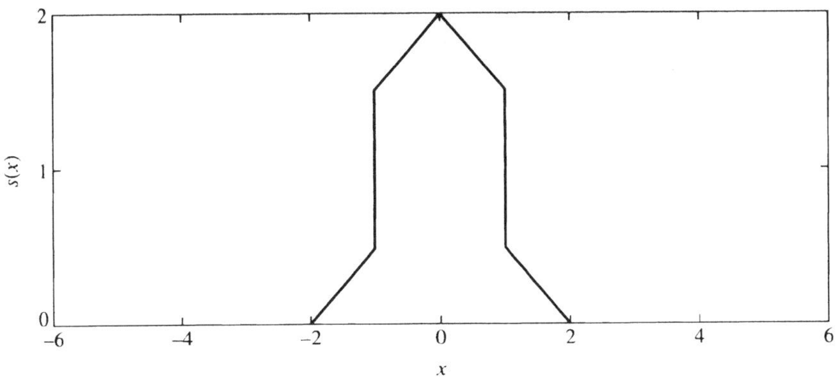

Figure A-3 Function whose Fourier transform is to be found.

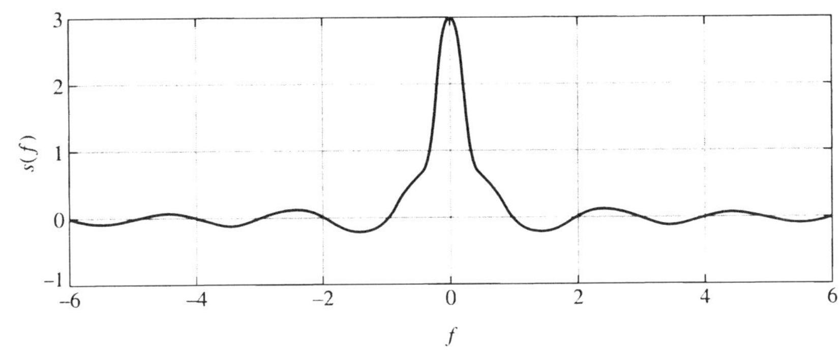

Figure A-4 Graph of the Fourier transform (A-19) of the function given by (A-18).

A–4 LINEAR SYSTEM THEORY[1]

Systems can be characterized in various ways: discrete time versus continous time, linear versus nonlinear, fixed versus time varying, causal versus noncausal, and stable versus unstable, among other classifications that are convenient. We denote the input–output relationship for a system by the symbolic notation

$$y(t) = \mathcal{H}[x(t)] \tag{A-20}$$

where $x(t)$ is the input, $y(t)$ is the output, and $\mathcal{H}[\cdot]$ denotes the operation that produces the output from the input.

[1]The subject matter of this section pertains mainly to Chapter 10. Consequently, the independent variable used will now be t to denote time.

A system is *linear* if superpostion holds. That is, let $x_1(t)$ and $x_2(t)$ be two inputs that produce the respective outputs $y_1(t)$ and $y_2(t)$. Then, for a linear system, the linear combination of inputs $\alpha_1 x_1(t) + \alpha_2 x_2(t)$ produces the output

$$\begin{aligned} y(t) &= \mathcal{H}[\alpha_1 x_1(t) + \alpha_2 x_2(t)] \\ &= \alpha_1 \mathcal{H}[x_1(t)] + \alpha_2 \mathcal{H}_2[x_2(t)] \\ &= \alpha_1 y_1(t) + \alpha_2 y_2(t) \end{aligned} \tag{A-21}$$

where α_1 and α_2 are two arbitrary constants.

A system is *fixed* if a constant time shift of the input produces the same constant time shift on the output. That is, for a fixed system, if $x(t)$ produces the output $y(t)$, a delayed input produces the output delayed by the same amount:

$$y_1(t) = \mathcal{H}[x(t - \tau)] = y(t - \tau) \tag{A-22}$$

A fixed linear system can be characterized in terms of its *impulse response, $h(t)$*, which is defined as the response of the system to a unit impulse funciton, $\delta(t)$, applied at time zero. That is,

$$h(t) = \mathcal{H}[\delta(t)] \tag{A-23}$$

A *causal system* is one that does not anticipate its input. Consequently, for a fixed linear system which is also causal, it follows that $h(t) = 0$ for $t < 0$.

Finally, a system is *stable* if every bounded input produces a bounded output. It can be shown that a fixed linear system is stable if and only if

$$\int_{-\infty}^{\infty} |h(t)|\, dt < \infty$$

For an arbitrary input, $x(t)$, to a fixed linear system the output can be obtained by means of the *superposition integral,* which is

$$y(t) = \int_{-\infty}^{\infty} h(\tau)x(t - \tau)d\tau = \int_{-\infty}^{\infty} h(t - \lambda)x(\lambda)d\lambda \tag{A-25}$$

where the first integral is obtained through application of the linearity property and definition of the unit impulse response, and the second integral follows from the first by the substitution $\tau = t - \lambda$.

The *transfer function* or *frequency response function* of a fixed linear system is defined as the Fourier transform of the impulse response. It can also be shown to be the ratio of the response of the system to a unit-amplitude complex exponential input, $e^{j\omega t}$, to that input by direct substitution into (A-25):

$$y(t)\big|_{x(t)=e^{j\omega t}} = \int_{-\infty}^{\infty} h(\tau)e^{j\omega(t-\tau)}d\tau = e^{j\omega t}\int_{-\infty}^{\infty} h(\tau)e^{j\omega(-\tau)}d\tau \tag{A-26a}$$

or

$$\frac{y(t)|_{x(t)=e^{j\omega t}}}{e^{j\omega t}} = \int_{-\infty}^{\infty} h(\tau)e^{j\omega(-\tau)}d\tau = \mathcal{F}\left[h(\tau)\right] = H(j\omega) = \tilde{H}(j2\pi f) \tag{A-26b}$$

where $\mathcal{F}[\cdot]$ denotes the Fourier transform integral. In (A-26b), two independent variables have been indicated: ω, or frequency in radians per second, and $f = \omega/2\pi$, or frequency in hertz. Whichever is most convenient for the situation at hand will be the one employed.

The discussion so far has assumed continuous-time systems. A discrete-time system is one for which the input and output signals are discrete time (i.e., signals that take on values only at discrete instants of time which, without loss of generality, can be taken to be integer values). For a discrete-time system, we can define the properties of linearity and time invariance similarly to a continuous-time system. The idea of an impulse response translates to that of a unit pulse response, which is the response of the system to the input $\delta(n) = 1, n = 0$, and 0 otherwise. That is, for a linear fixed discrete-time system, the unit pulse response is defined as

$$h(n) = \mathcal{H}[\delta(n)], \quad n = 0, \pm 1, \pm 2, \ldots \tag{A-27}$$

The properties of time invariance, causality, and stability are defined similarly to those for a continuous-time system.

For a linear fixed discrete-time system, the output can be obtained from the input by the *superposition sum,* which is

$$y(n) = \sum_{k=-\infty}^{\infty} h(k)x(n-k) = \sum_{k=-\infty}^{\infty} x(k)h(n-k) \tag{A-28}$$

where the second sum follows from the first by a change of the sum index.

Example A-3

Consider a linear fixed continuous-time system with impulse response $h(t) = u(t) - u(t - 10)$, where $u(t) = 1$ for $t \geq 0$, and 0 for $t < 0$ is the unit step. Let the input be $x(t) = e^{-t}u(t)$. Is this system causal? Is it stable? Find the output.

Solution

The system is causal because the impulse response is 0 for $t < 0$. To test for stability, we compute

$$\int_{-\infty}^{\infty} |h(t)|\,dt = \int_{0}^{10} dt = 10 < \infty$$

Therefore, this system is stable. To find the output, we use (A-25) to obtain

$$y(t) = \begin{cases} 0, & t < 0 \\ \displaystyle\int_0^t e^{-(t-\tau)}d\tau, & 0 \leq t \leq 10 \\ \displaystyle\int_0^{10} e^{-(t-\tau)}d\tau, & t > 10 \end{cases}$$

Table A-3 COMPUTATION OF THE SUPERPOSITION SUM FOR THE OUTPUT OF A FIXED LINEAR DISCRETE-TIME SYSTEM

n	$x(k) \rightarrow$ / $h(n-k)\downarrow$	$k=0$ 3	$k=1$ 2	$k=2$ 1	$k=3$ 0	$y(n)$
−1	0 1 1 1	0	0	0	0	0
0	0 0 1 1	1	0	0	0	3
1	0 0 0 1	1	1	0	0	5
2	0 0 0 0	1	1	1	0	6
3	0 0 0 0	0	1	1	1	3
4	0 0 0 0	0	0	1	1	1
5	0 0 0 0	0	0	0	1	0

$$= \begin{cases} 0, & t < 0 \\ 1 - e^{-t}, & 0 \le t \le 10 \\ (e^{10} - 1)e^{-t}, & t > 10 \end{cases}$$

where the first form of (A-25) has been used

Example A-4

Consider a linear fixed discrete-time system with $h(n) = 1, n = 0, 1, 2$, and 0 otherwise. Also, let $x(n) = 3, n = 0, 2, n = 1, 1, n = 2$, and 0 otherwise. The output can be found by using a tabular listing as shown in Table A-3, where the layout is suggestive of the convolution sum of the form

$$y(n) = \sum_{k=-\infty}^{\infty} x(k)h(n-k) \tag{A-30}$$

Example A-5

Find the output of a nonlinear system with input-output relationship, or transfer characteristic, given by

$$y(t) = \alpha_1 x(t) + \alpha_2 x^2(t) + \alpha_3 x^3(\mathrm{t}) \tag{A-31}$$

if the input is $x(t) = A\cos(\omega_0 t + \theta)$ where α_1, α_2, α_3, A, ω_0, and θ are constants. Such a system is called *zero memory* because its output depends on the input at the present time only and not on past or future values of the input.

Solution

We make use of the trigonometric identities

$$\cos^2\phi = \tfrac{1}{2} + \tfrac{1}{2}\cos 2\phi$$

$$\cos^3\phi = \tfrac{3}{4}\cos\phi + \tfrac{1}{4}\cos 3\phi$$

to obtain

$$y(t) = \alpha_1 A\cos(\omega_0 t + \theta) + \alpha_2 A^2\cos^2(\omega_0 t + \theta) + \alpha_3 A^3\cos^3(\omega_0 t + \theta) \tag{A-32}$$

$$= \frac{\alpha_2}{2}A_2 + (\alpha_1 A + \tfrac{3}{4}\alpha_3 A^3)\cos(\omega_0 t + \theta)$$
$$+ \frac{\alpha_2}{2}A^2\cos[2(\omega_0 t + \theta)] + \frac{\alpha_3}{4}A^3\cos[3(\omega_0 t + \theta)]$$

The nonlinear operation has generated terms at harmonics of the frequency of the input signal.

A–5 FURTHER READING

A good source for studying linear system theory is Ziemer et al. (1993).

B

MATLAB Hints and Overview

B–1 INTRODUCTION

MATLAB is a matrix or array-based language. We will give some examples of programs that make use of this feature in the following sections. In this section, several hints are given that should make running MATLAB programs handier.

Directory Management and Help

It is suggested that you create a special directory on your hard disk from which to run your MATLAB programs. This will give the fastest response time. If you are using a computer where you are not able to do this (e.g., in a open computer lab where this is prohibited), you can of course run the programs from a floppy disk, but it will make for slower saving and retrieving of programs from the floppy.

In either case you will want to change to the directory that your programs are in immediately when you activate MATLAB by double clicking on the MATLAB icon. This is easily done by typing "cd a:" (type only what is within the quotes, not the quotes themselves) to change to the a: floppy drive, or "cd c:\mystuff" at the » prompt in the MATLAB command window to change to the directory "mystuff" on the c: drive. You will be in this directory until you do this again when you change to another directory. If you want to see what programs are in that particular directory, you can type "ls" at the » prompt.

You might have three or more windows that you are working with when running MATLAB programs: the Command Window, the Notepad (perhaps more than one), and possibly a Plot Window. Each time you run a program from the Command Window, you can make the other windows into icons by clicking the left mouse button on the ∨ in the upper right-hand corner of that particular window. However, this soon becomes time consuming. It is suggested that you resize at least the Command Window and the Notepad(s) so that you can see all of them cascaded on the screen. You can then change to another window quickly by clicking the left mouse button on it.

MATLAB has extensive help capability. You can access it in two ways. The most obvious is by clicking on the "Help" button at the top of the Command Window. You can access all MATLAB functions through the "Index" on the drop-down menu under the "Help" button. An alternative way to access help on any function whose name you know is to type "help *function*" after the » prompt. You can access help on any operation by typing "help + " at the » prompt. This produces the following list (in fact, you can type "help" followed by any of the characters on this list):

```
» help +
Operators and special characters
Char Name                              HELP topic
+       Plus                           arith
-       Minus                          arith
*       Matrix multiplication          arith
.*      Array multiplication           arith
^       Matrix power                   arith
.^      Array power                    arith
\       Backslash or left division     slash
/       Slash or right division        slash
./      Array division                 slash
kron    Kronecker tensor product       kron
:       Colon                          colon
()      Parentheses                    paren
[]      Brackets                       paren
.       Decimal point                  punct
..      Parent directory               punct
...     Continuation                   punct
,       Comma                          punct
;       Semicolon                      punct
%       Comment                        punct
!       Exclamation point              punct
'       Transpose and quote            punct
+       Assignment                     punct
==      Equality                       relop
<>      Relational operators           relop
&       Logical AND                    relop
|       Logical OR                     relop
~       Logical NOT                    relop
xor     Logical EXCLUSIVE OR           xor
Logical characteristics
  exist   -Check if variables or functions are defined.
  any     -True if any element of vector is true.
```

```
all       -True if all elements of vector are true.
find      -Find indices of nonzero elements.
isnan     -True for Not-A-Number.
isinf     -True for infinite elements.
finte     -True for finite elements.
isempty   -True for empty matrix
isreal    -True for real matrix.
issparse  -True for sparse matrix.
isstr     -True for text string.
isglobal  -True for global variables.
```

Commenting, Saving, and Running Programs

It is advised that you save often when writing programs. In the Editor, you save by left-button clicking on the "File" button, clicking on "Save As" and typing the name of the program followed by an .m (note that MATLAB will not automatically attach the .m, but it is absolutely necessary). If you have not done a "cd c:\mystuff" in the Command Window, you will first have to identify the directory by clicking on "Drives" in the lower right center of the "Save As" window that comes up when you click on "Save As" under "Files."

If you are about to make major modifications in a program that already runs, save your modification to a new file, or save your old one to a backup file before modifying. This cannot be emphasized too strongly. You will probably learn the hard way, like most of us have, the first time you make major modifications to a perfectly good program that runs, and then wish you had the working version when your modified version gives you problems initially.

Another piece of advice: always begin your programs with a line or two of comments (a comment line is preceded by a % character) that tells what the program is for, what its input variables are, and what its outputs are before typing the program. You may think you will do this later, but you will probably forget until you bring up the program sometime later and have to spend time figuring out what it is for. You can access your comment lines at the beginning of a program from the Command Window by typing "help *program_name*" at the » prompt.

Note that programs are run from the Command Window by typing the program name *without the .m* at the » prompt. If you have not made a change of directory to where your programs are saved previously, MATLAB will not know the path until you do so. An alternative is to click on "File," then on "Run m-File," then on "Browse," then clicking the down arrow to display the drives, and finally, clicking on the program in the list displayed form the drive. Obviously, the first method takes the least time and mouse clicking.

B–2 SOME PROGRAM EXAMPLES

Introductory Examples

MATLAB is a vector- or array-based program. For example, if one wishes to generate and plot a sinusoid by means of MATLAB, the statements involved would be

```
t=0:.01:10;
x=sin(2*pi*t);
plot(t,x,'-w'),xlabel('t'),ylabel('x(t)'),grid
```

This can be stored in any conveniently named file, say *sinusoid* (note that the name *sin* would not be permissible because it is the name of a MATLAB function already defined). The first line generates a vector of values for the independent variable starting at 0, ending at 10, and spaced by 0.01; the second statement generates a vector of values for the dependent variable $x = \sin(2\pi t)$; and the third statement plots the vector x versus the vector t. The resulting plot is shown in Figure B-1. Note that the '-w' in the plot function is not necessary—it was included there to make a white line on the black plot window background, which made for better contrast in plotting Figure B-1.

In MATLAB, one has the options of running a program stored in an m-file, invoking the statements from the Command Window, or writing a function to perform the steps in producing a sine wave or other operation. For example, the Command Window option would be invoked as follows:

```
»t = 0:.-1:10;
»x = sin(2*pi*t);
» plot(t,x,'-w'),xlabel('t'),ylabel('x(t)'),grid
```

The Command Window prompt is » and each line is executed as it is typed and entered.

An example of a function implementation is provided by the generation of a unit step:

```
%     Function for generation of a unit step
function u = stepfn(t)
L = length(t);
u = zeros(size(t));
for i = 1:L
     if t(i) >= 0
          u(i) = 1;
     end
end
```

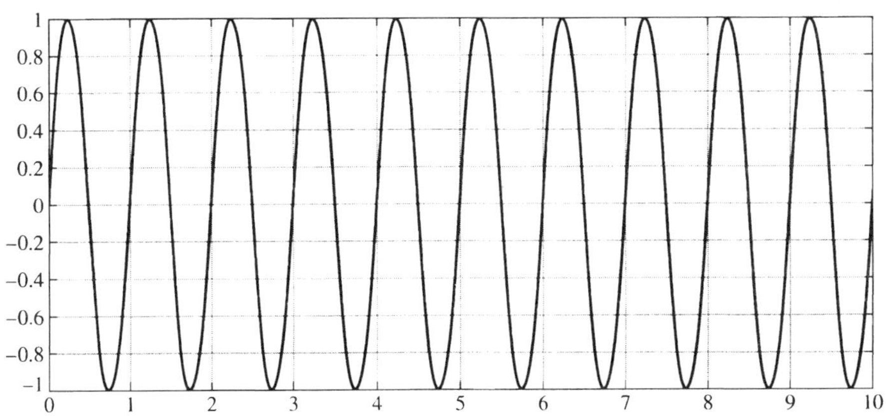

Figure B-1 Plot of sinusoid generated by the program listing above.

The Command Window statements for generation of a unit step starting at $t = 2$ are given below and a plot is provided in Figure B-2.

```
» t = -10:0.01:10;
» u = stepfn(t-2);
» plot(t,u,'-w'), xlabel('t'),ylabel('u(t)'),grid, title('unit step'),
axis([-10 10 -0.5 1.5])
```

Curve Fitting

MATLAB has several functions for fitting polynomials to data and a function for evaluation of a polynomial fit to the data. These functions include *table1, table2, spline,* and *polyfit.* The first one makes a linear fit to a set of data pairs, the second does a planar fit to data triples, and the third does a cubic fit to data pairs. The *polyfit* function does a least-mean-squared-error fit to data pairs. The use of these is illustrated by the program that follows.

```
%      Example of polynomial fitting to data
%      and plotting the resulting fit.
%
x = [0 1 2 3 4 5 6 7 8 9];
y = [0 20 60 68 77 110 113 120 140 135];
newx = 0:0.1:9;
newy = spline(x, y, newx);
for n = 1:5
     X = polyfit(x,y,n);
     f(:,n) = polyval(X,newx)';
     subplot(3,2,n+1),plot(newx,f(:,n),'w',x,y,'ow'),axis([0 10 0 150]),grid
end
subplot(3,2,1),plot(x,y,'w',newx,newy,'w',x,y,'ow'),axis([0 10 0 150]),grid
```

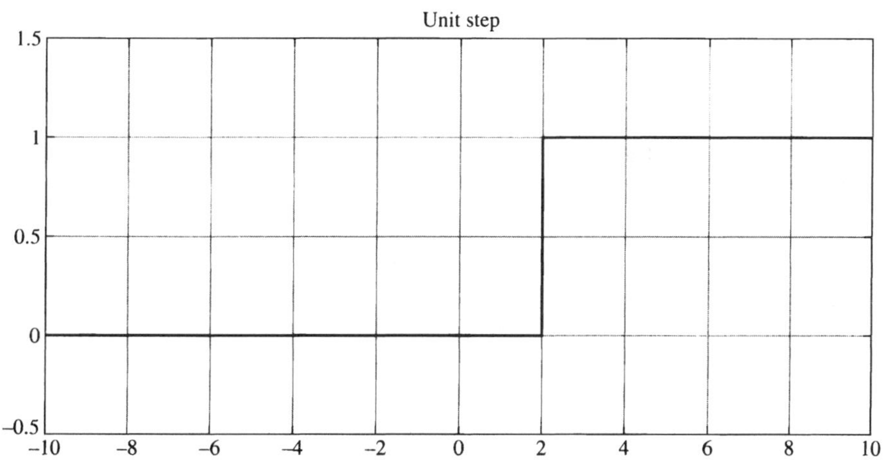

Figure B-2 Unit step starting at $t = 2$ generated by the step generation function given above.

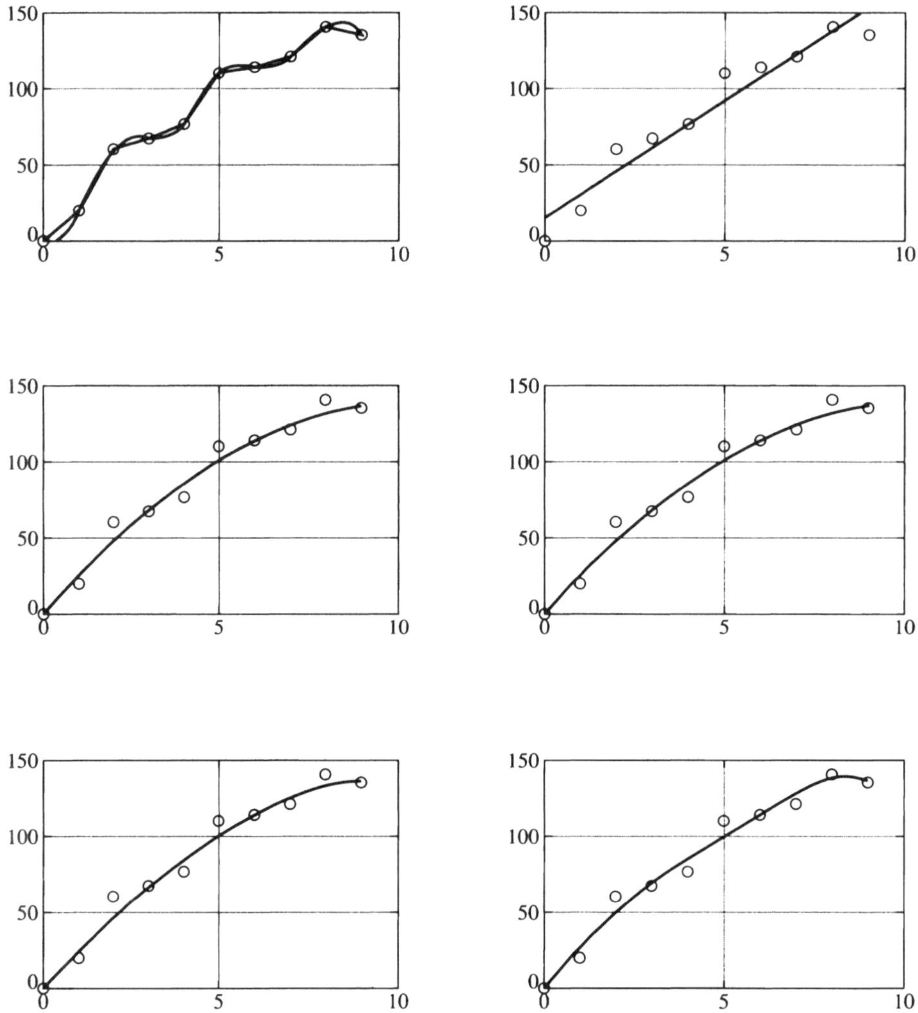

Figure B-3 Various fits to the data pairs shown by the circles. From left to right and top to bottom: linear and spline fits; linear least-squares fit; quadratic least-squares fit; cubic least-squares fit; quartic least-squares fit; fifth-order least-squares fit.

Plots of the various fits are shown in Figure B-3. In the program above, the linear interpolation is provided by the plotting routine itself, although the numerical value for the linear interpolation of a data point is provided by the statement *table1(x, y, x_0)*, where x_0 is the x-value for which an interpolated y-value is desired. The *polyfit* statement returns the coefficients of the least-squares-fit polynomial of specified degree n to the data pairs. For example, the coefficients of the fifth-order polynomial returned by *polyfit* are −0.0150, 0.3024,

−1.9988, 3.3400, 24.0124, and −1.1105 from highest to lowest degree. The *polyval* statement provides an evaluation of the polynomial at the element values of the vector *newx*.

Statistical Data Analysis

MATLAB has several *statistical data analysis functions*. Among these are random number generation, sample mean and standard deviation computation, histogram plotting, and correlation coefficient computation for pairs of random data. The following program illustrates several of these functions.

```
X = rand(1,5000);
Y = randn(1,5000);
mean_X = mean(X)
std_dev_X = std(X)
mean_Y = mean(Y)
std_dev_Y = std(Y)
rho = corrcoef(X,Y)
subplot(211), hist(X,[0 .1 .2 .3 .4 .5 .6 .7 .8 .9 1]),grid
subplot(212),hist(Y,15),grid
```

The computed values returned by the program (note that the semicolons are left off) are mean_X = 0.4997; std_dev_X = 0.2866; mean_Y = –0.0144; std_dev_Y = 0.9943;

rho = 1.0000 0.0088

0.0088 1.0000

The theoretical values are 0.5, 0.2887, 0, and 1, respectively, for the first four. The correlation coefficient matrix should have 1's on the main diagonal and 0's off the main diagonal. Histograms for the two cases of uniform and Gaussian variates are shown in Figure B-4. In the first plot statement, a vector giving the centers of the desired histogram bins is given. The two end values at 0 and 1 will have, on average, half the values in the other bins since the random numbers generated are uniform in $[0,1]$ (this illustrates that one must be careful when specifying the locations). In the second histogram plot statement, the number of bins is specified at 15.

B–3 FURTHER READING

The MATLAB user may consult the *Student Edition of MATLAB* and the *MATLAB User's Guide* (see the Bibliography).

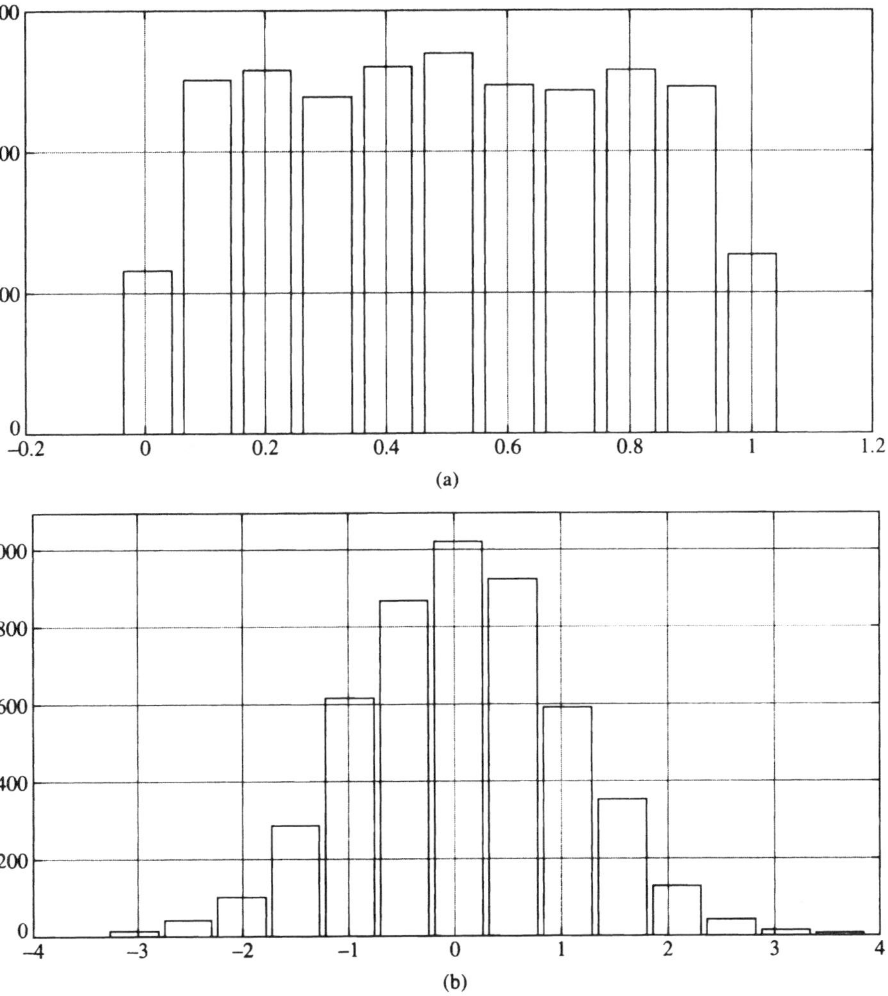

Figure B-4 Histograms for 5000 uniform pseudorandom numbers in [0,1] and 5000 Gaussian pseudorandom numbers with mean zero and variance 1. Note that the endpoints on the first histogram are centered at 0 and 1 for purposes of illustration and therefore give a false picture of the true histogram.

C

Tables and MATLAB Programs for Computation of Certain Probability Distributions

C–1 GAUSSIAN DISTRBUTIONS: THE Q-FUNCTION, GAUSSIAN CDF, AND ERROR FUNCTION

The Q-function was defined by (3-29) as

$$Q(x) = \frac{1}{\sqrt{2\pi}}\int_x^\infty e^{-u^2/2}du \tag{C-1}$$

The cdf of a zero-mean, unit-variance Gaussian random variable is $1 - Q(x)$. The error function was defined in (3-30) as

$$\mathrm{erf}(x) = \frac{2}{\sqrt{\pi}}\int_0^x e^{-u^2}du \tag{C-2}$$

The first is the area under one tail of a Gaussian pdf with zero mean and unit variance from x to ∞, and the latter is the area under the central hump of the same pdf from $-x$ to x. The complementary error function is

$$\mathrm{erfc}(x) = 1 - \mathrm{erf}(x) \tag{C-3}$$

By straightforward substitution, it is easily shown that

$$Q(x) = 0.5\ \mathrm{erfc}\left(\frac{x}{\sqrt{2}}\right) \tag{C-4}$$

A MATLAB function is available for computing the error function and complementary error function. Simply typing erf(*numerical value*) or erfc(*numerical value*) at the prompt in the

MATLAB command window will return the corresponding erf or erfc values. The argument can be a vector.

Values of the Q-function can be obtained by using a rational approximation as given in Abramowitz and Stegun (1972). A MATLAB m-file function is given in Table C-1 for computing the Q-function.

Table C-1 MATLAB M-FILE FUNCTION FOR COMPUTING THE Q-FUNCTION

```
%       This function computes the Gaussian Q-function using the rational
%       approximation 26.2.17 of Abramowitz and Stegun. For sufficiently large
%       arguments, the symptotoic expansion 26.2.12 is used. The rational
%       approximation is accurate to within 7.5e-8. Q is the value of the
%       Q-function (integral of Gaussian pdf of zero mean and unit variance from the
%       input quantity to infinity) and P = 1 − Q is the cumulative distribution
%       function. Vector arguments are allowed.
%
%       R.E. Ziemer
%       4/17/94; revised 12/27/95
%
Function [Q,P]=qfn(x)
format short e
N = length(x);
b(1)=0.31938153;
b(2)=-0.356563782;
b(3)=1.781477937;
b(4)=-1.821255978;
b(5)=1.330274429;
p=0.2316419;
y=abs(x);
T=1./(1+p*y);
TT=zeros(5,N);
for r=1:5
        TT(r,:)=T.^r;
end
Z = exp(-y.^2/2)/sqrt(2*pi);
C = b*TT;
for k = 1:N
        if y(k)<4
                Q1(k) = Z(k)*C(k);
        else
                Q1(k) = (Z(k)/y(k))*(1-1/y(k)^2+3/y(k)^4);
        end
end
for k = 1:N
```

```
        if x(k) > 0
                Q(k)=Q1(k);
        else
                Q(k) = 1-Q1(k);
        end
end
P=1-Q;
```

Values for the Q-function are given in Table C-2.

For large values of the argument, the asymptotic approximation

$$Q(x) \approx \frac{e^{-x^2\,2}}{x\sqrt{2\pi}}, \qquad x \gg 0$$

is useful. The error is about 10% for $x = 4$ and gets smaller rapidly as x gets larger.

We sometimes want the argument of the Q-function for a given value of the function. This can be obtained using the MATLAB m-file function given in Table C-3.

For example, entering the argument

```
P = [0.0500 0.1000 0.1500 0.2000 0.2500 0.3000 0.3500 0.4000 0.45000 .5000]
```

gives the result

```
>> X = invqfn(P)
x = [1.6452 1.2817 1.0364 0.8415 0.6742 0.5240 0.3849 0.2529 0.1254 0.0000]
```

Table C-2 Q-FUNCTION VALUES

x	Q(x)	x	Q(x)
0.0	0.5000	2.1	1.7864e-002
0.1	0.4602	2.2	1.3903e-002
0.2	0.4207	2.3	1.0724e-002
0.3	0.3821	2.4	8.1975e-003
0.4	0.3446	2.5	6.2097e-003
0.5	0.3085	2.6	4.6612e-003
0.6	0.2743	2.7	3.4670e-003
0.7	0.2420	2.8	2.5552e-003
0.8	0.2119	2.9	1.8659e-003
0.9	0.1841	3.0	1.3500e-003
1.0	0.1587	3.1	9.6767e-004
1.1	0.1357	3.2	6.8720e-004
1.2	0.1151	3.3	4.8348e-004
1.3	0.0968	3.4	3.3698e-004
1.4	0.0808	3.5	2.3267e-004
1.5	0.0668	3.6	1.5915e-004
1.6	0.0548	3.7	1.0783e-004
1.7	0.0446	3.8	7.2372e-005
1.8	0.0359	3.9	4.8116e-005
1.9	0.0287	4.0	3.1759e-005
2.0	0.0228		

Table C-3 MATLAB M-FILE FUNCTION FOR COMPUTING THE INVERSE Q-FUNCTION

```
%       Inverse Q-function: based on 26.2.23 of Abramowitz and Stegun.
%       Will take a vector as an argument; input arguments > 0 & < = 0.5.
%       The absolute value of the error is < 4.5e-4
%
%       R.E. Ziemer; 12/27/95
%
function xp = invqfn(p)
N = length(p);
t = sqrt(-2*log(p));
c(1) = 2.515517;
c(2) = 0.802853;
c(3) = 0.010328;
c(4) = 0;
d(1) = 0;
d(2) = 1.432788;
d(3) = 0.189269;
d(4) = 0.001308;
T = zeros (4,N);
for K = 1:4
        T(k,:) = t.^(k-1);
end
num = c*T;
den = ones(1,N) + d*T;
xp = t - num./den;
```

Note that the value of the Q-function input must be less than 0.5. For values larger than 0.5, the x value is negative. It can be shown that for $x < 0$, $Q(x) = 1 - Q(|x|)$.

C-2 CHI-SQUARE DISTRIBUTION

Finding confidence intervals for the variance or the mean with variance unknown required finding the value of χ^2 for a chi-square random variable V for a given probability [i.e., $P(V > \chi^2 = A)$, where A is specified]. The MATLAB m-file function in Table C-4 can be used to compute the cdf of a chi-square random variable of nu degrees of freedom, or the complementary cdf, or by using the MATLAB function fmins the abscissa value, χ^2, corresponding to the probability under the tail PP.

As an example, entering the following in the command window (abscissa value of 10 and degrees of freedom =5):

```
» [min, QX2,PX2]=chimin(10,5,0.95)
```

yields the returned values

```
min=
0.87476475385349
QX2 =
0.0752352414651
PX2 =
0.9247647385349
```

Table C-4 MATLAB M-FILE FUNCTION FOR COMPUTING THE CDF OF A CHI-SQUARE RANDOM VARIABLE

```
%       Function for use in finding abscissas for cum chi-square: PX2
%       is the chi-squre cdf, QX2 is the complementary chi-square cdf,
%       and pr_min is abs(QX2-PP) which is used to find the abscissa
%       corresponding to the tail probability PP using the MATLAB
%       function fmins. Based on 26.4.4 and 26.4.5 of Abramowitz and Stegun.
%
%       R.E. Ziemer
%       12/26/95
%
function [pr_min,QX2,PX2] = chimin(X2,nu,PP);
format long
dof=rem(nu,2);
if dof == 1
        NS=(nu-1)/2;
else
        NS=(nu-2)/2;
end
chi=sqrt(X2);
if nu < 30
Q=0.5*erfc(chi/sqrt(2));
Z=exp(-X2/2)/sqrt(2*pi);
        if NS > 0
                T=zeros(1,NS);
                if dof==1
                        T(1)=chi;
                        for r=1:NS - 1
                                T(r + 1)=X2*T(r)/(2*r + 1);
                        end
                        QX2=2*Q + 2*Z*sum(T);
                else
                        T(1)=chi^2/2;
                        for r=1:NS-1
                                T(r +1)=X2*T(r)/(2*(r + 1));
                        end
                        QX2=sqrt(2*pi)*Z*(1 + sum(T));
                end
        else
                if dof==1
                        QX2=2*Q;
                else
                        QX2=sqrt(2*pi)*Z;
                end
        end
        PX2=1-QX2;
```

```
else
        K1=(X2/nu)^(1/3;
        K2=1-2/(9*nu));
        K3=sqrt(2/9+nu));
        arg=(K1-K2)/K3;
        A=0.5*erfc(arg/sqrt(2));
        QX2=A;
        PX2=1-QX2;
end
if X2==0
        QX2=1;
        PX2=0;
end
pr_min =abs(QX2-PP);
```

To compute the abscissa values corresponding to specified tail probabilites, the MATLAB function given in Table C-5 can be used. Note that special cases are used for nu = 1 and nu = 2. In the first case,

$$P(V > v) = 2Q(\sqrt{v}) \tag{C-6}$$

which can be solved for $v = \chi^2$ using the inverse Q-function of the MATLAB function given in Table C-3. In the second case, for nu = 2,

$$P(V > v) = e^{-v/2} \tag{C-7}$$

If the program is run for a minimum nu = 1 and maximum nu = 15 with the tail probability vector equal to [0.9950 0.9900 0.9750 0.9500 0.7500 0.9000 0.2500 0.1000 0.0500 0.0250 0.0100 0.0050], we obtain the values listed in Table C-6.

C–3 STUDENT'S t-DISTRIBUTION

For confidence intervals for the variance with mean unknown, we need the Student's t-distribution. In particular, we need the abscissa values for the tail probability equal to a given value. These abscissa values can be computed using the rational approximation given by 26.7.5 of Abramowitz and Stegun (1972). Note that it makes use of the inverse Q-function, which was discussed earlier (see Table C-3 and 26.2.23 of Abramowitz and Stegun). A MATLAB program is given in Table C-7 for computing the abscissa values corresponding to a desired tail probability. Note that special cases are used for $n = 1$ (Cauchy) and $n = 2$. For these two cases, the pdf and cdf of the Student's t random variable are, respectively:

$$f_T(t) = \frac{1}{\pi(t^2 + 1)}$$

$$F_T(t) = 0.5 + \frac{\tan^{-1}(t)}{\pi}, \quad n = 1 \tag{C-8}$$

and

$$f_T(t) = \frac{1}{(t^2 + 2)^{3/2}}$$

$$F_T(t) = 0.5\left(1 + \frac{t}{\sqrt{t^2 + 2}}\right), \quad n = 2 \tag{C-9}$$

(Williams 1991, Appendix C)

Figure C-1 illustrates the Student's t pdf for $n = 10$ degrees of freedom. Also shown are two abscissa values, $A = 1.37$ and $B = -1.37$. Direct numerical integration shows that for $n = 10$,

$$P(T > A) = 0.1 \text{ and } P(T > B) = 1 - P(T > A) = 0.9 \tag{C-10}$$

Table C-5 MATLAB PROGRAM FOR FINDING THE ABSCISSA VALUE OF A CHI-SQUARE DISTRIBUTION CORRESPONDING TO A GIVEN TAIL PROBABILITY

```
%       Program to compute the abscissa for P = P(V > v)
%       for the chi-square random variable for a given P
%
nu_min=input('Enter starting value for degrees of freedom - integer ');
nu_max=input('Enter ending value for degrees of freedom ');
P = input('Input desired probability vector for P =P(V > v)');
Lp = length(P);
nu_values = zeros(1,nu_max-nu_min);
absc_values = zeros(nu-max-nu_min,Lp);
K = 1;
for nu = nu_min:nu_max
        for 1 = 1:Lp
        if nu ==1
                Y =(invqfn(P(1)/2))^2;
        elseif nu == 2
                Y = -2*log(P(1));
        else
                start = 0.7*(1-P(1))*nu;
                Y=fmins('chimin',start,[0,1.e-8,1.e-8],[],nu,P(1));
                format short
        end
                absc_values(k,1) = Y;
        end
        nu_values(k) =nu;
        k = k + 1;
end
disp('The values of P = P(V > v) are:')
disp(P)
A = [nu_values' absc_values];
disp('nu and the abscissa values, V, for chosen P vector are:')
disp(A)
```

Table C-6 ABSCISSA VALUES FOR VARIOUS TAIL PROBABILITIES AND NU = 1 to 15 FOR THE CHI-SQUARE DISTRIBUTION

	Tail Probability											
nu	0.9950	0.9900	0.9750	0.9500	0.9000	0.7500	0.2500	0.1000	0.0500	0.0250	0.0100	0.0050
1	0.0000	0.0002	0.0010	0.0039	0.0157	0.1013	1.3235	2.7067	3.8431	5.0259	6.6370	7.8815
2	0.0100	0.0201	0.0506	0.1026	0.2107	0.5754	2.7726	4.6052	5.9915	7.3778	9.2103	10.5966
3	0.0717	0.1148	0.2158	0.3518	0.5844	1.2125	4.1083	6.2514	7.8147	9.3484	11.3449	12.8382
4	0.2070	0.2971	0.4844	0.7107	1.0636	1.9226	5.3853	7.7794	9.4877	11.1433	13.2767	14.8603
5	0.4117	0.5543	0.8312	1.1455	1.6103	2.6746	6.6257	9.2364	11.0705	12.8325	15.0863	16.7496
6	0.6757	0.8721	1.2373	1.6354	2.2041	3.4546	7.8408	10.6446	12.5916	14.4494	16.8119	18.5476
7	0.9893	1.2390	1.6899	2.1673	2.8331	4.2549	9.0371	12.0170	14.0671	16.0128	18.4753	20.2777
8	1.3444	1.6465	2.1797	2.7326	3.4895	5.0706	10.2189	13.3616	15.5073	17.5345	20.0902	21.9550
9	1.7349	2.0879	2.7004	3.3251	4.1682	5.8988	11.3888	14.6837	16.9190	19.0228	21.6660	23.5894
10	2.1559	2.5582	3.2470	3.9403	4.8652	6.7372	12.5489	15.9872	18.3070	20.4832	23.2093	25.1882
11	2.6032	3.0535	3.8157	4.5748	5.5778	7.5841	13.7007	17.2750	19.6751	21.9200	24.7250	26.7568
12	3.0738	3.5706	4.4038	5.2260	6.3038	8.4384	14.8454	18.5493	21.0267	23.3367	26.2170	28.2995
13	3.5650	4.1069	5.0088	5.8919	7.0415	9.2991	15.9839	19.8119	22.3620	24.7356	27.6882	29.8195
14	4.0747	4.6604	5.6287	6.5706	7.7895	10.1653	17.1169	21.0641	23.6848	26.1189	29.1412	31.3193
15	4.6009	5.2293	6.2621	7.2609	8.5468	11.0365	18.2451	22.3071	24.9958	27.4884	30.5779	32.8013

Table C-7 MATLAB PROGRAM FOR COMPUTING A TABLE OF VALUES FOR THE ABSCISSA CORRESPONDING TO VARIOUS TAIL PROBABILITIES FOR THE STUDENT'S *T*-DISTRIBUTION

```
%       Table of abscissa values for Student's t-distribution
%
n=input('Enter minimum n value');
n_steps=input('Enter number of steps for n');
p=input ('Enter vector of probability values - tail area');
N=length(p);
n_values=zeros(1,n+n_steps-1);
absc_values=zeros(n+n_steps-1,N);
for K=1:n_steps
        n_values(k)=n+(k-1);
        absc_values(k,:)=stu_absc(p,n+k-1);
end
A=[n_values' absc_values];
disp(A)
%       This function computes the abscissa for the area under the tail
%       of the Student's t-distribution using the rational approximation 26.7.5
%       and 26.2.23 of Abramowitz and Stegun. Vector arguments for p are
%       allowed, but not for n.
%
%       R.E. Ziemer
%       12/26/95
%
function t=stu_absc(p,n)
format short
if n > 2
        xp =invqfn(p);
        g1 = (xp.^3+xp)/4;
        g2 = (5*xp.^5+16*xp.^3+3*xp)/96;
        g3 = (3*xp.^7+19*xp.^5+17*xp.^3-15*xp)/384;
        g4 = (79*xp.^9+776*xp.^7+1482*xp.^5-1920*xp.^3-945*xp)/92160;
        t = xp+g1/n+g2/n^2+g3/n^3+g4/n^4;
elseif n ==2
        K=1-2*p;
        t=sqrt(2)*K./sqrt(1-K.^2);
elseif n ==1
        t=tan(pi*(0.5-p));
end
```

Checking the abscissa values given in Table C-8 for $n = 10$ and $P(T > t)$ gives the same results. Values for the abscissas corresponding to various tail probabilities and values for the degrees of freedom (n) are given in Table C-8.

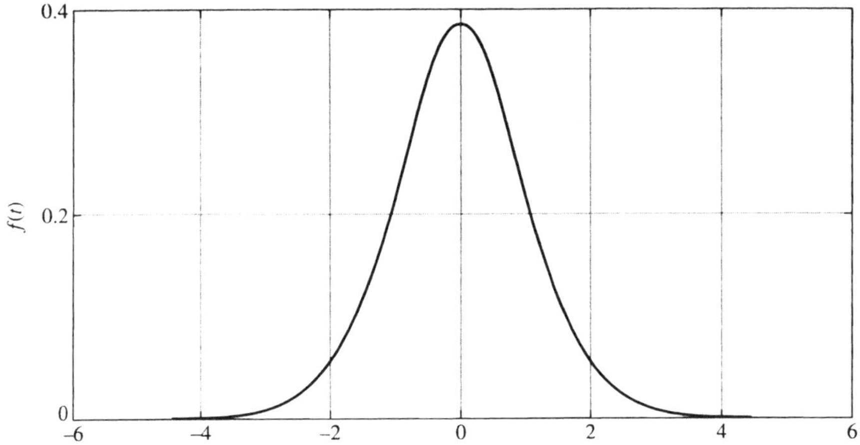

Figure C-1 Student's t pdf showing two abscissa values symmetrically spaced about $t = 0$ at $A = 1.37$ and $B = -1.37$. Due to symmetry, $P(T > A)$. Numerical integration and the algorithm given in Table C-7 give the same results.

C-4 FURTHER READING

Good resources for studying the computation of probability distributions are Abramowitz and Stegun (1972) and Williams (1991, Appendix C).

Table C-8 ABSCISSA VALUES FOR VARIOUS *N* VALUES (DEGREES OF FREEDOM) AND TAIL PROBABILITIES FOR STUDENT'S *t*-DISTRIBUTION

n	0.100	0.050	0.025	0.01	0.005
1	3.0777	6.3138	12.7062	31.8205	63.6567
2	1.8856	2.9200	4.3027	6.9646	9.9248
3	1.6377	2.3532	3.1799	4.5251	5.7973
4	1.5334	2.1323	2.7766	3.7444	4.5955
5	1.4761	2.0156	2.5711	3.3648	4.0302
6	1.4400	1.9437	2.4475	3.1431	3.7072
7	1.4151	1.8951	2.3653	2.9986	3.4998
8	1.3970	1.8600	2.3066	2.8971	3.3559
9	1.3832	1.8336	2.2628	2.8221	3.2504
10	1.3724	1.8129	2.2287	2.7644	3.1699
11	1.3636	1.7963	2.2016	2.7187	3.1064
12	1.3564	1.7827	2.1794	2.6816	3.0551
13	1.3504	1.7714	2.1609	2.6509	3.0129
14	1.3452	1.7617	2.1453	2.6251	2.9774
15	1.3408	1.7535	2.1320	2.6031	2.9473
16	1.3370	1.7463	2.1204	2.5841	2.9213
17	1.3336	1.7400	2.1103	2.5675	2.8988
18	1.3306	1.7345	2.1014	2.5529	2.8790
19	1.3279	1.7295	2.0935	2.5400	2.8615
20	1.3255	1.7251	2.0865	2.5285	2.8459
21	1.3234	1.7211	2.0801	2.5182	2.8319
22	1.3214	1.7175	2.0744	2.5089	2.8193
23	1.3196	1.7143	2.0692	2.5004	2.8078
24	1.3180	1.7113	2.0644	2.4927	2.7974
25	1.3165	1.7085	2.0600	2.4856	2.7879
26	1.3152	1.7060	2.0560	2.4791	2.7792
27	1.3139	1.7037	2.0523	2.4732	2.7712
28	1.3127	1.7015	2.0489	2.4677	2.7638
29	1.3116	1.6995	2.0457	2.4625	2.7569
30	1.3106	1.6976	2.0428	2.4578	2.7505

D

Mathematical Formulas

D–1 TRIGONOMETRIC IDENTITIES

Euler's theorem: $e^{\pm ju} = \cos(u) \pm j\sin(u)$

$\cos(u) = \frac{1}{2}(e^{ju} + e^{-ju}))$

$\sin(u) = (e^{ju} - e^{-ju})/2j$

$\sin^2(u) + \cos^2(u) = 1$

$\cos^2(u) - \sin^2(u) = \cos(2u)$

$2\sin(u)\cos(u) = \sin(2u)$

$\cos^2(u) = \frac{1}{2}[1 + \cos(2u)]$

$\sin^2(u) = \frac{1}{2}[1 - \cos(2u)]$

$\sin(u \pm v) = \sin(u)\cos(v) \pm \cos(u)\sin(v)$

$\cos(u \pm v) = \cos(u)\cos(v) \mp \sin(u)\sin(v)$

$\sin(u)\sin(v) = \frac{1}{2}[\cos(u - v) - \cos(u + v)]$

$\cos(u)\cos(v) = \frac{1}{2}[\cos(u - v) + \cos(u + v)]$

$\sin(u)\cos(v) = \frac{1}{2}[\sin(u - v) + \sin(u + v)]$

D–2 INDEFINITE INTEGRALS

$$\int \sin(ax)\,dx = -\frac{1}{a}\cos(ax)$$

$$\int \cos(ax)\,dx = \frac{1}{a}\sin(ax)$$

$$\int \sin^2(ax)\,dx = \frac{x}{2} - \frac{\sin(2ax)}{4a}$$

$$\int \cos^2(ax)\,dx = \frac{x}{2} + \frac{\sin(2ax)}{4a}$$

$$\int x\sin(ax)dx = \frac{\sin(ax) - ax\cos(ax)}{a^2}$$

$$\int x\cos(ax)dx = \frac{\cos(ax) + ax\sin(ax)}{a^2}$$

$$\int x^m \sin(x)dx = -x^m\cos(x) + m\int x^{m-1}\cos(x)dx$$

$$\int x^m \cos(x)dx = x^m\sin(x) - m\int x^{m-1}\sin(x)dx$$

$$\int \sin(ax)\sin(bx)dx = \frac{\sin(a-b)x}{2(a-b)} - \frac{\sin(a+b)x}{2(a+b)}, \quad a^2 \neq b^2$$

$$\int \sin(ax)\cos(bx)dx = -\frac{\cos(a-b)x}{2(a-b)} - \frac{\cos(a+b)x}{2(a+b)}, \quad a^2 \neq b^2$$

$$\int \cos(ax)\cos(bx)dx = \frac{\sin(a-b)x}{2(a-b)} + \frac{\sin(a+b)x}{2(a+b)}, \quad a^2 \neq b^2$$

$$\int e^{ax}dx = e^{ax}/a$$

$$\int x^m e^{ax}dx = \frac{x^m e^{ax}}{a} - \frac{m}{a}\int x^{m-1}e^{ax}dx$$

$$\int e^{ax}\sin(bx)dx = \frac{e^{ax}}{a^2 + b^2}\left[a\sin(bx) - b\cos(bx)\right]$$

$$\int e^{ax}\cos(bx)dx = \frac{e^{ax}}{a^2 + b^2}[a\cos(bx) + b\sin(bx)]$$

D–3 DEFINITE INTEGRALS

$$\int_0^\infty \frac{a\,dx}{a^2 + x^2} = \pi/2, \quad a > 0$$

$$\int_0^{\pi/2} \sin^n(x)dx = \int_0^{\pi/2} \cos^n(x)dx = \begin{cases} \dfrac{1 \cdot 3 \cdot 5 \cdots (n-1)}{2 \cdot 4 \cdot 6 \cdots (n)} \dfrac{\pi}{2}, & n \text{ even, } n \text{ an integer} \\ \dfrac{2 \cdot 4 \cdot 6 \cdots (n-1)}{1 \cdot 3 \cdot 5 \cdots (n)}, & n \text{ odd} \end{cases}$$

$$\int_0^\pi \sin^2(nx)dx = \int_0^\pi \cos^2(nx)dx = \pi/2, \qquad n \text{ an integer}$$

$$\int_0^\pi \sin(mx)\sin(nx)dx = \int_0^\pi \cos(xm)\cos(nx)dx = 0, \quad m \neq n, \quad m \text{ and } n \text{ integer}$$

$$\int_0^\pi \sin(mx)\cos(nx)dx = \begin{cases} 2n/(m^2 - n), & m + n \text{ odd}, m^2 \neq n^2 \\ 0, & m + n \text{ even} \end{cases}$$

$$\int_0^\infty \frac{\sin(ax)}{x} dx = \frac{\pi}{2}, a > 0$$

$$\int_0^\infty \frac{\sin^2 x}{x^2} dx = \frac{\pi}{2}$$

$$\int_0^\infty e^{-a^2x^2} dx = \frac{\sqrt{\pi}}{2a}, \qquad a > 0$$

$$\int_0^\infty x^n e^{-ax} dx = \frac{n!}{a^{n+1}}, \ n \text{ an integer and } a > 0$$

$$\int_0^\infty x^{2n} e^{-ax^2} dx = \frac{1 \cdot 3 \cdot 5 \cdots (2n-1)}{2^{n+1} a^n} \sqrt{\frac{\pi}{a}}$$

$$\int_0^\infty e^{-ax} \cos(bx) dx = \frac{a}{a^2 + b^2}, \ a > 0$$

$$\int_0^\infty e^{-ax} \sin(bx) dx = \frac{b}{a^2 + b^2}, \ a > 0$$

$$\int_0^\infty e^{-a^2x^2} \cos(bx) dx = \frac{\sqrt{\pi}}{2a} e^{-b^2/4a^2}$$

E

Answers to Selected Problems

CHAPTER 1

1-1. Sample mean of X = 5.5; sample mean of Y = 23.3. **1-4.** Decade from -10 to -1 has -2 in its row; decade from 10 to 19 has 5 6 6 in its row; decade from 20 to 29 has 0 1 4 5 in its row; decade from 30 to 39 has 8 in its row; decade from 60 to 69 has 1 in its row; all the rest just show the decades with nothing in their respective rows.

CHAPTER 2[1]

2-1. (a) Personal; (c) relative frequency or possibly equally likely outcomes; (e) relative frequency. **2-2.** (a) $P(\text{red}) = 1/3$; $P(\text{W3}) = 1/15$; (c) $P(\text{2 or 3 spots up}) = 1/3$. **2-3.** $P(\text{ace of spades}) = 1/52$; (c) $P(\text{red ace}) = 1/26$; (e) $P(\text{any face card}) = 2/13$; (f) $P(\text{any ace}) = 1/221$. **2-4.** $P(A) = 45/360$; (c) $P(A \cap B) = 1/16$. **2-6.** $A \cup B = \{0, 1, 2, 3, 4, 5, 6, 7, 8\}$; (c) $A \cup B \cup C \cup D = S$; (e) $A \cap B = \varnothing$; (f) $A \cup B \cap C \cup D = \{3, 4, 5, 6\} = D$. **2-7.** $A^c = \{0,2,4,6,8,10\}$; $C^c = \{0, 1, 2, 3, 4, 5, 6, 7, 8, 9\}$. **2-9.** (a) $P(A) = 3/8$; (c) $P(A \cup B = 3/4$. **2-10.** (a) $P(X) = 372/1372$; (c) $P(A \cap X) = 57/1372$. **2-11.** (a) number of persons reading 1 paper only = 32,000; (c) number of persons reading no newspapers = 55,000. **2-12.** (b) $P(X + Y > 1/2) = 1 - 1/8 = 7/8$. **2-13.** (a) $P(B) = 0.6$; (d) $P(A \mid B) = 1/3$. **2-14.** Either $P(A) = 0$ or $P(B) = 0$ or both—your challenge is to explain why. **2-15.** $P(A \mid B) = 0.25$; (c) not statistically independent—why? **2-16.** (a) $P(A \mid X) = 113/372$;

[1] The notation for complement used here will be superscript c.

(c) $P(B \mid X) = 57/372$; (e) $P(C \mid X) = 202/372$; (g) $P(A \mid Y) = 207/406$; (i) $P(B \mid Y) = 116/406$; (k) $P(C \mid Y) = 83/406$; (m) $P(A \mid Z) = 342/594$; (o) $P(B \mid Z) = 175/594$; (q) $P(C \mid Z) = 77/594$. **2-17.** (a) $P(E_5) = 1/2$; (c) $P(E_5 \cap E_6) = 0$. **2-18.** Not statistically independent—why? **2-19.** (b) $P(Y \mid C) = 83/362$. **2-20.** (a) $P(\text{nickel}) = 7/216$; (c) $P(\text{quarter}) = 31/72$. **2-21.** (b) $P(S = 1 \mid R = 1) = 0.99$; (d) $P(S = 1 \mid R = 0) = 0.005$. **2-22.** (b) $P(\text{cloudy} \mid \text{no rain}) = 2/9$. **2-23.** (a) $P(X = 2 \mid Y = 2) = 0.968$; (c) $P(X = 1 \mid Y = 2) = 0.005$. **2-24.** $P(D^c \mid F^c) = 0.9901$. **2-25.** $P(U \mid P^c) = 0.0011$. **2-26.** P (show goat | goat revealed) $= 1/n$ if host uses prior knowledge. What is it for the case of host randomly selecting? **2-27.** For 23 persons in the room, the probability of two or more with same birthday is 0.51. Derive general case and show this. **2-28.** Probability of four aces is 1.847×10^{-5}. **2-29.** Probability of all hearts in a 13-card hand is 1.575×10^{-12}. **2-31.** If you guess on 8 out of 10 questions in order to pass the test, the probability of passing is 0.055.

CHAPTER 3

3-1. $P(X = 0) = 1/6$; $P(X = 4) = 1/18$; $P(X = -2) = 1/9$; $P(X = -5) = 1/36$. **3-2.** P (2 heads) = 3/8. **3-3.** $P(X = 5) = 0.046$; $P(X = 9) = 0.092$; $P(X = 13) = 0.194$. **3-4.** (b) No–why? **3-7.** (a) $P(5 \leq X \leq 7) = 0.1213$; (c) $P(X > 3) = 0.549$. **3-9.** (a) $P(2 \leq X \leq 4) = 0.4214$; (c) $P(X < 2) = 0.5$. **3.12.** (a) $P(> 3 \text{ errors}) = 0.0802$; (b) $P(> 3 \text{ errors}) = 0.0803$ by Poisson approximation. **3-14.** (a) $A = a$. **3-15.** For $\lambda = 50$ per second, $P(W > 10^{-2}\ \text{s}) = 0.6065$. **3-16.** $P(W > 1\ \text{min}) = 0.1889$ for λ $100/60 = 1.667$ events per minute. **3-17.** Apply the geometric distribution; P(success at any trial up to and including 5) = 0.4095. **3-18.** (b) Apply the Pascal distribution: $P(X = 3) = 0.0018$. **3-19.** Apply the hypergeometric distribution: $P(\geq 1 \text{ defective}) = 0.005$. **3-21.** (b) $f_Y(y) = 0.5\delta(y) + 1.5e^{-5y}u(y)$. **3-25.** (b) $E(X) = 3.7$ and $E(X^2) = 15.7$. **3-26.** Mode at $x = 1$; median $= 2^{-1/2}$; mean $= 2/3$. **3-27.** $E(X) = 21/6$ and $E(X^2) = 91/6$. **3-28.** $E(X) = 6$. **3-30.** (a) $E(Y) = 50.75$; variance = 1138.58. **3-31.** $E(X) = 3.7$; variance = 21.61. **3-32.** Characteristic function is $9/(9 + \nu^2)$. **3-34.** $E(X) = a$ and $E(X^2) = a + a^2$. **3-35.** The bound for the desired probability is 0.25; (c) the bound for this case is e^{-1}. **3-38.** $E(X) = 1/6$. **3-39.** The transformation is $V = [-2\sigma^2 \ln(1 - U)]^{1/2}$, where U is uniform. **3-40.** $X = \alpha \tan[\pi(U - 0.5)]$, where U is uniform.

CHAPTER 4

4-1. (a) $A = 2$; (d) the desired probability is 2/15. **4-3.** $f_X(x) = [2/(2x + 1)^2]u(x)$; $f_Y(y) = [1/(y + 1)^2]u(y)$. **4-5.** (b) $f_X(x) = e^{-x}u(x)$; $f_Y(y) = [1/(y + 1)^2]u(y)$. **4-6.** The probability of me arriving before my friend is 5/6. **4-8.** $P(999.5 < X \leq 1000.5 \mid 999 < X \leq 1001) = 0.7311$. **4-9.** If $A = s/2$, then the probability of a miss and false alarm for $s/\sigma = 1$ are 0.309 and 0.309, respectively. For $s/\sigma = 4$, they are 0.023 and 0.023. Derive general result and fill in a table of values. **4-10.** (a) Yes, they are independent. **4-11.** No—why not? **4-12.** (b) The desired expectation is 0.8. **4-13.** (a) The mnth moments are zero for m or n or both odd. For m and n even the result is $4m!n!/(2^{m+1}2^{n+1})$. The central moments are the same since the means are zero. (c) $\rho_{XY} = 0$. **4-14.** (a) $C = 4/\pi$. (c) The random variables are uncorrelated—why? **4-15.** Some cases are: $m = n = 1$, joint moment $= 7/4$; $m = 2$ and $n = 1$; joint moment $= 7/4$; $m = 1$ and $n = 2$; joint moment $= 91/12$. **4-17.** 1.43 projectiles, or 2 projectiles since we cannot have a fraction of a projectile. **4-23.** The mean of the measurement is 52,800 feet and its variance is 9292.8 ft^2. **4-24.** Variance is 11.4.

CHAPTER 5

5-1. $E(\bar{x}) = m$. **5-3.** $s_x = 15.904$. **5-9.** Sample means of $\{x\}$ and $\{y\}$ are 2.34 and 6.37, respectively. Sample standard deviations of $\{x\}$ and $\{y\}$ are 1.52 and 3.18, respectively. The sample correlation coefficient of $\{x\}$ and $\{y\}$ are 0.02. **5-13.** UCL $= 107.27$ and LCL $= 87.73$.

CHAPTER 6

6-1. Median $= 10.3$ and sample mean $= 8.99$. **6-2.** (b) Variance $= 1$. **6-3.** n 100,000 samples. **6-8.** Neither is totally satisfactory; the median is biased toward the lower end and the mean is biased toward the upper end. **6-9.** The maximum likelihood estimate is k/T. **6-10.** (a) Use the binomial distribution to show that the estimate for the probability is the ratio of the number of heads observed to the total number of tosses; (c) the sample standard deviation for this case is 0.06697. **6-12.** The variance of the estimate for p is $p(1 - p)/100$. **6.15.** (a) The sample mean is 15.89. **6-16.** The 95% confidence interval for the mean in this case is [14.04, 17.74]. **6-17.** The 95% confidence interval for the variance in this case is [19.13, 120.65]. **6-18.** The 95% confidence interval for the variance in this case is [20.18, 142.16] **6-19.** The 95% confidence interval for the mean in this case is [11.22, 20.56]. **6-20.** (a) The 95% confidence interval for p is [0.477, 0.564]. **6-23.** (b) The MMSE estimate is 5.625 feet. **6-25.** The estimator is $\theta(X) = -5X + 11$.

CHAPTER 7

7-1. The test is if $Z > 6.355$, decide H_1; otherwise, decide H_0. **7-2.** The average cost of a decision is 0.028. **7-3.** 7-3. For $Z < -6.941$, or >6.941, choose H_1; for Z in $[-6.941, -1.059)$ or $[1.059, 6.941)$, choose H_0; for $[-1.059, 1.059)$, choose H_1; the average cost for making a decision is 0.375. **7-4.** The critical region boundary is 1.96. **7-5.** (a) Accept the hypothesis that the bar is 60 oz. **7-6.** Reject the hypothesis that the bar is 60 oz. **7-8.** The probability of error averaged over all possible numbers is 0.605.

CHAPTER 8

8-2. $\lambda = 0.1386$ yr^{-1}. **8-3.** (b) The mean time to failure is 10 years. **8-4.** The mean time to failure is 5.7×10^4 yr. **8-6.** (b) The mean time to failure is $0.667T$. **8-7.** (a) The system reliability is $1 - r^2f$; (c) the system reliability is r^3; (e) the system reliability is $(1 - f^3)(1 - f^2)$. **8-8.** Find that $r = 0.9$ gives the desired overall reliability. **8-9.** $n = 3$.

8-10. The mean time to failure is 200 days. **8-11.** The reliability for 5000 h is 0.9695. **8-12.** (b) The mean time to failure is 8862 s.

CHAPTER 9

9-2. (a) Case 3: continuous independent variable and discrete dependent variable. **9-3.** (b) Case 1: discrete independent variable and discrete dependent variable. **9-4.** (a) The ensemble-average mean is $4A[\sin(\omega_0 t + \pi/4) - \sin(\omega_0 t)]/\pi$; (b) the time-average mean of a single sample function is 0; (c) neither stationary nor ergodic. **9-7.** (a) Suitable; (c) not suitable—its Fourier transform is not nonnegative. **9-8.** (a) The mean-square value is A^2; (c) not a periodic process. **9-9.** The mean and variance are 1 and 1, respectively. From these, the Gaussian pdf at any time can be written down. **9-11.** (a) Average powers are both 1 W; (c) average power of their sum is 3.732 W. **9-13.** The covariance is 0.5, the mean is 0, and the variance is 4; from these, the joint pdf can be written down. **9-16.** $\alpha = 0.6$, so $r_X(M) = 0.6^{|M|}$.

CHAPTER 10

10-1. $c/\sigma = 0.7$ is close, but not exact. **10-3.** $P_n = 0.207A^2$. **10.4** (a) $m_x = 5$; (c) $m_Y = 5$. **10-8.** The variance of $Y(t)$ is 45.5 and the mean is zero. From these the first-order pdf can be written down. The correlation coefficient is approximately 0, so the process joint pdf at times separated by 1 s is the product of two marginal Gaussian pdfs. The same is true for a separation of o.ls. **10-9.** The autocorrelation funciton is $N_0 B \operatorname{sinc}(2B\tau)$. **10-11.** $B_N = 5$ Hz. **10-12.** (b) 2.667 Hz. **10-13.** (a) 10.45 μV. **10-14.** Overall noise figure is 2 dB. **10-15.** $G_1 = 19.42$ dB. **10.17.** $e_{min} = B/(4B/N_0 + 1)^{1/2}$.

Bibliography

M. Abramowitz and L. Stegun, eds., *Handbook of Mathematical Functions.* New York: Dover, 1972 (originally published in 1964 as part of the National Bureau of Standards Applied Mathematics Series 55.).

Carol Ash, *The Probability Tutoring Book.* New York: IEEE Press, 1993.

Donald R. Barr and Peter W. Zehna, *Probability: Modeling and Uncertainty.* Reading, MA: Addison-Wesley, 1983.

George E. P. Box, William G. Hunter, and J. Stuart Hunter, *Statistics for Experimenters.* New York: Wiley, 1978.

Arthur M. Breipohl, *Probabilistic Systems Analysis.* New York: Wiley 1970.

Harald Cramér, *Mathematical Methods of Statistics.* Princeton, NJ: Princeton University Press, 1946.

William Feller, *An Introduction to Probability Theory and Its Applications,* Vol. I, 2nd ed. New York: Wiley, 1957.

William Feller, *An Introduction to Probability Theory and Its Applications,* Vol. II, New York: Wiley, 1966.

Thomas S. Ferguson, *Mathematical Statistics.* San Diego, CA: Academic Press, 1967.

Carl W. Helstrom, *Probability and Stochastic Processes for Engineers,* 2nd ed. Reading, MA: Addison-Wesley, 1992.

Ernest J. Henley and Hironmitsu Kumamoto, *Probabilistic Risk Assessment.* New York: IEEE Press, 1992.

Richard A. Johnson, *Miller and Freund's Probability and Statistics for Engineers.* Upper Saddle River, NJ: Prentice Hall, 1994.

John B. Kennedy and Adam M. Neville, *Basic Statistical Methods for Engineers and Scientists,* 3rd ed. New York: Harper & Row, 1986.

Edwin Kreyszig, *Advanced Engineering Mathematics,* 6th ed. New York: Wiley, 1988.

Alberto Leon-Garcia, *Probability and Random Processes for Electrical Engineering,* 2nd ed. Reading, MA: Addison-Wesley, 1994.

MATLAB User's Guide. Natick, MA: Mathworks, 1993

D.C. Montgomery and G.C. Runger, *Applied Statistics and Probability for Engineers.* New York: Wiley, 1994.

Michael O'Flynn, *Probabilities, Random Variables, and Random Processes.* New York: Harper & Row, 1982.

R. Lyman Ott, *An Introduction to Statistical Methods and Data Analysis.* Belmont, CA: Wadsworth, 1993.

Athanasios Papoulis, *Probability and Statistics.* Upper Saddle River, NJ: Prentice Hall, 1990.

Athanasios Papoulis, *Probability, Random Variables, and Stochastic Processes,* 3rd ed. New York: McGraw-Hill, 1993.

Peyton Z. Peebles, *Probability, Random Variables, and Random Signal Principles,* 2nd ed. New York: McGraw-Hill, 1987.

Richard A. Roberts, *An Introduction to Applied Probability.* Reading, MA: Addison-Wesley, 1992.

Sheldon M. Ross, *Introduction to Probability and Statistics for Engineers and Scientists.* New York: Wiley, 1987.

Sheldon M. Ross, *A First Course in Probability,* 4th ed. Upper Saddle River, NJ: Prentice Hall, 1994.

Mischa Schwartz and Leonard Shaw, *Signal Processing: Discrete Spectral Analysis, Detection, and Estimation.* New York: McGraw-Hill, 1975.

Murray R. Spiegel, *Theory and Problems of Probability and Statistics.* New York: McGraw-Hill, 1975.

Henry Stark and John W. Woods, *Probability, Random Processes, and Estimation Theory for Engineers,* 2nd ed. Upper Saddle River, NJ: Prentice Hall, 1994.

Student Edition of MATLAB. Upper Saddle River, NJ: Prentice Hall, 1995.

John B. Thomas, *Introduction to Probability.* New York: Springer-Verlag, 1986.

Harry L. van Trees, *Detection, Estimation, and Modulation Theory,* Part I, New York: Wiley, 1968.

Stephen B. Vardeman, *Statistics for Engineering Problem Solving.* Boston: PWS, 1994.

Ronald E. Walpole and Raymond H. Meyers, *Probability and Statistics for Engineers and Scientists,* 5th ed. New York: Macmillan, 1993.

Richard H. Williams, *Electrical Engineering Probability.* St. Paul, MN: West, 1991.

Rodger E. Ziemer and William H. Tranter, *Principles of Communications: Systems, Modulation, and Noise,* 4th ed. New York: Wiley,1995.

R. E. Ziemer, W. H. Tranter, and D. R. Fannin, *Linear Systems: Continuous and Discrete.* Upper Saddle River, NJ: Prentice Hall, 1993.

Index

D

E

H

I

L

S